NMR 23

Basic Principles and Progress

Editors: P. Diehl E. Fluck H. Günther
R. Kosfeld J. Seelig

Deuterium and Shift Calculation

With contributions by
U. Fleischer, W. Kutzelnigg, H.-H. Limbach,
G. J. Martin, M. L. Martin, M. Schindler

With 92 Figures

Springer-Verlag
Berlin Heidelberg New York London
Paris Tokyo Hong Kong Barcelona

ISBN 3-540-52949-7 Springer-Verlag Berlin Heidelberg NewYork
ISBN 0-387-52949-7 Springer-Verlag NewYork Berlin Heidelberg

Library of Congress Cataloging-in-Publication Data
Deuterium and shift calculation / with contributions by U. Fleischer ... [et al.].
p. cm. -- (NMR ; 23) Includes index.
ISBN 3-540-52949-7 (Springer-Verlag Berlin Heidelberg New York). -
ISBN 0-387-52949-7 (Springer-Verlag New York Berlin Heidelberg)
1. Nuclear magnetic resonance spectroscopy. 2. Deuterium. 3. Isotope shifts. I. Fleischer, U.
II. Series: NMR (Series) ; v. 23.
[QC490.N2 vol. 23] [QD96.N8] 538'.362 s--dc20 [538'.362] 90-10273 CIP

Printed in Germany

Typesetting Thomson Press (India) Ltd, New Delhi
Offsetprinting: Colordruck Dorfi GmbH, Berlin
Bookbinding: Lüderitz & Bauer, Berlin
2151/3020-543210 - Printed on acid-free paper

Editorial

The preceding volume of this series was devoted to NMR isotope effects and the first two contributions of the present publication continue to deal with this topic. The emphasis is, however, shifted from the structural aspect of analysis and reactivity.

M.L. Martin and G.J. Martin report on site-specific natural isotope fractionation of deuterium as studied by quantitative ^{2}H-NMR. They give a detailed account of this new and important experimental technique which they discovered in 1981 and treat its application in various areas. Deuterium isotope effects on kinetic and thermodynamic parameters are discussed as well as the origin of natural site-specific deuterium labelling and its potential for the elucidation of reaction mechanisms and quality control of synthetic and natural products.

Deuterium isotope effects on reaction rates as determined from lineshape analyses of NMR spectra are reviewed by H.-H. Limbach. The author develops general theories and strategies for the application of dynamic NMR methods to the field of isotopic reaction kinetics. He also discusses experimental aspects of such measurements and demonstrates with examples from the liquid and the solid state, the sensitivity of NMR for the detection of small differences in rate constants.

Finally, the chemical shift itself is the topic of the last chapter by W. Kutzelnigg, U. Fleischer and M. Schindler who present a detailed review of the IGLO-method which has been developed in recent years by the Bochum group for the calculation of shielding constants and magnetic susceptibilities. A thorough comparison between calculated and observed chemical shifts for the elements of the first two rows of the Periodic Table convincingly demonstrates the progress which has been made and encourages the use of these types of calculation in the field of structural chemistry.

Siegen, September 1990 H. Günther

Table of Contents

Deuterium NMR in the Study of Site-Specific Natural Isotope Fractionation (SNIF-NMR)

Maryvonne L. Martin and Gérard J. Martin

Université de Nantes-URA-CNRS 472, 2 rue de la Houssinière 44072 Nantes Cedex-France

Table of Contents

NMR Basic Principles and Progress, Vol. 23

The isotope content and in particular the ratio of deuterium and hydrogen atoms in a given molecular species, D/H, are not constant on Earth. On the basis of the overall molecular isotope contents accessible by mass spectrometry techniques, important information has been obtained on biochemical and physiological aspects of biosyntheses under natural conditions. The original approaches provided by the NMR study of *Site-Specific Natural Isotope Fractionation* of hydrogen must be seen in this context.

The advantages and requirements of quantitative deuterium NMR are recalled and experimental techniques appropriate to precise, accurate, and eventually automatic, analyses of site-specific isotope ratios are described. It is shown, on a theoretical basis, how several primary and secondary kinetic isotope effects can be determined in a single experiment without the need for isotope labelling, and applications are reviewed.

The new approaches derived for determining thermodynamic fractionation factors in exchange equilibria, for detecting natural chirality associated with methylene groups, or more generally, for investigating chemical and biochemical mechanisms, are discussed in the light of the traditional applications of NMR to mechanistic studies involving deuterium labelling.

Of major interest is the powerful capability of SNIF-NMR to infer properties of raw materials, which have since disappeared, from the isotope parameters measured on end products. The different kinds of information concerning the roles played by the metabolic pathway and by the environmental conditions of the biosynthesis are reviewed and discussed.

Finally, various applications of practical interest for quality assessment of food products are summarized.

Terms and Symbols

BCR:	Bureau Commun de References (Community Bureau of References B-1049 Brussels)
C_3:	Photosynthetic mechanism of plants in which the carboxylation product is a compound with a 3-carbon skeleton
C_4:	Photosynthetic mechanism of plants in which a first carboxylation step produces a compound with a 4-carbon skeleton
CAM:	Photosynthetic mechanism of plants with a Crassulacean Acid Metabolism
δ-scale:	Relative scale of isotope ratios expressed in ‰ (see Eq. (2) [and (1)] for definition)
DEPT:	Distorsionless Enhancement by Polarisation Transfer
GISP	Greenland Ice Sheet Precipitation
IAEA:	International Atomic Energy Agency (P.O. Box 100 A-1400 Vienna)
INEPT:	Insentive Nuclei Enhanced by Polarisation Transfer
ISO-NORMS:	NORMS from the International Standard Organization (CH-1211 Geneva 20, Switzerland)
OIV:	Office International de la Vigne et du Vin (International Office of Vine and Wine)
PDB	Peedee Belemnite (Rf. Table1)
SLAP:	Standard Light Antartic Precipitation
SMOW:	Standard Mean Ocean Water
SNIF:	Site-specific Natural Isotope Fractionation
V.SMOW:	Vienna Standard Mean Ocean Water (Reference sample, for hydrogen and oxygen isotope ratios, available from IAEA in Vienna)
WALTZ:	Wideband Alternating Phase Low-power Technique for Zero residue splitting

1 The Distribution of Isotopes in Nature

In natural products the main atoms of hydrogen, ^{1}H, carbon, ^{12}C, nitrogen, ^{14}N, and oxygen, ^{16}O, for example, are accompanied by their stable isotopes, ^{2}H, ^{13}C, ^{15}N, and ^{17}O, ^{18}O, respectively. These heavier atoms are present only in very small quantities. The number of deuterium atoms with respect to that of protium atoms, D/H, is about 156×10^{-6} (or 156 ppm) in water from the oceans [1, 2]. In the case of carbon the $^{13}C/^{12}C$ ratio is about 11×10^{-3} [3] and the corresponding isotope ratio is about 20×10^{-4} for oxygen-18 [1] and 37×10^{-4} for nitrogen-15 [4] (Table 1).

In fact, the isotope contents are not constant among the different natural products and even among pure samples of a given molecular species from different raw materials.

In our galaxy and solar system, the isotope contents may exhibit considerable variations. The D/H ratio varies from about 0.01 ppm in outer space to 900 ppm on Mars [10] and 16000 ppm on Venus [11]. On the Earth, relatively small but reproducible variations in the isotope contents of biogenic materials are observed as a result of isotope discrimination due to physical processes or biochemical reactions. Whereas the deuterium content of natural water reaches about 155 ppm in the Ivory Coast, it is only 89 ppm in the Antarctic region [2, 5] (Fig. 1). The large variations in the ^{2}H and ^{18}O contents of meteoric waters may be understood in terms of evaporation and condensation effects in the clouds [12–15]. The isotope ratios of precipitation therefore depend on complex atmospheric phenomena. Relationships have been found either with geographical parameters such as latitude, altitude, distance from the sea, or with meteorological parameters such as temperature and amount of precipitation [13, 16]. As illustrated in Fig. 1, water is increasingly depleted in deuterium when going from the equator towards the poles. A depletion is also observed

Table 1. Natural isotope contents of hydrogen, carbon, nitrogen, and oxygen in international standards. V.SMOW denotes Vienna Standard Mean Ocean Water [1, 2, 5–7] and PDB refers to Peedee Belemnite, the carbonate skeleton of a fossil cephalopod (*Belemnitella americana*) from the Cretaceous Peedee formation of South Carolina [3, 8, 9]. Since its isotope content is relatively homogeneous, atmospheric nitrogen can be used as a standard [4, 250]. R is defined in equation (1) and $F = A_H/(A_H + A_L)$

Isotopic Ratio	$\frac{^{2}H}{^{1}H}$	$\frac{^{13}C}{^{12}C}$	$\frac{^{15}N}{^{14}N}$	$\frac{^{18}O}{^{16}O}$
R (ppm)	155.76 ($\pm$0.5)	11237.2	3676.5	2005.2 ($\pm$0.4)
Relative abundance F (ppm)	155.74	11123.3	3663	2001.2
Standard	H_2O V.SMOW	$CaCO_3$ PDB	N_2 Air	H_2O V.SMOW

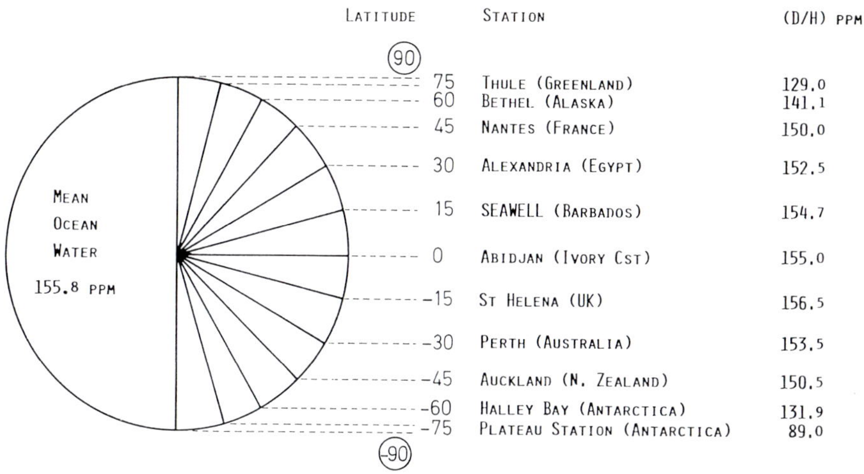

Fig. 1. Variations in the hydrogen isotope ratio of atmospheric water samples as a function of latitude. The D/H values, expressed in ppm correspond in most cases to samples collected in meteorological stations of the IAEA network [16]. The monthly values have been weighted by the amounts of precipitation

from coast to inland and, within a given region, the heavy isotope content tends to decrease with altitude [17].

The isotope contents in plants and animals depend on the biochemical and environmental factors which preside over the biosynthesis in auto- or heterotrophic conditions. During the last thirty years, numerous investigations of the *carbon* isotope contents in natural compounds have been carried out [18, 19]. It has been observed that the Earth's biomass as a whole is depleted in the carbon-13 isotope relative to the inorganic pool of the ocean-atmosphere system. This isotopic discrimination has been mainly attributed to the key step of carbon assimilation. In the principal mechanism of photosynthesis, CO_2 is directly introduced by the ribulose 1,5-biphosphate carboxylase into the Calvin cycle of reactions, also termed "C_3 cycle" because the carboxylation product is a compound with a 3-carbon skeleton, phosphoglycerate. A significant fractionation effect occurring in this CO_2-fixing reaction is considered to be responsible for the depletion in carbon-13 observed in C_3 plants which constitute the majority of the vegetal species [9, 20–23]. In contrast, only small fractionation effects with respect to atmospheric CO_2 occur in C_4 plants where carbon assimilation involves, as a first step, the synthesis by phosphoenolpyruvate carboxylase of a 4-carbon compound, oxaloacetate [24–26]. Succulent plants which exhibit crassulacean acid metabolism (CAM) may fix atmospheric carbon in a manner in which both the C_3 and C_4 pathways participate and the range of their $^{13}C/^{12}C$ ratios overlaps those of the C_4 and C_3 plants [19, 27–30]. Figure 2 illustrates the ranges of variations in the carbon-13 contents of natural compounds.

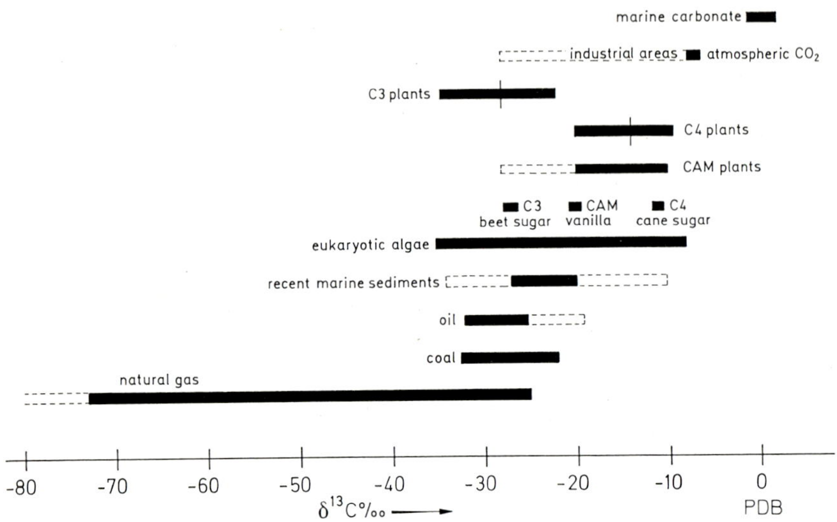

Fig. 2. Ranges of variations in the ^{13}C content of natural products. The isotope parameter $\delta\ ^{13}C$‰ is defined in Eq. (2). It enables the isotope ratio, $^{13}C/^{12}C$, of the considered compound to be compared to that of the international standard PDB (Table 1). In industrial areas, the ^{13}C content of atmospheric CO_2 may be significantly smaller because of combustion of highly depleted fossil materials [18]. In organic constituents of recent marine sediments, 90% of the samples are found between -20 and -27‰ [31]. Strong depletions are observed in biogenic gases [32]

Although the overall *deuterium* content of organic molecules has been the subject of fewer investigations than that of ^{13}C, it has been shown that relatively large variations can be observed among plants with different photosynthetic metabolism [33–35]. Moreover, even for a given plant such as potato, large differences are exhibited by the hydrogen isotope ratios of the various metabolites [36] (Fig. 3).

Until recently only an average deuterium content of the investigated molecule could be obtained directly since the mass spectrometry method, usually employed, requires a combustion of the sample before measurement either of the carbon isotope ratio in CO_2, or of the hydrogen isotope ratio on molecular hydrogen resulting from the reduction of water [4]. In 1981, we showed that large deviations with respect to a statistical distribution of deuterium among the different sites of a given molecule can be directly evidenced by deuterium Nuclear Magnetic Resonance [42, 43]. The Site-specific Natural Isotope Fractionation as studied by NMR (SNIF-NMR) provides a very powerful tool for the investigation of biochemical mechanisms in natural conditions. It enables isotope labelling to be investigated without the need for enrichment [44] and has greatly widened the possibilities of origin inference of natural oz synthetic products.

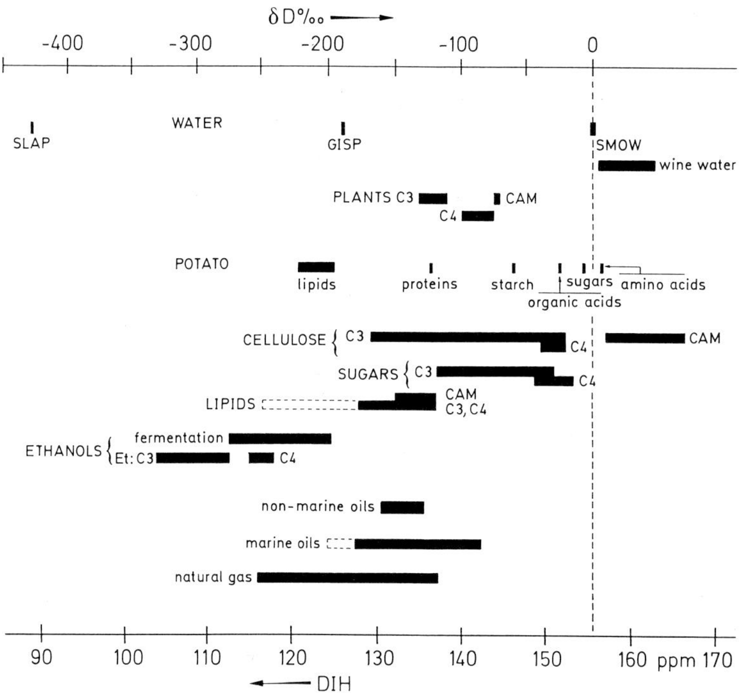

Fig. 3. Ranges of variations in the deuterium content of natural products. The isotope parameter δD‰ is defined in Eq. (2). The diagram is built from literature results referring to: plants with different metabolism [34, 37], marine and non-marine oils [38], different components extracted from potato [36], lipids from some C3, C4, and CAM [35], cellulose [33, 39, 35], sugars [37, 40], fermentation ethanols [37, 40, 41], ethyl group of ethanol from C_3 and C_4 plants [40]. GISP denotes a water sample of Greenland Ice Sheet Precipitation used for intercalibration [41]

2 Site-Specific Natural Isotope Fractionation of Hydrogen

2.1 The Non-Random Distribution of Deuterium

If we consider the natural abundance deuterium NMR spectra of two samples of ethylacetate, A and B, (Fig. 4), it is immediately obvious that the distribution of deuterium among the different molecular sites is far from being statistical [42, 43]. The methyl site I of B, for example, is significantly enriched as compared with the other methyl site III and the deuterium contents deviate greatly from the ratios 2:3:3 which would characterize a random repartition of ^2H among the three sites II, I, III. Evidently the chemist should be aware of this phenomenon if he intends to complement ^1H NMR by ^2H NMR to estimate relative concentrations of the components of a mixture. More importantly, the direct evidence of a non-statistical distribution of deuterium and the

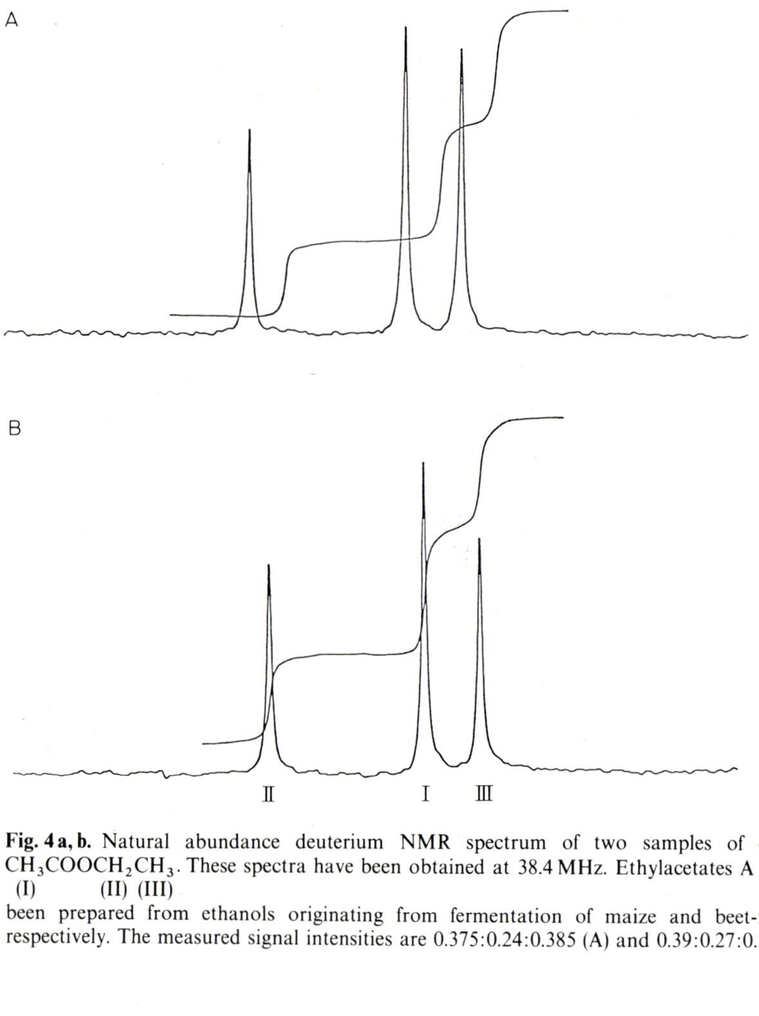

Fig. 4 a, b. Natural abundance deuterium NMR spectrum of two samples of ethylacetate, $CH_3COOCH_2CH_3$. These spectra have been obtained at 38.4 MHz. Ethylacetates A and B have
(I) (II) (III)
been prepared from ethanols originating from fermentation of maize and beet-root sugars, respectively. The measured signal intensities are 0.375:0.24:0.385 (A) and 0.39:0.27:0.34 (B)

demonstration of reproducible behaviors in this distribution makes the isotopic fingerprint a very faithful probe of the chemical, biochemical, and technological history of the product [45, 46].

As a result of the small deuterium content at the natural abundance level (Sect. 1), only monodeuterated species are observable in the conventional ^2H-NMR spectrum. In the presence of proton decoupling the spectrum is composed of single lines associated with all the monodeuterated isotopomers corresponding to diastereotopic deuterium environments. In the example of ethylacetate (Fig. 4), the three signals are assigned to the three isotopomers: $CH_2DCO_2CH_2CH_3$ (I), $CH_3CO_2CHDCH_3$ (II), and $CH_3CO_2CH_2CH_2D$ (III). The molar fractions, f_i, can be obtained from area measurements and compared to the molar fractions, F_i, expected on the basis of a statistical distribution:

		I	II	III
	F =	0.37_5	0.25	0.37_5
A	$f_i =$	0.37_5	0.24	0.38_5
B	$f_i =$	0.39	0.27	0.34

This comparison illustrates the existence and the quantification of significant relative deuterium depletions (site III of B and II of A) or enrichments (sites I and II of B).

2.2 *Site-Specific Isotope Ratios; Definitions*

There are several ways of defining an isotope content but the most obvious is to simply consider the ratio, R, of the number of heavy isotopes, A_H, to that of light isotopes, A_L:

$$R = \frac{A_H}{A_L}\left(\frac{^2H}{^1H}, \frac{^{13}C}{^{12}C}, \frac{^{15}N}{^{14}N}, \frac{^{18}O}{^{16}O} \cdots\right) \tag{1}$$

R is usually expressed in percent for carbon and in ppm for hydrogen. This parameter is well-suited to NMR spectroscopy since this technique gives a response which is directly proportional to the number of nuclei in resonance.

Mass spectroscopists who pioneered the field of natural abundance isotope ratios, used to report their results on the δ-scale[1] as defined by the International Atomic Energy Agency (IAEA) [41, 47]. In this scale, the mean isotope ratio of a given chemical species, S, is referred to that of a reference compound, Ref, and the ratio, δS is expressed in ‰:

$$\delta S‰ = \frac{R_S - R_{Ref}}{R_{Ref}} 1000 \tag{2}$$

The reference compounds for the main biogenic elements are defined in Table 1 which also gives the values of their absolute isotope ratios, R_{Ref}. In the case of hydrogen, the primary standard is the Vienna Standard Mean Ocean Water (V.SMOW) [1] which has a D/H value of 155.8 ppm [2, 5]. When the value of R_S is very different from that of the V.SMOW standard, the extrapolation involved in the use of Eq. (2) may induce some inaccuracy. A secondary reference, the Standard Light Antartic Precipitation, SLAP, has been proposed following

[1] The reader should be aware of the difference between "δ" as defined by Eq. (2) and the δ-scale used in NMR spectroscopy

calibration work carried out in an intercomparison study [2, 5]. This secondary standard is characterized by an isotope ratio $R_{SLAP} = 89.05$ ppm and the value $\delta D‰$ of the considered species, S, can be interpolated [6] from Eq. (3).[1]

$$\delta D_S‰ = \frac{R_S^D - R_{SMOW}^D}{R_{SLAP}^D - R_{SMOW}^D} 428 \tag{3}$$

Site-specific isotopic parameters can be defined in the same way as the mean isotopic ratios. Thus the specific parameter, R_i, represents the ratio of the number of deuterium atoms, D_i, to the number of protium atoms, H_i, associated with site i.

$$R_i = \left(\frac{D}{H}\right)_i = \frac{D_i}{H_i} \tag{4}$$

D_i is approximately equal to the number of monodeuterated isotopomers of type i, whereas H_i is approximately defined as:

$$H_i \approx P_i N_H \tag{5}$$

where P_i is the number of equivalent i positions of the hydrogen atoms and N_H is the number of fully protonated molecules.

Similarly, a $\delta D_i‰$ parameter characterizing the molecular site i can be defined as:

$$\delta D_i‰ = \frac{(D/H)_i - (D/H)_{SMOW}}{(D/H)_{SMOW}} 1000 \tag{6}$$

An interpolation equation similar to Eq. (3) can also be used.

2.3 *Theoretical Aspects*

Taking into account the low concentration of ^2H in natural products, the fractional abundances of deuterium, $[^2H]/([^2H] + [^1H])$, may be approximated to the corresponding absolute isotope ratio, R. Consequently, if we consider a reaction in which a starting compound, I, is completely transformed into a mole fraction x of product II and a mole fraction $1 - x$ of product III, the average isotope ratios, $\overline{(D/H)}$, of II and III are related to that of I by the mass balance Eq. (7):

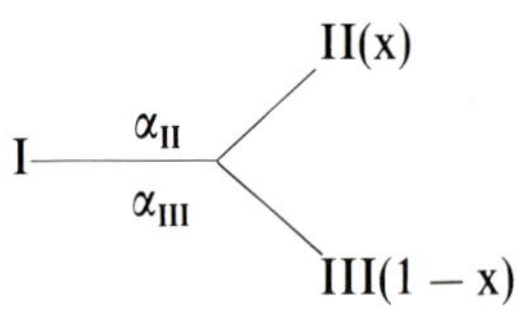

$$\left(\frac{\bar{D}}{H}\right)_I \approx x\left(\frac{\bar{D}}{H}\right)_{II} + (1-x)\left(\frac{\bar{D}}{H}\right)_{III} \tag{7}$$

Product fractionation factors α_{II} and α_{III} describing the disproportionation of isotopes from I to II and III can be defined according to:

$$\alpha_{II}R_{II} = \alpha_{III}R_{III} = R_I \tag{8}$$

For a molecular species containing P hydrogen atoms and characterized by an average isotope parameter $(\overline{D/H})$, the site-specific isotope ratio $(D/H)_i$ associated with isotopomer i, can be calculated from:

$$\left(\frac{D}{H}\right)_i = \frac{f_i}{F_i}\left(\frac{\bar{D}}{H}\right) \tag{9}$$

where f_i is the molar fraction of i obtained from the ^{2}H-NMR spectrum as discussed in section 2.1 and F_i is the statistical molar fraction computed over the n isotopomers of the considered species:

$$F_i = \frac{P_i}{P} \tag{10}$$

Conversely, the mean isotope ratio $(\overline{D/H})$ can be calculated from the site-specific isotope ratios $(D/H)_i$:

$$\left(\frac{\bar{D}}{H}\right) = \sum_{i=1}^{n} F_i\left(\frac{D}{H}\right)_i \tag{11}$$

Equation (7) is very useful for quantifying the isotopic repartition between different chemical species derived from the same starting material on the condition that the transformation is complete. It may be generalized to a reaction involving distinct starting, S, and end, Q, products.

$$S_A + S_B + \cdots \rightarrow Q_A + Q_B + \cdots$$

The deuterium content, $(D/H)_j^Q$, in every site, j, of an end product Q can be approximately described in terms of the deuterium contents in the different sites, i, of the starting materials S by a set of linear equations:

$$\left(\frac{D}{H}\right)_j^Q \approx a_{j1}\left(\frac{D}{H}\right)_1^S + \cdots + a_{ji}\left(\frac{D}{H}\right)_i^S + \cdots + a_{jm}\left(\frac{D}{H}\right)_m^S \tag{12}$$

or in matrix notation:

$$\vec{D}^Q = [A]\vec{D}^S \tag{13}$$

where $\vec{D}^Q$ and $\vec{D}^S$ are the column vectors of the deuterium contents in the end and starting molecules, respectively. [A] is the redistribution matrix built from the coefficients which relate a specific isotopic ratio in the products to that of a given site in the starting materials.

For a reaction such as the fermentation of glucose into ethanol, the matrix [A] reflects the technical and mechanistic aspects of the bioconversion [40, 48]. Thus, provided that this matrix has been previously determined by a series of

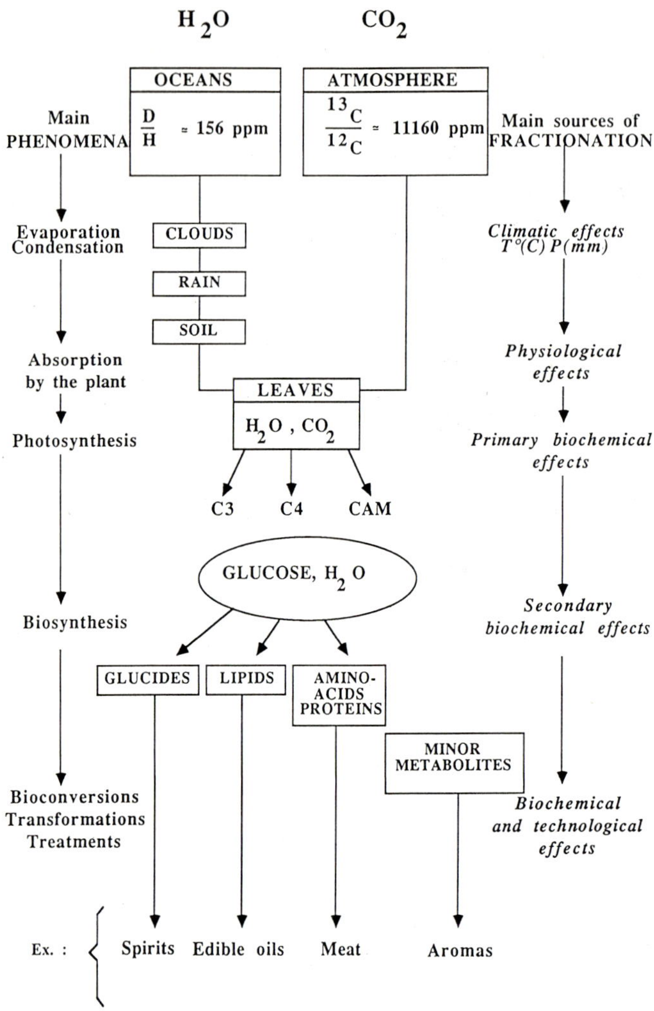

Fig. 5. Isotopic filiation in natural products and main sources of fractionation

calibration experiments, it becomes possible to produce from the vector $\vec{D}^{Q}$, deduced from SNIF-NMR measurements, a fingerprint of the deuterium contents of the starting materials. This point is of fundamental importance in the study of natural products since it means that the isotope distribution in materials which have disappeared can be inferred from measurements performed on their end derivatives.

In the case of incomplete reactions, the isotope ratios of the end products and of the remaining reactants depend on the fractional amount of conversion and on the primary and secondary kinetic isotope effects as discussed in Sect. 4.

On the other hand, isotopic exchanges may occur between reagents in chemical equilibrium and the product fractionation factors then depend on the equilibrium constants of the reactions (see Sect. 4).

2.4 The Sources of Natural Isotope Fractionation of Hydrogen

If we consider a molecule such as ethanol, which will be shown in Sects. 6 and 7 to constitute a very good probe for origin recognition, a great number of effects are likely to have contributed to the site-specifc deuterium contents. For a given environment, these effects, which may be physical, biochemical, or technological in nature, have been proved to be very reproducible [46]. The main filiation pathways are described in the diagram of Fig. 5 which summarizes the key steps responsible for fractionation effects from water and carbon dioxide to the primary products of photosynthesis and then to secondary metabolites and eventually to commercial bioconverted products.

In light of the considerable complexity of the pathways followed by the hydrogen isotopes in the course of the biosynthesis and the fact that isotopic discrimination may intervene all along these pathways, it is obvious that much caution must be exercised in the interpretation of variations in the isotope ratios of natural products.

3 Deuterium Nuclear Magnetic Resonance and the Determination of Site-Specific Isotope Ratios

3.1 Nuclear Magnetic Resonance of ^2H

In spite of technical difficulties, high-resolution nuclear magnetic resonance spectra of deuterium were obtained as early as 1964 [49]. However, in 1972 there were only about 90 papers which dealt with ^2H NMR [50]. Until now deuterium has occupied a rather marginal position in NMR not only with respect to ^1H and ^{13}C but also with respect to other heteroatoms such as ^{15}N and ^{31}P. This rather limited development of ^2H NMR is explained by the strong superiority of the proton isotope in structural analysis which constitutes the main field of application of high-resolution NMR.

Observation of the ^2H resonance at the natural abundance level [51] suffers from sensitivity limitations due to both an intrinsic low receptivity of ^2H and a low natural abundance (Table 2). This renders ^2H NMR less sensitive than proton NMR by a factor of about 1.5×10^{-6}.

Since deuterium is a nucleus with a spin quantum number different from 1/2 ($I = 1$), it has an electric quadrupolar moment (Table 2). Its nuclear relaxation

Table 2. Properties of the two stable isotopes of hydrogen ^1H and ^2H. **a** μ_N is the nuclear Bohr magneton $\mu_N = 5.05095 \times 10^{-27}$ J.T^{-1} **b** Resonance frequency for a field ($B_o \approx 2.35$ T) in which the proton frequency is exactly 100 MHz. **c** Relative receptivity in the same field for an equal number of nuclei

	^1H	^2H
Natural abundance	99.985×10^{-2}	1.5×10^{-4}
Nuclear spin	1/2	1
Nuclear magnetic moment (in units of μ_N) (a)	4.8372	1.2126
Electric quadrupole moment (10^{-28} m^2)	—	2.8×10^{-3}
Gyromagnetic ratios (10^7 rad. s^{-1} T^{-1})	26.7522	4.1066
Resonance frequency (b) (MHz)	100	15.351
Intrinsic relative receptivity (c)	1.0	9.6×10^{-3}
Receptivity vs ^{13}C	5.68×10^3	8.2×10^{-3}

is therefore dominated by the relatively efficient quadrupolar mechanism, and nuclear Overhauser effects may be neglected (Sect. 3.2). These properties have important repercussions on the technical conditions of spectrum acquisition and on the methods of quantitative analysis (Sects. 3.3 and 3.4). In addition, since the predominance of the quadrupolar relaxation mechanism greatly facilitates the interpretation of the relaxation times in terms of molecular motion (Sect. 3.2), ^2H NMR is particularly suited for the study of rotational diffusion [52, 53].

It must be emphasized that chemical shift discrimination expressed in frequency units is 6.5 times lower in ^2H NMR than in proton NMR at the same magnetic field. At a proton nominal frequency of 400 MHz, for example, deuterium NMR has therefore a resolution power somewhat analogous to that of ^1H NMR at 60 MHz. As regards scalar coupling, the J(^1H, ^2H) values are 6.51 times smaller than the corresponding J(^1H, ^1H) values. The J(^2H, ^2H) coupling constants are generally negligible since they reach only 2.4% of the corresponding J(^1H, ^1H) values. Moreover, ^2H, ^2H coupling can be ignored at the natural abundance level of deuterium since the dideuterated species are very diluted ($\sim 2 \times 10^{-8}$).

Except for usually small intrinsic isotope shifts [54–56], the deuterium chemical shifts expressed in ppm are identical to the corresponding proton chemical shifts and all J(X, ^2H) values are to a good approximation predictable from the relationship:[1]

$$J(X, {}^2H) = \frac{\gamma_D}{\gamma_H} J(X, {}^1H) \qquad (14)$$

[1] For a discussion of primary isotope effects on scalar spin–spin coupling constants see N.M. Sergeyev, NMR – Basic Principles and Progress, *22*, 31 (1990).

From the point of view of structural analysis, the information provided by deuterium spectra of isotropic media is therefore largely redundant with that more easily obtained in proton spectroscopy. In this respect deuterium labelling at specific positions is sometimes exploited for estimating the corresponding proton chemical shifts and therefore for facilitating the analysis of complicated proton spectra. Such an approach has been used, for instance, in analyses of the spectra of nicotine [57] or in separate observations of the positions α to the carbonyl group in *trans*-decalone-1 [58]. Similarly, Eq. [14] enables the unobservable $J(^1H, ^1H)$ coupling constant between magnetically equivalent hydrogens to be calculated from the $J(^1H, ^2H)$ value determined for the ^2H-substituted compound. However, this coupling constant is usually more accurately measured on the proton spectrum. For example, a value $J(^1H, ^1H) = (-)\ 15$ Hz has been calculated in this way from the double quantum spectrum of an allenic ketone [59].

Deuterium labelling is also commonly employed for investigating chemical or biochemical reaction mechanisms and ^2H NMR is very convenient for directly observing the fate of the labels [52, 60] (Sect. 5).

Other applications more typical to deuterium NMR can be found in studies of systems undergoing relatively slow anisotropic motions [50, 52–54, 61–63]. In oriented fluids such as thermotropic liquid crystals or biological membranes, the ^2H spectra of partially deuterated molecules exhibit quadrupolar splittings whose magnitude depends on the nuclear quadrupole coupling constant and on an order parameter, S, which characterizes the restricted motion of the C–D bond. Appropriate analyses of the splitting patterns then enable segmental motions of the oriented molecules to be defined [63–67]. Orientation parameters of fatty acid chains in phospholipid structures, for example, have thus been precisely characterized by using ^2H spectra of selectively enriched species.

Finally, it should be mentioned that $^2H/^1H$ isotope effects on chemical shifts can be indirectly utilized to derive structural information from changes in the chemical shift of ^{13}C or other nuclei produced by deuterium substitution [68].

In the SNIF-NMR method, emphasis is laid on the quantitative dimension of deuterium NMR and the experimental techniques are largely dependent on the relaxation properties of the ^2H nucleus.

3.2 Relaxation Properties of ^2H and Nuclear Overhauser Effect

Although the quadrupole moment of the deuterium nucleus (Table 2) is weaker than that of most nuclei with a spin quantum number $I > 1/2$, the longitudinal or transversal relaxation rate constant, $1/T_1$ or $1/T_2$, is practically governed by the quadrupolar mechanism. For nuclei undergoing isotropic motion characterized by a correlation time τ_c the values of T_{1Q}^{-1} and T_{2Q}^{-1} are given by Eqs. (15) and (16):

$$\frac{1}{T_{1Q}} = \frac{6}{100}\pi^2\left(\frac{e^2q_{zz}Q}{h}\right)^2\left(1+\frac{\eta^2}{3}\right)\frac{2I+3}{I^2(2I-1)}\left(\frac{\tau_c}{1+\omega^2\tau_c^2}+\frac{4\tau_c}{1+4\omega^2\tau_c^2}\right) \tag{15}$$

$$\frac{1}{T_{2Q}} = \frac{3}{100}\pi^2\left(\frac{e^2q_{zz}Q}{h}\right)^2\left(1+\frac{\eta^2}{3}\right)\frac{2I+3}{I^2(2I-1)} \times\left(3\tau_c+\frac{5\tau_c}{1+\omega^2\tau_c^2}+\frac{2\tau_c}{1+4\omega^2\tau_c^2}\right) \tag{16}$$

$\chi = (e^2q_{zz}Q/h)$ is the quadrupole coupling constant expressed in frequency units. It characterizes the interaction between the nuclear electric quadrupole moment, Q, assumed to lie at a position of axial symmetry, and the single electric field gradient q_{zz} in its environment. The asymmetry parameter, η, which characterizes the lack of axial symmetry is negligible for $\geq$C–D fragments when z is the direction of the bond. The quadrupole coupling constant of ^{2}H takes values between 150 and 250 KHz.

Figure 6 represents the variation of the longitudinal relaxation time T_{1Q} as a function of τ_c for two typical values of the quadrupole coupling constant and for several magnetic field strengths corresponding to nominal proton frequencies of the spectrometer ranging from 200 to 500 MHz. For small molecules

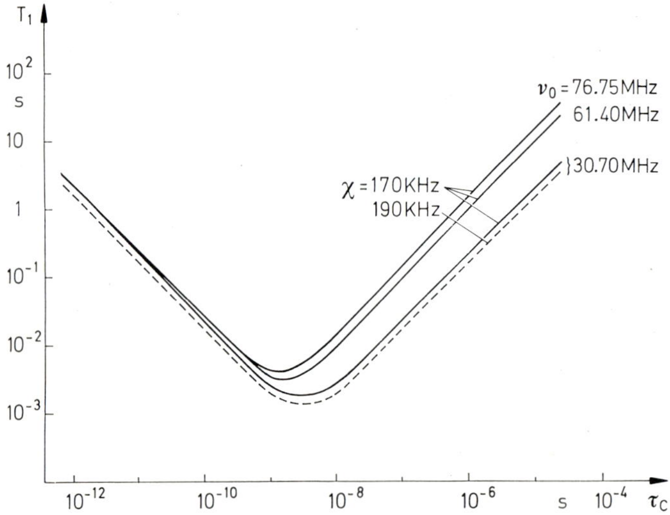

Fig. 6. Variation in the quadrupolar relaxation time of ^{2}H as a function of the correlation time τ_c. The results are presented on a log/log scale. The *solid* and *dashed* lines correspond to values of the quadrupolar coupling constant $\chi = 170$ kHz and $\chi = 190$ kHz, respectively. In the case $\chi = 170$ kHz, three values of the nominal spectrometer frequency are considered: $\nu_o = 200$ MHz (^{1}H)/30.7 MHz (^{2}H), $\nu_o = 400$ MHz (^{1}H)/61.4 MHz (^{2}H) and $\nu_o = 500$ MHz (^{1}H)/176.7 MHz (^{2}H). The curve for $\chi = 190$ kHz is associated with $\nu_o = 200$ MHz (^{1}H)/30.7 MHz (^{2}H)

reorienting relatively rapidly, molecular tumbling is characterized by τ_c values of the order of 10 ps. The relaxation time is then long enough to ensure convenient sampling of the free induction decay.

The influence of the magnetic field strength on the transverse relaxation time, T_{2Q}, corresponding to a typical value of the quadrupole coupling constant, $\chi = 170$ kHz, is represented in Fig. 7. The influence of an increase in the value of χ is also illustrated. From these curves it appears that the linewidths of the deuterium signals remain relatively moderate for small mobile molecules with tumbling in the range $\tau_c = 1$ to 10 ps. However, the signal rapidly broadens with the increase in the reorientational correlation time and therefore with an increase in the size of the molecule. Typical values of the quadrupole coupling constant χ [53] and of the longitudinal relaxation time T_1 are given in Table 3. It is worth noting that the intrinsic ^2H linewidths of molecules with a molecular weight less than 200 may not exceed 0.5 Hz. Broader signals are observed in hydrogen-bonded species and the lines pertaining to exchangeable deuterium atoms such as those of hydroxyl groups are particularly broadened.

Since the magnetic moment of the ^2H nucleus is significantly smaller than that of ^1H (Table 2), the dipolar interactions between a deuterium nucleus and the surrounding protons are expected to be significantly attenuated with respect to the corresponding $^1H \sim {}^1H$ interactions. More precisely, as the dipole-dipole interactions depend on the square of the product of the gyromagnetic ratios,

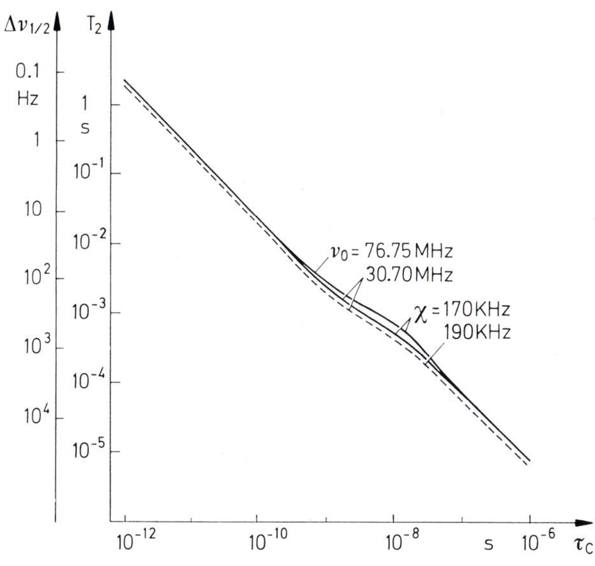

Fig. 7. Variation of the transverse relaxation time of ^2H and of the half-height linewidth, $\Delta\nu_{1/2}$, as a function of the correlation time τ_c. The results are presented on a log/log scale. The *solid lines* are calculated for a value of the quadrupolar coupling constant $\chi = 170$ kHz and for three values of the nominal spectrometer frequency: $\nu_o = 200$ MHz (^1H)/30.7 MHz (^2H), $\nu_o = 400$ MHz (^1H)/61.4 MHz (^2H) and $\nu_o = 500$ MHz (^1H)/76.7 MHz (^2H). The dashed line corresponds to $\chi = 190$ MHz and $\nu_o = 200$ MHz (^1H)/30.7 MHz (^2H)

Table 3. Nuclear quadrupole coupling constants, χ, and spin-lattice relaxation times, T_1, of deuterium in typical molecules. The experimental conditions for the measurement of χ are indicated in Refs. [52, 53]. Large differences are frequently exhibited in the published values. The retained values are the means over several determinations. The T_1 values have been determined at 61.4 MHz and 306 K by the IRFT method on neat compounds at the natural abundance level. Typical values of the correlation times are 0.8 to 1×10^{-12} s for CH_3CN and $(CH_3)_2CO$, 4×10^{-12} s for $(CH_3)_2SO$, and 0.4 to 0.8×10^{-11} s for H_2O [69]

Molecule		χ(kHz)	T_1(s)
CH_3CN		169	6.5
CH_3NO_2		164	5.0
$(CH_3)_2CO$		163	4.1
CH_2Cl_2		164	2.3
C_6H_{12}		174	1.8
C_6H_6		190	1.6
$(CH_3)_2SO$			1.0
$\overset{1}{C}H_3\overset{2}{O}H$	1	141	5.1
	2	192	0.4
$\overset{1}{C}H_3\overset{1}{C}H_2\overset{3}{O}H$	1	175	1.1
	2		1.25
	3		0.15
H_2O		215	0.5

the $^2H \sim {}^1H$ interactions are about 42 times weaker than the corresponding $^1H \sim {}^1H$ interactions. This limited efficiency of the dipolar mechanism compared to the fast quadrupolar relaxation usually justifies the use of Eqs. (15) and (16) for directly calculating local correlation times from measured values of the relaxation times. More complete formulas apply to the general case of anisotropic motions [69, 70] and relaxation measurements of several nuclei in the same molecule then enable the principal components of the rotational diffusion tensor to be estimated.

Another important consequence of the low relative efficiency of dipolar relaxation is the usual absence of NOE enhancements in proton-decoupled ^2H spectra. This property is highly beneficial for quantitative analysis of deuterium spectra. It avoids the need to resort to time-consuming quantitative techniques such as gated decoupling sequences.

3.3 Quntitative Deuterium NMR. Principle and Methods

The SNIF-NMR method exploits either relative or absolute values of the site-specific isotope ratios. When the isotope content in a given molecular site, A, is compared to that of a reference site, ref., in the same molecule, the method simply involves an accurate comparison of the area of signals A and ref. in a spectrum obtained under "quantitative conditions". For example, a relative parameter, r, can be defined from the intensities I_I and I_{II} of the CH_2D and CHD signals in the ^2H spectrum of ethanol:

$$r_{II} = 3\,\frac{I_{II}}{I_I} \tag{17}$$

This parameter represents the deuterium content in the methylene site, the methyl site being chosen as reference and arbitrarily given the statistical abundance factor 3. A random distribution of deuterium would be characterized by a relative methylene factor $r_{II} = 2$.

Absolute values of the specific ratio of site i in compound A, $(D/H)_i^A$, can be obtained either by using an intra- or intermolecular reference of known isotope content or by measuring independently the overall average isotope ratio of the compound, $\overline{(D/H)}^A$, by isotope mass spectrometry [71]. The first approach makes use of a working standard WS, the $(D/H)^{WS}$ value of which has been accurately determined on the V.SMOW/SLAP scale [41]. The working standard may be either introduced into a hermetic cell coaxial to the sample cell or weighed and dissolved into the sample.

In these methods the intensity, I_i^A, of the signal of site i in compound A is compared to that of the reference signal, I^{WS}:

$$\left(\frac{D}{H}\right)_i^A = \frac{I_i^A P^{WS} N_H^{WS}}{I^{WS} P_i^A N_H^A}\left(\frac{D}{H}\right)^{WS} \tag{18}$$

where P^A and P^{WS} are the stoichiometric numbers of hydrogens in sites i and WS, and N_H^A, N_H^{WS} are the numbers of fully protonated molecules of compounds A and WS in the active volume of the coil. When the coaxial arrangement is used, the isotope content in SMOW can also be referred to that of WS by using the same cell. Experimental parameters, T, are measured which represent the ratios of the deuterium contents in site i of A or in SMOW with respect to WS:

$$T_i^A = \frac{I_i^A}{I^{WS}} \quad \text{and} \quad T^{SMOW} = \frac{I^{SMOW}}{I^{WS}} \tag{19}$$

Since the active volumes of A and SMOW in these experiments are identical, the $(D/H)_i^A$ ratio can be calculated from the values of the specific weights d^A, d^{SMOW} and of the molecular weights M^A, M^{SMOW} [71]:

$$\left(\frac{D}{H}\right)_i^A = \frac{T_i^A}{T^{SMOW}}\,\frac{P^{SMOW}}{P_i^A}\,\frac{d^{SMOW}}{d^A}\,\frac{M^A}{M^{SMOW}}\left(\frac{D}{H}\right)^{SMOW} \tag{20}$$

When the working standard is mixed with sample A, the masses m^{WS} and m^A or the volumes V^{WS} and V^A must be precisely known:

$$\left(\frac{D}{H}\right)_i^A = \frac{P^{WS}}{P^A}\,\frac{m^{WS}}{m^A}\,\frac{M^A}{M^{WS}}\,T_i^A\left(\frac{D}{H}\right)^{WS} \tag{21}$$

A complementary measurement of $(D/H)^{WS}$ on the V.SMOW scale is then required. This can be achieved by ^{2}H NMR using a known mixture of WS and V.SMOW.

In certain cases it may also be valuable to use an intramolecular working standard introduced by reacting the investigated molecule with an appropriate reference substance. For example, when ethanol is transformed into acetate by reacting it with acetic anhydride, the isotope content in the ethyl sites from ethanol can be referred to that of the reference methyl group from acetic anhydride. In this "intramolecular" referencing method, possible kinetic isotope effects must be either negligible or precisely known. Denoting $(D/H)^{ref}$ the isotope ratio of the "internal" reference and $T^{A}_{i/ref}$ the ratio of the signal intensities of site i and ref., the isotope ratio of site i is then expressed as:

$$\left(\frac{D}{H}\right)^{A}_{i} = \frac{P^{ref}}{P^{A}_{i}} T^{A}_{i/ref} \left(\frac{D}{H}\right)^{ref} \tag{22}$$

Recourse to a reference NMR signal pertaining to a fragment of known isotope ratio can be avoided if the average isotope ratio $\overline{(D/H)}^{A}$ of the considered molecule, A, is independently measured by isotopic mass spectrometry. In this technique the sample is burned and the resulting water reduced to hydrogen gas. Only an average isotope ratio of A is therefore obtained and the site-specific parameters are computed from Eq. (9).

3.4 Experimental Techniques. Automation of the Analytical Chain

Obtaining accurate values of the isotope ratios requires very careful experimental procedures [72a].

At moderate field strengths corresponding to nominal ^{1}H frequencies in the range of 200–300 MHz, field drift occurring in the absence of field-frequency locking usually remains very limited and does not introduce too much broadening of the signals. At a deuterium frequency of 61.4 MHz, drift is more awkward and a ^{19}F locking device is desirable. In all cases, the implementation of an automatic sample changer on the spectrometer requires efficient field-frequency locking. Broad-band proton decoupling is applied to remove line-splittings and field homogeneity must be carefully adjusted. Low power decoupling techniques such as the WALTZ pulse sequence [247] generate less heat and temperature gradients and thus ensure a satisfactory reproducibility of the results, especially in large sample tubes (10 mm). In some instances, a Dante Water Eliminated Fourier Transform pulse sequence DASWEFT [248] can be used for attenuating the signal intensity of the solvent. The acquisition time should be sufficiently long in order to guarantee good peak definition. Phase adjustment is also particularly crucial. Owing to the small frequency range of the deuterium resonances, a frequency window of 1200 Hz is sufficient at 61.4 MHz and a 16 K memory size then corresponds to a 6.8 s acquisition time, i.e., a digital resolution of 0.15 Hz. This value of the acquisition time is higher than 3 to 5 times the longitudinal relaxation times of most deuterium nuclei and is usually sufficient for ensuring complete relaxation between successive pulses. For example, for a simple of 95% ethanol investigated at 46.1 MHz in a 10-mm tube, a signal-to-noise ratio of at least 140 can be reached after NS = 300 accumulations. The experiment is repeated up to ten times when very precise results are desired. Exponential multiplication of the free induction decay associated with a line-broadening of 1 Hz or 2 Hz significantly increases the signal-to-noise ratio and is usually beneficial to the accuracy of the quantitative determinations.

The isotopic parameters exhibit a great sensitivity to experimental settings. For example, the value of the relative intensity ratio of ethanol, r_{II}, expressed in terms of signal heights [Eq. (17)] is strongly dependent on the adjustment of the phase and on the decoupling parameters; r_{II} varies from 2.747 to 2.751 when the decoupling offset is changed by 2000 Hz (at 61.4 MHz) and from 2.753 to 2.733 when the decoupling power is reduced by 0.5 W from the optimum value. A change of $-10°$ from the optimum zeroth order phase correction is associated with a variation of r_{II} from 2.737 to 2.754.

Since the SNIF-NMR method is likely to be used routinely in control laboratories, efforts have been devoted to automating the technical procedures and the analyses. A sample-changing robot can be used and a dedicated PASCAL software used with the DISNMR program of Bruker spectrometers, for example, directly supplies the values of the statistical parameters in terms of mean specific isotope ratios, standard deviations, Student estimations for a given confidence interval etc. These NMR data are transferred, along with all technical parameters concerning the sample, to a host computer which also holds a local data bank. A dedicated program, ISOLOG enables a statistical analysis of the isotopic data to be carried out [73]. A diagnosis procedure involves the following computerized steps: data collection and transfer, search for typical sub-sets of authentic samples from similar origin, check for environmental consistency, and discriminant analysis of the results [74].

When more complex molecules than ethanol are studied, i.e., sugars, terpenes etc., it is recommended that a careful analysis of the signals be carried out using dedicated software to obtain accurate quantitative data.

3.5 Precision and Accuracy

Since the SNIF-NMR method has now been adopted by the International Office of Vines and Wine (OIV) and by the EEC as an official method for determining the enrichment of wines with sugar, it is essential that reasonable repeatability and reproducibility of the NMR results are achieved. An intercomparison exercise was organized in 1986 under the responsibility of the Community Bureau of References (BCR) in Brussels [75].

The repeatability, r_i, of the measurement corresponding to the ith molecular site is defined as:

$$r_i = 2.83\, Sr_i \tag{23}$$

the variance of the repeatability Sr_i^2 being computed from the variance S_{ij}^2 of the $j = 1$ to m series of measurements:

$$S_{r_i}^2 \frac{1}{m} \sum_{j=1}^{m} S_{ij}^2 \tag{24}$$

with

$$S_{ij}^2 = \frac{1}{N-1} \sum_{k=1}^{n} (T_{ijk} - \bar{T}_{ij})^2 \tag{25}$$

In the present inter-laboratory comparison, m is the number of experiments performed on the same sample in the same laboratory over a short period of time and N is the number of spectra run in each experiment. T_{ijk} is the value of the intensity parameter [Eq. (19)] of the ith site, obtained from the kth spectrum in the jth series of experiments, and $\bar{T}_{ij}$ is the mean value over the $k = 1$ to N spectra in this j^{th} series. Sr_i can be considered as the mean of the intra-laboratory variance.

The reproducibility, R_i, is defined from both intra- and inter-laboratory variance parameters:

$$R_i = 2.83\, (S_{Li}^2 + S_{r_i}^2)^{1/2} \tag{26}$$

The corrected inter-laboratory variance S_{Li}^2 is computed from the inter-laboratory variance, S_{Loi}^2, normally calculated for the $j = 1$ to q participants:

$$S_{Li}^2 = S_{Loi}^2 - S_{r_i}^2/N \tag{27}$$

$$S_{Loi}^2 = \frac{1}{q-1} \sum_{j=1}^{q} (T_{ij} - \bar{\bar{T}}_i)^2 \tag{28}$$

$\bar{\bar{T}}_i$ denotes the mean value over the $j = 1$ to q series of experiments performed in the different laboratories.

Sixteen different spectrometer systems installed in fifteen laboratories in Belgium, Denmark, France, Germany, Ireland, Netherlands, and the United Kingdom participated in this collaborative study. Three ethanol samples originating from beet-root, grape and maize were distributed and three isotopic parameters were considered: the relative parameter r_{II}^A [Eq. (17)] and the experimental ratios T_I and T_{II} [Eq. (19)] corresponding to the methyl and methylene sites respectively. The results selected in Table 4 show that confidence intervals of the order of 0.25% can be achieved in the determination of relative intensity ratios. The repeatability defined according to the ISO-NORMS [Eqs. (23–25)] can be better than 1%. Even with different spectrometer configurations and different operators the reproducibility [Eqs. (26–28)] remains quite satisfactory.

Table 4. Absolute (Abs) and relative (%) values of the repeatability, r_i, and reproducibility, R_i, of the isotopic parameters measured for a sample of maize ethanol in a given laboratory (A) and in 14 different laboratories (all). **a** The repeatability was not determined in every laboratory. It is in fact included in the reproducibility determinations. **b** The reproducibility $\mathbf{R_i}$ was calculated in two situations—for series of experiments performed in the same laboratory A over a long period of time (2 years) and for series of experiments performed in different laboratories [75]

		r_{II}		T_I		T_{II}	
		A	All	A	All	A	All
Means		2.318	2.317	0.954	0.954	0.737	0.737
Confidence interval		0.006	0.009	0.002	0.004	0.002	0.002
Repeatability[a] r_i	Abs	0.024	a)	0.009	a)	0.008	a)
	%	1.0		0.9		1.0	
Reproductibility[b] R_i	Abs	0.029	0.058	0.010	0.020	0.010	0.018
	%	1.2	2.5	1.0	2.0	1.3	2.4

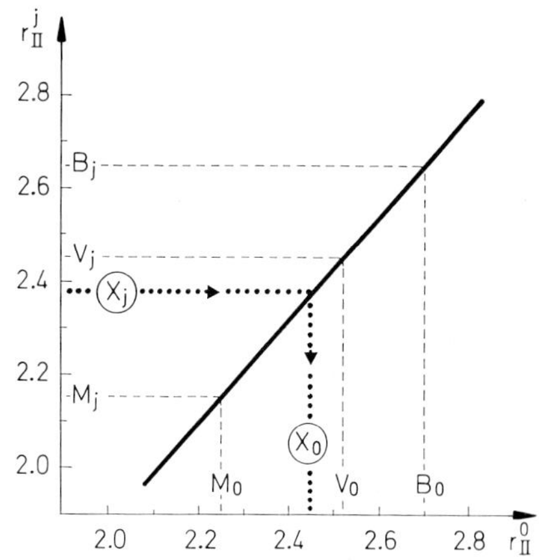

Fig. 8. Standardization of the isotope ratios measured on an unknown sample X in a given laboratory j. B and M and V are three reference samples the isotopic parameters of which (B_o, M_o, and V_o) have been accurately determined in a collaborative study and are certified by the Community Bureau of Reference. B_j, M_j, and V_j are the values measured in laboratory j: The "true" value for the unknown sample, X_o, is obtained from the measured value X_j by resorting to the interpolation line. The diagram is for an example involving the isotope parameter r_{II} [Eq. (17)]

It must be emphasized that the precision reached in routine analysis in a given laboratory using defined protocols, is high enough to detect the presence of less than 5% v/v of ethanol issued from beet-sugar in a sample of pure grape ethanol (Sect. 7). A standardization of the isotope ratios measured on an unknown sample X in a given laboratory, j, can be realized by resorting to the interpolation procedure defined in Fig. 8 on the condition that reference materials, such as B, M, and V have been previously certified.

4 Site-Specific Natural Isotope Fractionation in the Determination of Kinetic and Thermodynamic Isotope Effects

Site-specific Natural Isotope Fractionation as studied by NMR provides a very attractive approach for determining isotope effects. This method presents two important advantages with respect to conventional techniques. It avoids the recourse to selectively enriched materials and it enables all types of primary and secondary isotope effects to be determined in a one-pot experiment. The method is very easy to manage and we may expect it to become standard usage in laboratories concerned with the study of reaction mechanisms exploiting the information provided by deuterium isotope effects.

The general equations of kinetics, as summarized by Melander, for instance [76, 77], readily apply to the conditions of the SNIF-NMR experiments. In order to illustrate the advantages of the method, we shall consider the general case of reactions involving the substitution or the elimination of a hydrogen atom from a group of n equivalent hydrogens. Such reactions have already been discussed by Melander from a kinetic point of view [76] and we shall only recall some basic equations in order to give a better appraisal of the facilities and complementary information provided by SNIF-NMR.

4.1 *Natural Isotope Labelling and Rate Equations*

When a hydrogen is abstracted from a set of n equivalent hydrogens in a reactant AH_nBH_m possessing a group of m equivalent hydrogens in a position α to the substituted site, several isotope effects simultaneously intervene as shown in the following scheme:

$$\underset{(A_H)}{AH_nBH_m} + \cdots \xrightarrow{nk_H} \underset{(Y_H)}{YH_{n-1}BH_m} + ZH \qquad (a)$$

$$\underset{(A_D)}{AH_{n-1}DBH_m} + \cdots \xrightarrow{(n-1)k_H^{\alpha}} \underset{(Y_D)}{YH_{n-2}DH_m} + ZH \qquad (b)$$

$$\underset{(A_D)}{AH_{n-1}DBH_m} + \cdots \xrightarrow{k_D} \underset{(Y_D^*)}{YH_{n-1}BH_m} + ZD \qquad (c)$$

$$\underset{(A'_D)}{AH_nBH_{m-1}D} + \cdots \xrightarrow{nk_H^{\beta}} \underset{(Y'_D)}{YH_{n-1}BH_{m-1}D} + ZH \qquad (d)$$

In conventional experiments only the competitive reactions **a–c** involving the fully protonated species, A_H, and a selectively deuterated species A_D, are usually concerned. These reactions are characterized by kinetic rate constants, k_H for the substitution of a hydrogen in the fully protonated species, k_D for the substitution of the corresponding deuterium atom, and k_H^α for the substitution of a hydrogen atom in the monodeuterated molecule A_D.

In the SNIF experiment, various types of monodeuterated species are naturally present in trace amount in the reaction medium. Thus, we have specified the case of the A'_D species which contains a deuterium in a position α to the site of substitution (reaction **d**). In such molecules the deuterium atom may introduce a secondary isotope effect characterized by a rate constant k_H^β. More generally, since all possibilities of monodeuteration naturally occur in the medium, all kinds of secondary isotope effects can be simultaneously investigated in the same experiment.

As an example, if we consider the substitution of a methylenic hydrogen in a compound A of type CH_3CH_2R giving the product CH_3CHYR denoted Y_H, the reacting medium naturally contains trace amounts of four monodeuterated species: CH_3CHDR (A_D), CH_2DCH_2R (A'_D), CH_3CDYR (Y_D), and CH_2DCHYR (Y'_D). Reactions **a** to **d** are then characterized by the rate constants $2k_H$ for **a**, k_H^α for **b**, k_D for **c**, and k_H^β for **d**. The determination of the kinetic isotope effects will be fully developed in this case but the treatment can be easily generalized.

Since NMR signal area benefits from the convenient property of being simply proportional to the number of nuclei engaged in the resonance, they are easily related to the numbers of molecules, N, of the corresponding species.

4.2 Kinetic Equations

4.2.1 Intermolecular Competition

In the considered reactions, the reactant molecules A_H, A_D, and A'_D are subject to intermolecular competition and the rates of the reactions can be written as:

$$-\frac{dN_H^A}{dt} = nk_H N_H^A f(N_X \ldots) \tag{29}$$

$$-\frac{dN_D^A}{dt} = [k_D + (n-1)k_H^\alpha]\, N_D^A f(N_X \ldots) \tag{30}$$

$$-\frac{dN_D^{A'}}{dt} = nk_H^\beta N_D^{A'} f(N_X \ldots) \tag{31}$$

The numbers of product molecules N_H^Y, N_D^Y, and $N_D^{Y'}$ can be expressed in terms of the initial numbers, N_H^{Ao}, N_D^{Ao}, $N_D^{A'o}$ and of the remaining numbers N_H^A, N_D^A, $N_D^{A'}$, of the reactant molecules by:

$$N_H^Y = N_H^{Ao} - N_H^A; \quad N_D^Y = N_D^{Ao} - N_D^A; \quad N_D^{Y'} = N_D^{A'o} - N_D^{A'} \tag{32}$$

In tracer conditions the fraction of fully protonated product, Y_H, represents the fraction of reaction ρ:

$$\frac{N_H^Y}{N_H^{A_0}} = \frac{N_H^{A_0} - N_H^A}{N_H^{A_0}} = \rho \quad \text{and} \quad \frac{N_H^A}{N_H^{A_0}} = 1 - \rho \tag{33}$$

From Eqs. (29) and (31) it follows that:

$$\frac{1}{nk_H}\frac{dN_H^A}{N_H^A} = \frac{1}{k_D + (n-1)k_H^\alpha}\frac{dN_D^A}{N_D^A} = \frac{1}{nk_H^\beta}\frac{dN_D^{A'}}{N_D^{A'}} \tag{34}$$

and by integrating, taking into account the initial conditions, we obtain:

$$\frac{k_D + (n-1)k_H^\alpha}{nk_H} = \frac{\mathrm{Ln}(N_D^A/N_D^{A_0})}{\mathrm{Ln}(1-\rho)} \tag{35}$$

$$\frac{k_H^\beta}{k_H} = \frac{\mathrm{Ln}(N_D^{A'}/N_D^{A'_0})}{\mathrm{Ln}(1-\rho)} = \frac{\mathrm{Ln}[(N_D^{A'_0} - N_D^{Y'})/N_D^{A'_0}]}{\mathrm{Ln}(1-\rho)} \tag{36}$$

From these equations we derive the kinetic laws describing the variations in the concentrations as a function of the fraction of reaction for the remaining monodeuterated reactant molecules A_D and A'_D and for the product molecules, Y'_D, deuterated in α-position to the substituted site:

$$N_D^A/N_D^{A_0} = (1-\rho)^{1/n[(k_D/k_H) + ((n-1)k_H^\alpha/k_H)]} \tag{37}$$

$$N_D^{A'}/N_D^{A'_0} = (1-\rho)^{k_H^\beta/k_H} \tag{38}$$

$$N_D^{Y'}/N_D^{A'_0} = [1 - (1-\rho)^{k_H^\beta/k_H}] \tag{39}$$

4.2.2 Intramolecular Competition

When the reaction site bears more than one hydrogen, the presence of both hydrogen and deuterium atoms in the monodeuterated species is the origin of a simultaneous intramolecular competition:

$$\frac{dN_D^Y}{dt} = (n-1)k_H^\alpha N_D^A f(N_X\ldots) \tag{40}$$

$$\frac{dN_H^{Y*}}{dt} = k_D N_D^A f(N_X\ldots) \tag{41}$$

In these conditions the ratio between the concentrations of the products issued from the competing reactions is constant over the time and:

$$\frac{(n-1)k_H^\alpha}{k_D} = \frac{N_D^Y}{N_H^{Y*}} \tag{42}$$

Since: $N_H^{Y*} = N_D^{A_0} - N_D^A - N_D^Y$ (43)

$$\frac{k_D + (n-1)k_H^\alpha}{(n-1)k_H^\alpha} = \frac{N_D^{A_0} - N_D^A}{N_D^Y} \tag{44}$$

and:
$$\frac{N_D^Y}{N_D^{A_0}} = \frac{(n-1)k_H^\alpha}{k_D + (n-1)k_H^\alpha}\left[1 - (1-\rho)^{1/n((k_D/k_H)+(n-1)(k_H^\alpha/k_H))}\right] \tag{45}$$

By combining Eqs. (34) and (42), the kinetic isotope effects k_H/k_D and k_H/k_H^α can be expressed separately:

$$\left(\frac{k_H}{k_H^\alpha}\right)^{-1} = \frac{n}{(n-1)}\frac{N_D^Y}{N_D^{A_0} - N_D^A}\frac{\mathrm{Ln}(N_D^A/N_D^{A_0})}{\mathrm{Ln}(1-\rho)} \tag{46}$$

$$\left(\frac{k_H}{k_D}\right)^{-1} = n\frac{\mathrm{Ln}(N_D^A/N_D^{A_0})}{\mathrm{Ln}(1-\rho)}\left[1 - \frac{N_D^Y}{N_D^{A_0} - N_D^A}\right] \tag{47}$$

4.3 Relationships Between the Site-Specific Isotope Ratios $(D/H)_i$ and the Kinetic Parameters

In a SNIF-NMR experiment we determine absolute or relative values of the site-specific isotope ratios $(D/H)_i$ by using the most appropriate referencing procedure. When the reactant molecule contains one or several sites, sufficiently remote from the reaction site not to be subjected to any detectable secondary isotope effect, such sites can be used as an intramolecular reference. In contrast, when the isotope ratios of all the sites are likely to undergo non-negligible variations, the determination of the deuterium contents requires the use of an intermolecular reference either contained in an external capillary or added to the considered medium according to the procedures described in Sect. 3.3.

In order to quantify the variations in the isotope ratios as a function of the fraction of reaction and to specify the relationship which enables the kinetic isotope effects to be derived from the NMR spectra, we only need to express the $(D/H)_i$ ratios as a function of the mole numbers. Remembering that (Sect. 3.3):

$$\frac{(D/H)_i^A}{(D/H)_j^B} = \frac{P_j^B}{P_i^A}\frac{N_H^B}{N_H^A}\frac{N_D^A}{N_D^B} \tag{48}$$

the kinetic Eqs. (34–39) and (42–47) can be written in terms of the site-specific isotope ratios.

Thus for the reactions considered above the kinetic isotope parameters can be deduced from the three Eqs. (46, 47 and 36), respectively:

$$\left(\frac{k_H}{k_D}\right)^{-1} = \left\{1 + \frac{\mathrm{Ln}[(D/H)_A/(D/H)_{A_0}]}{\mathrm{Ln}(1-\rho)}\right\} \times \left\{n - \frac{(n-1)\rho(D/H)_Y/(D/H)_{A_0}}{1-(1-\rho)[(D/H)_A/(D/H)_{A_0}]}\right\} \tag{49}$$

$$\left(\frac{k_H}{k_H^\alpha}\right)^{-1} = \left\{1 + \frac{\mathrm{Ln}\,[(D/H)_A/(D/H)_{A_0}]}{\mathrm{Ln}\,(1-\rho)}\right\} \times \left\{\rho\,\frac{(D/H)_Y/(D/H)_{A_0}}{1-(1-\rho)[(D/H)_A/(D/H)_{A_0}]}\right\} \tag{50}$$

$$\left(\frac{k_H}{k_H^\beta}\right)^{-1} = 1 + \frac{\mathrm{Ln}\,[(D/H)_{A'}/(D/H)_{A'_0}]}{\mathrm{Ln}\,(1-\rho)} = \frac{\mathrm{Ln}\,\{1-\rho[(D/H)_{Y'}/(D/H)_{A'_0}]\}}{\mathrm{Ln}\,(1-\rho)} \tag{51}$$

4.4 *Evolution of the Site-Specific Isotope Ratios in the Course of the Reaction*

The variations in the site-specific deuterium contents of the reactants and of the products as a function of the fraction of reaction are easily expressed from the kinetic Eqs. (37–39) and (45) taking into account the relationship between the isotope ratios and the concentrations of the monodeuterated and fully protonated molecules [Eq. (48)]. Thus if we consider the reacting site of the remaining reactant, A, its isotope ratio referred to that in the initial state, A_0, depends on the primary and secondary α isotope effects according to:

$$\frac{(D/H)_A}{(D/H)_{A_0}} = (1-\rho)^{[(1/n)(k_D/k_H)+((n-1)/n)(k_H^\alpha/k_H)-1]} \tag{52}$$

This behavior is illustrated in Fig. 9a for a reaction involving the substitution of an hydrogen in a methylene group (n = 2). The variation of the methylene isotope ratio of the remaining reactant is represented for different combinations of the primary, k_H/k_D, and secondary, k_H/k_H^α, isotope effects. We shall consider, for example, the case of a compound A, with a methylenic group characterized by a deuterium content of 130 ppm, undergoing a chemical transformation subjected to a primary kinetic isotope effect of 4 and to a secondary effect of 1.2. At a 20% conversion, the isotope content of this compound is already enhanced to 144 ppm.

Similarly, the isotope ratio of the substituted site in the product, Y, varies according to Eq. (53):

$$\frac{(D/H)_A}{(D/H)_{A_0}} = \frac{n}{\rho}\,\frac{k_H^\alpha/k_H}{(k_D/k_H)+(n-1)k_H^\alpha/k_H}\left[1-(1-\rho)^{((1/n)(k_D/k_H)+((n-1)/n)(k_H^\alpha/k_H))}\right] \tag{53}$$

The evolution of the isotope ratio in the product as a function of the fraction of reaction is given in Fig. 9b for the considered example (n = 2). At low conversions and in the absence of secondary kinetic isotope effects, the isotope

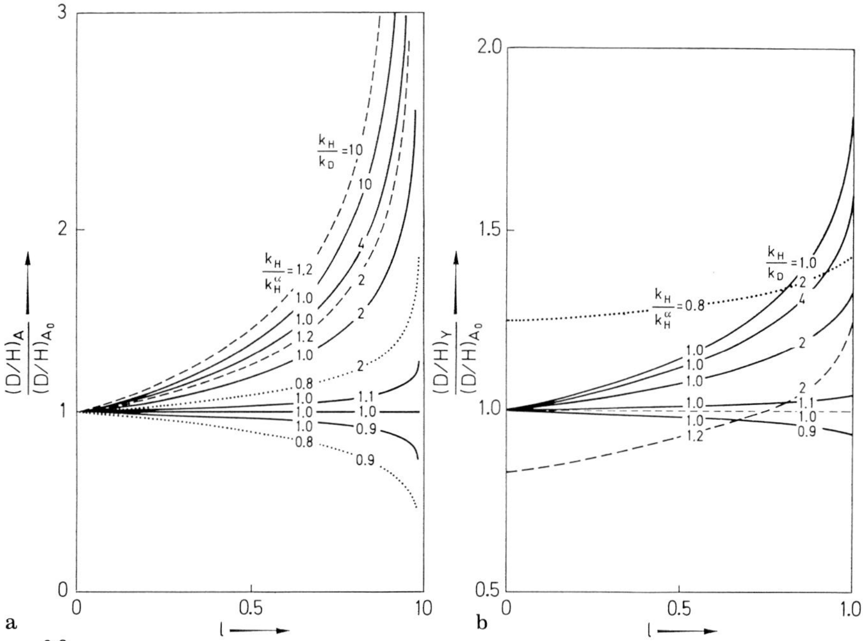

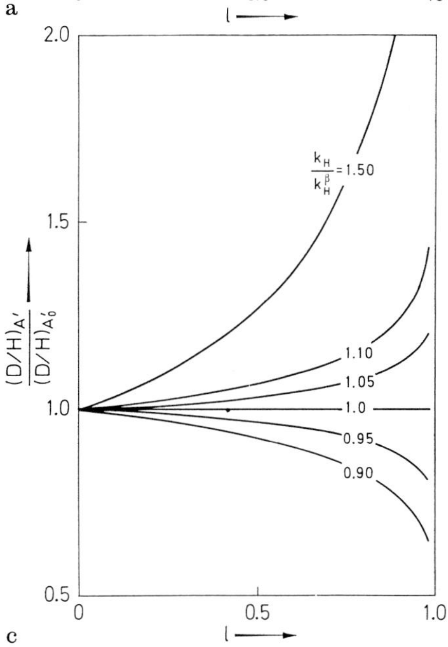

Fig. 9. Dependency of hydrogen isotope ratios on the fraction of reaction, ρ, for a reaction involving the substitution of a methylenic hydrogen. Different combinations of the kinetic isotope effects k_H/k_D, k_H/k_H^α, and k_H/k_H^β are considered. **a** Isotope ratio of the methylene site in the remaining reactant. **b** Isotope ratio of the substituted site in the product. **c** Isotope ratio of the site adjacent to the substituted group in the remaining reactant

content of the substituted product remains very close to that of the reactant whatever the value of the primary kinetic isotope effect. In contrast, the isotope ratio of the product is very sensitive to the secondary isotope effect. For a starting methylene group with an isotope content of 130 ppm, subjected to a secondary isotope effect $k_H/k_H^2 = 1.2$, the isotope ratio of the product is lowered to values varying from 108 to 114 ppm when the amount of reaction increases from 0 to 20% and when the value of the primary kinetic isotope parameter is equal to 4. The variation remains very similar for a primary isotope effect of 8 since it ranges from 108 to 115 ppm in the same conditions.

On the contrary, the primary isotope effect exerts a strong influence on the isotope ratio of the substituted product when the reaction approaches completion.

As regards the molecular group adjacent to the reacting site in the remaining reactant A′ or in the product Y′, no intramolecular competition is involved and the isotope contents only depend on a secondary β isotope effect, k_H/k_H^β:

$$\frac{(D/H)_{A'}}{(D/H)_{A'_0}} = (1-\rho)^{((k_H^\beta/k_H)-1)} \tag{54}$$

$$\frac{(D/H)_{Y'}}{(D/H)_{A'_0}} = \frac{1}{\rho}[1-(1-\rho)^{(k_H^\beta/k_H)}] \tag{55}$$

These isotope ratios exhibit the simple evolution represented in Fig. 9c for several values of the secondary β kinetic isotope effect in the case of the remaining reactant. Significant variations in the isotope ratio are observed at conversions rates higher than 50% when the secondary isotope effect is different from unity. For example, we may consider the isotopic behavior of a methyl group characterized by an isotope content of 110 ppm and adjacent to the substitution site of a molecule reacting with a secondary kinetic isotope effect of 1.05. When the degree of conversion is lower than 20%, the isotopic variation remains below 1 ppm but the isotope ratio reaches 115 ppm at a degree of conversion of 60% and 119 ppm at 80%.

4.5 *Applications of SNIF-NMR to the Determination of Kinetic Isotope Effects*

The theoretical methodology described in the previous section for the case of the abstraction of a hydrogen from a set of 2 equivalent hydrogens is readily extended to other types of reactions. Once the rate equations have been derived for all kinds of fully protonated and monodeuterated species present in the medium, the concentrations of the various molecular species can be directly translated in terms of isotope ratios, as shown above, by using Eq. (48).

If we consider the halogenation of the ethyl fragment in a compound of type CH_3CH_2R, the determination of the kinetic isotope effects can be performed on the basis of absolute or relative values of the isotope contents measured on the reactant before starting the reaction (A_0) and then at one or several stages of the reaction. Thus, the deuterium spectra of CH_3CH_2R enables the initial values of the isotope ratios of the methylene and methyl sites, $(D/H)_{CH_3^0} = (D/H)_{A_0}$, and $(D/H)_{CH_3^0} = (D/H)_{A'_0}$, to be determined as described in Sect. 4.4. Then values of the parameters $(D/H)_A$, $(D/H)_{A'}$ corresponding to the methylene and methyl site of the remaining reactant and/or $(D/H)_Y$, $(D/H)_{Y'}$, corresponding to the methine and methyl sites of the product, $CH_3CHYR(Y)$, can be measured at a given fractional amount of conversion, ρ. The primary, k_H/k_D, and secondary k_H/k_H^α, k_H/k_H^β, kinetic isotope effects are simultaneously obtained from the same set of experiments by applying Eqs. (49–51). The accuracy of the kinetic parameters depends on that of the isotope ratios and it is advisable to make use of mean values derived from several spectra of the same species and preferably from several measurements corresponding to different values of the fraction of reaction. The possibility of measuring the isotope contents in two reaction products, Y and Z, obviously constitutes a favorable situation.

The differences exhibited by the kinetic properties of compounds containing different hydrogen isotopes are mainly due to the difference in their masses which in turn influences the vibrational motions and the zero-point vibrational energies [76]. The investigation of kinetic hydrogen isotope effects using selectively labelled starting materials has therefore been largely exploited to gain information on the mechanisms of chemical or biochemical reactions and in particular to investigate the structure of the transition states [76–79]. Unfortunately, the conventional methods for the determination of kinetic isotope effects are usually very time-consuming since they require the preliminary synthesis of selectively labelled materials. Moreover, several experiments are necessary in order to estimate the various kinetic isotope effects. A number of isotope effects have thus been investigated by performing independent rate measurements for the reactions of the fully protonated and of a selectively labelled reactant. However, since it is not easy to reproduce identical experimental conditions for successive kinetic runs, the accuracy of such absolute rate measurements is low. More accurate values of the kinetic isotope effects associated with a single label have been achieved by resorting to an intermolecular competition realized by mixing the fully protonated and a selectively labelled reactant. Another successful approach consists in preparing appropriate mixtures of mono-, di-, tri-, and non-deuterated species. The phase-transfer-catalyzed chlorination of toluene, for example, has been investigated in this way [80]. A large value (5.90) of the primary kinetic isotope effect and low values (1.03) of the secondary effects were calculated from measurements performed on mixtures of $PhCH_3/PhCH_2D$ or $PhCH_2D/PhCHD_2$, $PhCHD_2/PhCD_3$. Such results suggest extensive breaking of the reacting C–H bond in the transition state and relatively little rehybridization and benzylic resonance stabilization at the reacting carbon.

The NMR methodology which is based, as described above, on the determination of site-specific isotope ratios at the natural abundance level, obviously presents definite advantages with respect to the traditional techniques of enrichment and it applies readily to the study of chemical [44] or biochemical [40] reaction mechanisms. The discriminating character of the natural deuterium distribution is the basis for an attractive labelling technique which avoids the need for selective enrichment [44] and provides a simultaneous access to the various primary and secondary isotope effects.

A very good agreement has been found between the natural abundance ^{2}H-NMR determinations and the results of conventional methods [81]. Thus, in a study of the insertion of a carbon into a C–H bond of cyclohexane, Pascal et al. have confirmed a value $k_H/k_D = 2.1(\pm 0.1)$ for the isotope effect associated with the cleavage of a C–H bond [81, 82]. Similarly, the value $k_H/k_D = 7.9(\pm 0.8)$ calculated for the dehydrobromination of (2-bromoethyl)-benzene [82] is in excellent agreement with the literature value [83]).

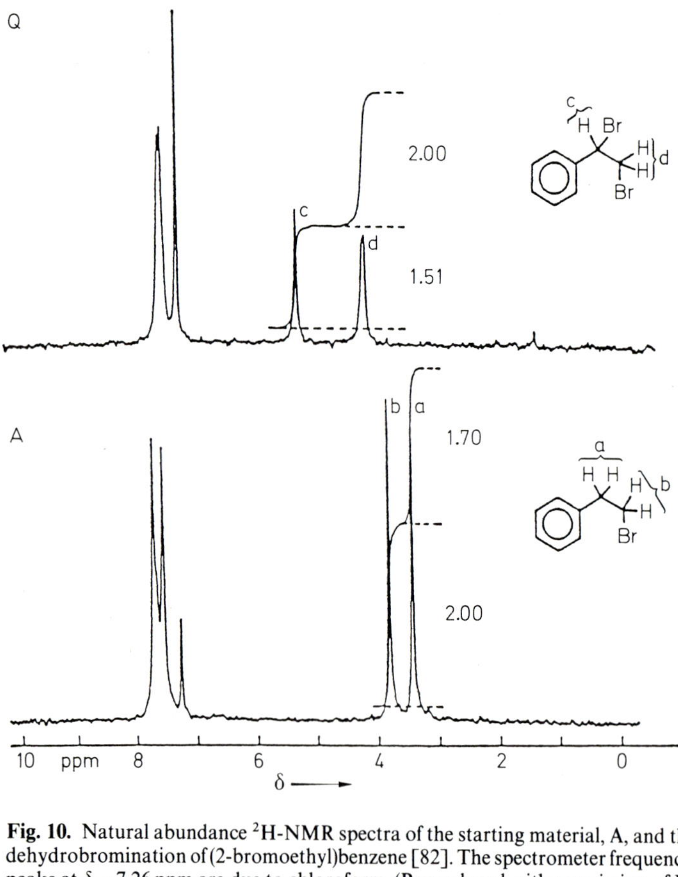

Fig. 10. Natural abundance ^{2}H-NMR spectra of the starting material, A, and the products, Q, of the dehydrobromination of (2-bromoethyl)benzene [82]. The spectrometer frequency was 38.4 MHz. The peaks at $\delta = 7.26$ ppm are due to chloroform. (Reproduced with permission of Dr. R.A. Pascal et al.)

$$\underset{(A)}{C_6H_5CH_2CH_2Br} \xrightarrow{tBuOK} C_6H_5CH=CH_2 \xrightarrow{Br_2} \underset{(Q)}{C_6H_5CHBrCH_2Br}$$

The deuterium signal of the site CH_2Br in the starting compound, A, and in the end product, Q, is used as a reference for the quantitative determinations and the kinetic isotope ratio is directly calculated from the area of the deuterium signals pertaining to the sites CH_2 of A and CHBr of Q (Fig. 10).

Various other reactions such the hydrolysis of Grignard reagents [84], the conversion of glucose into ethanol [40], the dimerization of ketene, and the biosynthesis of pinenes [82] and limonene [85] have been investigated.

Both primary and secondary kinetic isotope effects could be calculated by Zhang et al. in reactions such as the bromination of ethylbenzene, the chlorination of benzene, the oxidation of benzylalcohol and the chlorination of 1-2 dichloroethane [86, 87]. For example, in the study of the bromination of ethyl benzene, the deuterium signal of the aromatic ring could be used as an internal reference. Measurements of the area of the methyl and methylene signals of starting and remaining ethylbenzene and of the methyl and methine signals of the α bromobenzene product enabled the following values of the kinetic isotope effects to be calculated (uncertainty in parenthesis): $k_H/k_D = 2.39$ (< 1), $k_H/k_H^{\alpha} = 0.84$ (0.03), $k_H/k_H^{\beta} = 1.0$ (0.02) [87].

The various examples already considered are fairly illustrative of the wide interest in this approach which may constitute an efficient substitute to the traditional labelling techniques when sufficiently resolved deuterium spectra are obtainable.

4.6 Thermodynamic Fractionation Factors in Exchange Equilibria

Isotope effects are known to introduce differences in the isotopic composition of molecular species engaged in exchange equilibria. Thus, in aqueous solutions of alcohols, deviations with respect to a statistical distribution of deuterium between water and the alcoholic hydroxyl have been detected. Deuterium exhibits a significant preference for the hydroxyl bond and an equilibrium constant $K_1 = 1.09$ (± 0.02) has been determined, for example, for the methanol-water system [88] at 298 K:

$$2CH_3OH + D_2O \overset{K_1}{\rightleftharpoons} 2CH_3OD + H_2O$$

Equilibrium isotope effects reflect differences between the local force fields at the site of isotopic substitution in the exchanging molecules and a precise knowledge of the partitioning of deuterium and protium at equilibrium may be useful for a detailed interpretation of reaction mechanisms in systems containing exchangeable groups.

Several experimental methods have been used for investigating thermodynamic isotope effects. These methods do not provide a direct and simultaneous observation of the exchanging species and they frequently require tedious and time-consuming measurements. Thus a great deal of effort has been devoted to the determination of equilibrium constants, by using a combination of independent mass spectrometry measurements of isotope separation factors between hydrogen gas and liquid water or liquid alcohol (89), or by deriving these factors from various-vapor-pressure [90] calorimetric [91] distillation-density [92] and infrared [93] measurements, or from statistical mechanics calculations [94].

NMR methods have also been used for determining isotope effects on exchange equilibria [95]. Thus, the fractionation factor can be obtained by monitoring the chemical shift δ_H of the exchangeable protons referred to an external reference as a function of the mole fraction f_L or f_H of solute, respectively, in light water or in heavy water of known deuterium content [96–100]. The fractionation factor is then calculated from the slopes of the lines relating δ_H to f_L and f_H. However, caution must be exercised in interpreting such results due to possible variations in magnetic susceptibility effects intervening with external referencing [101]. In another procedure which has been proposed for investigating rapid protium-deuterium exchanges, the equilibrium isotope effect is derived from the measurement of the difference between a proton frequency and the corresponding deuterium frequency in the deuterated analogue [102]. Other interesting approaches take advantage of the isotope effects on carbon-13 chemical shifts. Such effects enable the protonated and deuterated isotopomers to be differentiated in the ^{13}C spectrum. In rapidly exchanging systems, the observed chemical shifts are weighted means over the parameters of the individual exchanging species and the method involves measurements of differences in chemical shift between dilute solutions in H_2O and D_2O contained in concentric tubes, and between solutions in H_2O and in H_2O/D_2O mixtures [103]. This procedure eventually provides a simultaneous determination of the fractionation factors associated with different exchangeable sites present in the investigated molecule. In the case of 2-aminoethanol, for instance, values of the equilibrium constant for the isotopic exchange with water of $K = 1.16$ and $K = 1.12$ have been simultaneously measured at 308 K for the OH and NH_2 group, respectively. More direct methods apply under conditions of slow exchange [104–106]. In the isomerization of the cyanobicyclo [4.2.0]octatriene-d_6, a significant preference of deuterium for the bridgehead position with respect to the vinyl position ($K = 1.55$) is unambiguously proved by a comparison of the area of the proton signals [105].

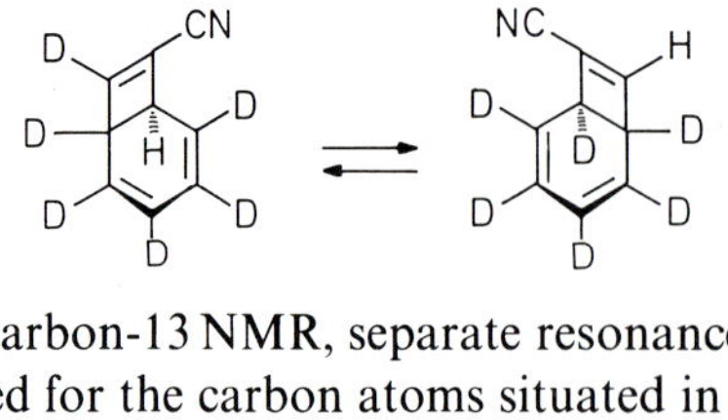

In carbon-13 NMR, separate resonances due to the $^1H/^2H$ isotopomers are observed for the carbon atoms situated in the vicinity of the exchangeable sites

[106]. The magnitude of the isotope shift which frequently lies in a range 0.05 to 0.2 ppm is usually large enough to enable reasonably accurate values of the signal area of the individual isotopomers to be obtained. The fractionation factors $\alpha_X = R_{substance}/R_{water}$, where R is the deuterium/protium ratio, are directly derived from signal area measured in dilute solutions in D_2O/H_2O mixtures of known composition. In non-aqueous media where slow exchange conditions are more likely to be realized, the isotope ratio of the investigated sample can be referred to that of a reference compound such as methanol, the fractionation factor of which is known or has been previously determined.

In principle, ^2H NMR at the natural abundance level is well suited for a direct determination of the fractionation factors in slowly exchanging systems. Since no isotopic enrichment is carried out, only the monodeuterated isotopomers are concerned in the observed exchange equilibria. Thus, in contrast with other procedures which resort to deuterated reactants, the natural abundance method provides a pure and direct access to the equilibrium constants for the following exchanges which only involve singly deuterated water molecules:

$$RAH + HOD \xrightleftharpoons{K_X} RAD + HOH$$

If $(D/H)_X$ and $(D/H)_W$ represent, respectively, the isotope ratios of compound RAH and of water, the equilibrium constant K_X is given by:

$$K_X = \frac{(D/H)_X}{2(D/H)_W} \tag{56}$$

and the fractionation factor α_X is defined as $\alpha_X = 2K_X$.

Since the mole fractions of the compound RAH (a) and of water $(1-a)$ can be very accurately determined, α_X is simply computed from the ratio of the RAD and HOD signal area measured in the deuterium spectrum:

$$\alpha_X = 2K_X = \frac{S_{RAD}}{S_{HOD}} \frac{(1-a)}{a} \tag{57}$$

If suitable, the ratio of the mole fractions of RAH and water can be derived from the area of the -AH and H_2O signals measured in the proton spectrum of the mixture. An indirect procedure making use of an intermediate reference compound can be carried out when a direct investigation of the aqueous solution is not possible due to solubility limitations or fast exchange conditions. The equilibrium constant K_{ref} of the exchange reaction between the investigated compound and the reference is then calculated from the area measured on the ^2H spectrum and from the value of the fractionation factor of the reference, α_{ref}. This value is either taken from the literature or determined by ^2H NMR in an independent experiment. The accuracy of the fractionation factor, α_X, either measured directly or estimated from the product of K_{ref} and α_{ref}, is obviously conditioned by that of area determinations in the deuterium spectrum[†]. The standard deviation of α_X therefore depends, as discussed in Sect. 3, on the relative magnitude of the linewidths and of the chemical shift difference between the considered signals. In favorable cases α_X can be determined at ± 0.02.

From a statistical treatment of a large number of experiments performed on the water-ethanol equilibrium, we have determined a value $K = 0.55 \pm 0.03$ at 302 °C for the equilibrium constant and therefore a value of 1.10 for the fractionation factor associated with the exchange [107]. This value is in reasonable agreement with the results obtained by other techniques [91, 103].

The method is also applicable to the determination of isotope effects on equilibrium constants for enzyme-catalyzed reactions. We have investigated, for example, the stereospecific exchange of the pro-R site of ethanol with water in the presence of the yeast *Saccharomyces cerevisiae*:

$$CH_3CH_2OH + HOD \rightleftharpoons CH_3CHDOH + HOH$$

This exchange involves two redox reactions which produce hydrogen transfers between the pro-R site of ethanol and nicotinamide adenine dinucleotide, NAD, on one hand, and between the pro-S site of NADH and flavine mononucleotide, FMN, on the other. The reaction is rendered possible by the rapid exchange between two NH sites of the reduced flavine mononucleotide, $FMNH_2$, and water [108]. Denoting by an asterisk the sites which may exchange deuterium and protium, the mechanism is described by the following scheme:

$$FMNH_2 + {}^*H_2O \rightleftharpoons FMN{}^*H_2 + H_2O$$

$$FMNH_2{}^* + NAD \xrightleftharpoons{\text{diaphorase}} FMN + (4S)[4\text{-}H^*]NADH + H^+$$

$$(4S)[4\text{-}H^*]NADH + CH_3CHO + H^+ \xrightleftharpoons{\text{ADH}}$$

$$CH_3CH_2OH + [4\text{-}H^*]NAD$$

$$[4\text{-}H^*]NAD + FMNH_2{}^* \xrightleftharpoons{\text{diaphorase}} [4\text{-}H_2{}^*]NADH + FMN + H^+$$

$$[4\text{-}H_2{}^*]NADH + CH_3CHO \xrightleftharpoons{\text{ADH}}$$

$$(1R)[1\text{-}H^*]CH_3CH_2OH + NAD$$

When a diastereotopicity of the two methylenic positions is introduced as described in Sect. 5.3, the hydrogen exchange can be quantitatively investigated.

Variations in the isotope content of the water medium are shown to be stereospecifically transmitted to the pro-R site of ethanol as expected from the above mechanism [109, 110]. Apparent values of the fractionation factor α can

[†] [For example, S_{RAD} and S_{HOD} of Eq. (57)].

be directly determined either for the enzyme system alcohol dehydrogenase-diaphorase or for the yeast *Saccharomyces cerevisiae*. Due to incomplete irreversibility and to glycolytic breakdown of endogeneous materials of the yeast [111], the measured α values are lower than predicted on the basis of pure equilibrium isotope effects [112, 113].

5 Natural Fractionation of Hydrogen or Labelling Without Enrichment Applied to the Study of Reaction Mechanisms

5.1 The Study of Deuterium Labelling by NMR

NMR is a non-destructive technique easily applicable to the investigation of chemical or biochemical mechanisms involving isotopically labelled molecules. In numerous studies of reaction pathways, the aim has been to follow the fate of hydrogen atoms from the precursors to the products of interest and only in some appropriate instances have the kinetic isotope effects of rate-determining steps been quantitatively measured as considered in Sect. 4.

Indirect detection of ^{2}H labelling by ^{1}H or ^{13}C NMR can be advantageous from the point of view of both sensitivity and frequency resolution. As an example, the enolization of organic substrates in strong aqueous acidic solutions is easily studied by following the evolution of ^{1}H peak intensities in the course of the substitution of enol-exchangeable hydrogens by deuterium [114]. Such studies now benefit from the possibility of implementing appropriate versions of modern pulse sequences [115–117]. Thus, J-modulation of ^{13}C spin-echo signals can be used for distinguishing mono-, di-, tri-, and non-deuterated carbons in mixtures of partially labelled species and for eliminating ^{13}C signals from deuterated solvents [118, 119]. Pulse sequences exploiting polarization transfer such as $^{13}C\{^{2}H\}$ INEPT and DEPT can also be applied if a possibility of phase shifting is implemented on the ^{2}H decoupler [120–122]. Limitations due to fast deuterium relaxation have been made less stringent and sensitivity has been optimized by using composite pulse sequences of ^{13}C editing which involve both ^{1}H-DEPT and ^{2}H spin–echo [123–125]. In practice the relative advantages and disadvantages of the polarization transfer methods and of the spin echo methods for editing ^{13}C spectra of CD_nH_m groups depend on the complexity of the mixture to be analysed [121].

Specially adapted two-dimensional $^{1}H/^{2}H$ chemical shift correlation experiments (COSY) are also helpful in revealing the location of deuterium labels [126]. $^{2}H/^{13}C$ shift correlation experiments are well suited to the study of deuterated compounds with a relatively low degree of deuterium incorporation. The assignment then benefits from the large dispersion of the signals in the carbon frequency dimension [127]. Finally, it should be mentioned that selective $^{1}H/^{13}C$ shift correlation experiments with deuterium decoupling enable stereospecific labelling at diastereotopic methylene hydrogens to be detected [128]. Last but not least, ^{2}H NMR is frequently well suited for directly identifying the sites of labelling in organic reactions and for quantifying the extent of enrichment while simultaneously providing stereochemical information [52, 58, 62, 129–132].

Due to the low natural abundance of deuterium, the incorporation of labelled precursors in biosynthetic pathways can be detected even after strong dilution of the precursor in the metabolic pool. If we consider that a twofold enrichment with respect to the natural abundance level is easily detected in ^{2}H or ^{13}C spectroscopy, the incorporation of singly labelled molecules is detectable up to dilutions about 70 times higher for ^{2}H than for ^{13}C.

A number of biosynthetic processes have been investigated by using deuterated materials [60]. For example, the introduction of ^{2}H into the methoxy group and other specific positions of griseofulvin and the microbial hydroxylation of this compound were readily followed by ^{2}H NMR [133]. Similarly, mechanistic routes of terpene biosynthesis involving the formation of rosenonolactone [134], tetrahymanol [135], or ovalicin [136] could be further interpreted. In the same respect, the incorporation of deuterium from CD_3CO_2H into mellein was elucidated by ^{2}H NMR and information was obtained about the stereospecificity of the reactions involving the two diastereotopic methylenic positions at C-4 [124]. In conjunction with proton NMR, deuterium NMR studies of the catechol dioxygenases were also a source of information on the active-structure in metalloproteins [137]. From a related point of view, it is worth noting that retention or transfer of deuterium can be indirectly detected in ^{13}C NMR spectroscopy by following the integrity of a doubly labelled $^{13}C-^{2}H$ fragment introduced into precursors of biological reactions [60].

5.2 Natural Abundance Labelling in ^{2}H NMR or Labelling Without Enrichment

An efficient approach for investigating or characterizing reaction mechanisms consists in exploiting the very large deviations with respect to a statistical distribution of deuterium atoms evidenced within natural or synthetic molecules [42, 43]. The Site-Specific Natural Isotope Fractionation (SNIF) creates in fact a situation of selective labelling which may avoid the need for cumbersome selective enrichment [44]. The NMR method therefore enables the fate of naturally labelled reactants to be followed in the course of chemical or biochemical transformations. Such investigations can be performed on a purely qualitative basis or they can resort to quantitative determinations of kinetic isotope effects as described in Sect. 4.

We shall consider, for example, the simple esterification reaction performed with a given pool of acetic anhydride and with ethanol samples from different origins:

$$CH_3CH_2OH + (CH_3CO)_2O \rightarrow CH_3COOCH_2CH_3 + CH_3COOH$$

If the reaction is complete, and in the absence of secondary isotope effects, the isotope ratios of the three hydrogenated sites of ethylacetate must be equal to those of the corresponding sites in the starting materials. The deuterium contents of the methyl and methylene sites of the ethyl fragment therefore constitute faithful witnesses of the site-specific deuterium contents of ethanol. They provide good criteria for inferring the origin of the ethanol precursor [40]. In the same respect, it was demonstrated that fractionation effects occurring in the course of acid or enzymatic hydrolyses of starch followed by yeast fermentation do not alter the answer of the glucose or ethanol probes to isotopic variations in starch [138].

When comparing the isotope ratios in starting and end products, the observation of eventual variations in the deuterium contents specific to certain molecular sites may be a source of valuable information on the reaction pathways and on their branchings. From a more quantitative point of view, the analysis of kinetic isotope effects, calculated as discussed in Sect. 4, plays an important role in the elucidation of the rate-determining step in organic reactions. The interpretation is less easy in the case of enzymic reactions which are more complex and usually involve multi-step equilibria. However, SNIF-NMR determinations present the advantage of involving trace amounts of purely monodeuterated isotopomers which frequently offer a possibility of intra-molecular competition on the reaction site. Although complications may arise from the participation of relatively slow dissociation of enzyme-substrate complexes, this situation is favorable for estimating the relative participation of bond breaking steps in biosynthetic reactions.

The potential of the SNIF-NMR method can be illustrated by the study of the complex multi-step reaction of alcoholic fermentation [40, 48]. Several series of fermentation reactions have been performed in the presence of yeast with glucose and water samples from different origins characterized by different deuterium distributions. Since the technical and biochemical aspects of the bioconversion may be assumed to be invariant in standardized conditions, a redistribution matrix [A] relates the vectors $\vec{D}^Q$ of the site-specific deuterium contents in the end products, ethanol and water, to the vector $\vec{D}^S$ associated with the starting materials, glucose and water (Sect. 2.3). Once the matrix [A] has been determined by a series of calibration experiments, it becomes possible to produce from any vector $\vec{D}^Q$ deduced from NMR-measurements performed on a fermented medium, a fingerprint of the deuterium content of disappeared starting materials. Such a methodology is obviously of prime importance for inferring the isotopic distribution in inaccessible precursors of products grown in natural conditions. We have demonstrated, for example, that the vector $\vec{D}^Q$ built from the site-specific isotope ratios of ethanol and end water provides a very reliable probe for studying the biochemical and physiological factors which govern the hydrogen cycle in photosynthetic processes occurring in natural environments (Sect. 6). Although a number of biochemical steps are involved in this bioconversion, the coefficient of the [A] matrix which relates the deuterium content in the methylene site of ethanol to that of the starting water is shown to represent to a large extent fractionation phenomena intervening between water and methylene hydrogens. Neither of the pro-R or pro-S methylenic sites exhibits significant direct connection with glucose [40]. Further oxidation of ethanol into acetic acid by *Gluconobacter suboxydans* proceeds with fractionation effects which depend on the yield of the transformation as discussed in Sect. 6 and result from a small secondary kinetic isotope effect. Standardized conditions of fermentation and bacterial oxidation therefore successively translate isotopic variations occurring in sugars to the ethanol and acetic acid molecules [139]. In the same respect, an investigation of the isotopic filiation between glucose and mannitol enables information on the metabolisms peculiar to several types of algae to be inferred from the isotopic distribution determined in pentaacetate derivatives of extracted mannitols [140].

Natural abundance labelling also provides interesting criteria for the understanding of monoterpene biosynthesis [82, 85, 141]. Thus, the isopropenyl methyl hydrogens, a, of *R*(+)-limonene were shown by Grant et al. [85] to exhibit a noticeable deuterium depletion whereas the vinyl hydrogens, b, are enhanced with respect to the methyl hydrogens c which are assumed to remain unperturbed throughout the biosynthesis (Fig. 11):

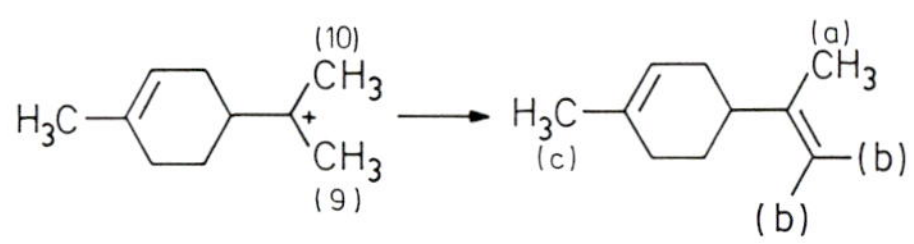

intermediate: α-terpinyl cation limonene

This behavior excludes the selective loss of a proton from C-10 of the α-terpinyl cation intermediate and is consistent with the abstraction of a proton from carbon 9 derived from the methyl group of mevalonic acid.

5.3 Natural Labelling and Chirality

It is worth emphasing that molecular species usually considered as achiral but possessing a methylene group may in fact present a "natural" enantiomeric imbalance. Taking into account the presence of the monodeurated isotopomers the question arises, for example, of the enantiomeric purity of ethanol, i.e., of the relative proportions of the *R* and *S*, CH_3CHDOH enantiomers. Since chemical synthesis usually leads to racemic mixtures whereas biosynthesis frequently involves stereospecific reaction steps, enantiomeric enrichments at methylene sites are likely to occur in the latter case. On this basis, ^{2}H-NMR determination of the relative concentrations of naturally present mono-deuterated methylenic enantiomers, pertaining to otherwise achiral compounds, eventually provides an absolute method for identifying biosynthetically produced compounds [142]. This method avoids the need for comparing the

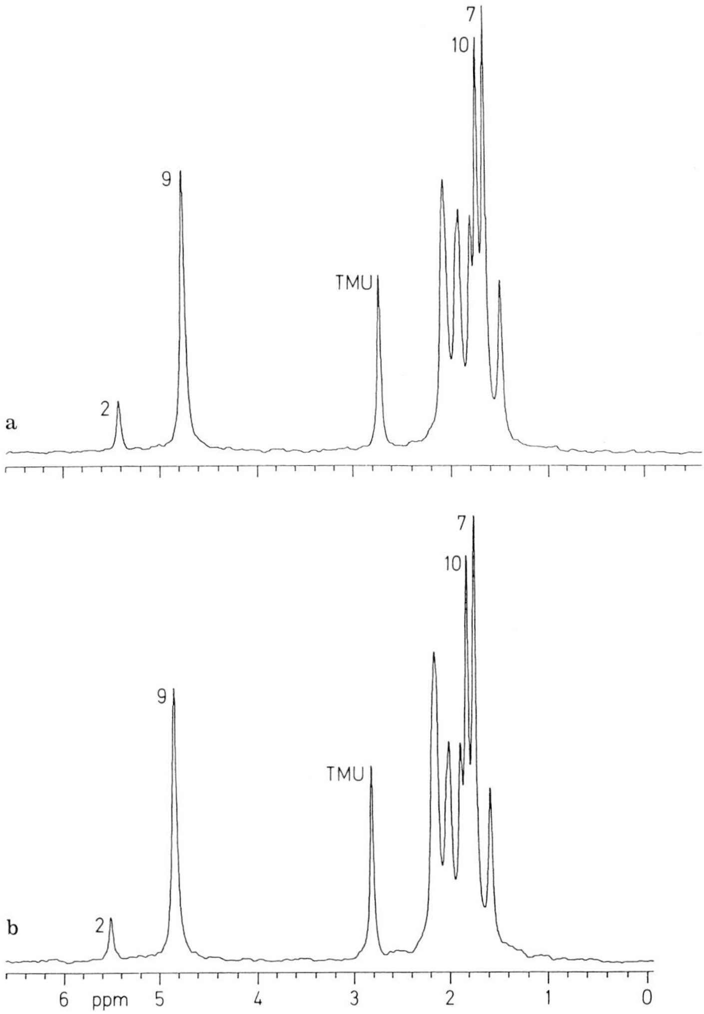

Fig. 11 a, b. Natural abundance ^{2}H-NMR spectra of *R*(+) (**a**) and *S*(−) (**b**) limonenes.

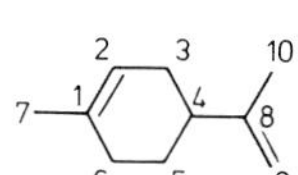

The samples are from different commerical origin. The spectra have been obtained at 61.4 MHz. An exponential multiplication (2 Hz line-broadening) has been applied for improving the signal-to-noise ratio. Tetramethylurea (TMU) has been added to the sample for internal referencing (Sect. 3.3). The strong deuterium depletion of the ethylenic site 2 as compared to the vinylic position 9 is remarkable. Mechanistic features concerning the biosynthesis of limonene are discussed by Grant et al. in Ref. [85]

isotope ratios of the unknowns to those of reference samples from an authenticated origin as described in Sect. 7. However, conversely, the lack of enantiomeric enrichment must not be considered as a criterium for certifying a synthetic origin.

From a more fundamental point of view, quantification of the enantiomeric purity also constitutes a new source of information about the mechanisms and stereospecificity of reactions involving methylenic hydrogens. A prerequisite to the method is the existence or the creation of a diastereotopicity of the enantiomeric methylenic deuterium nuclei. In principle, such a diastereotopicity is an inherent property in dissymmetric compounds. However, several conditions must be fulfilled in practice. First, the magnitude of the chemical shift difference between the methylenic ^2H signals must be large enough for quantitative measurements to be performed in satisfactory conditions. This shift is predictable from the proton spectrum by taking into account the values of the ^1H and ^2H gyromagnetic ratios and of the spectrometer frequencies (Sect. 3.1). Secondly, for the methylenic enantiomeric purity to be readily accessible, the dissymmetric molecule must exist either in a pure *R* or *S* configuration or at least as a mixture strongly enriched in one of the enantiomers.

Thus, in the case of aspartic acid $HO_2CCH_2CH(NH_2)CO_2H$, the methylenic hydrogens adjacent to the chiral center exhibit a chemical shift difference of 0.29 ppm ($pH \approx 13.1$, $T = 298$ K) which corresponds to 11.1 Hz at a deuterium frequency of 38.4 MHz. It is shown in this case that both a racemic sample and a D-sample obtained by chemical synthesis have identical deuterium content at the two configurations of the chiral methylene center. In contrast, the deuterium content at the two methylenic sites of L-samples obtained by fermentation is not identical and has a ratio of about 1.17 [142)].

The method is of particular interest for investigating the stereospecificity of reactions involving the methylenic hydrogens of compounds devoid of an asymmetric centre. However, it is necessary in this case to introduce a diasterotopicity of the methylene sites. This can be achieved by adding a chiral shift reagent [143–144] or by reacting the sample with an appropriate optically active derivative [145–148]. In the case of ethanol, for example, the prochiral methylene protons are differentiated in the ester resulting from the reaction with (−) camphanoyl chloride [145]. However, in order to obtain sufficient separation of the ^2H signals, it is convenient to add the shift reagent $Eu(dpm)_3$ at a suitable concentration. For a solution of about 0.6 M in CCl_4 containing 32 mol% of the europium complex, a chemical shift difference of 0.34 ppm is reached. Then it could be directly shown that ethanol produced by fermentation of beet-sugar in usual conditions contains a nearly racemic mixture of the monodeuterated enantiomers [142]. At equilibrium the fractionation effects associated with the stereospecific reactions involving the individual pro-*R* and pro-*S* sites are therefore rather similar. The method can be applied to the investigation of the stereospecific exchange between the pro-*S* site of ethanol and water governed by the enzymic system, alcohol dehydrogenase-diaphorase,

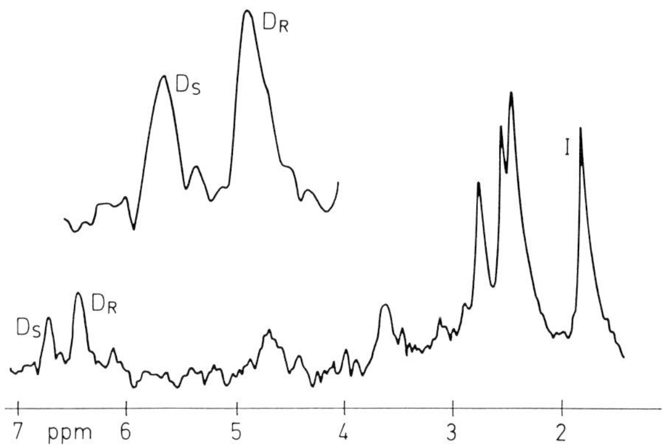

Fig. 12. ^{2}H-NMR spectrum of camphanate obtained by reacting (-)camphanoylchloride with an ethanol sample previously incubated with yeast in slightly enriched water (760 ppm) [139]. The higher isotope ratio of the R methylenic site can be interpreted by a stereospecific exchange mediated by the linked alcohol dehydrogenase-diaphorase system present in the yeast. The spectrum was obtained at a ^{2}H frequency of 61.4 MHz in the presence of Eu(dpm)$_3$

present in the yeast sacchoromyces cereoisiae. Thus, the deuterium spectrum (Fig. 12) of the camphanate prepared from ethanol exchanged in a water medium slightly enriched in deuterium illustrates the selective enrichment of the *R* configuration which can be explained on the basis of the reaction scheme described in Sect. 4.6. Unfortunately, the accuracy of these quantitative measurements is restricted by line-broadening and sensitivity limitations caused by the paramagnetic shift reagent. Epimeric mixtures of organophosphorus derivatives bearing an enantiomerically pure chiral substituent may exhibit a high discriminating power [109] and appropriate ^{2}H spectra are also obtained with mandelate derivatives [110]. The synthesis of such diamagnetic chiral products is expected to greatly improve the capability of the method to detect small enantiomeric imbalances.

In the experiments described above, the enantiomeric purity has been investigated by direct measurement of the concentrations of each isotopomer. Another kind of application of the SNIF-NMR method can be based on possible correlations between the values of certain site-specific deuterium contents and optical activity. Thus in the case of α-pinenes, the isotope ratio of the ethylenic site, e, in particular, exhibits large differences between the *R* and *S* configurations and $(D/H)_e$ is linearly related to the measured optical deviation α [141] (Sect. 6.2):

$$(D/H)_e = 105 + 0.97\alpha \qquad (58)$$

This property offers an easy way of estimating the optical purity of enantiomeric mixtures of given origins. The method avoids the need for adding a chiral reagent [72b] and preserves the sample from contamination. In addition, the isotopic distribution measured in each enantiomer is a source of information about the mechanisms which determine the synthesis of the two configurations.

6 Influence of Chemical, Biochemical, and Environmental Factors on Site-Specific Isotope Fractionation

6.1 Isotope Ratios in the Characterization of Chemical or Biochemical Synthesis

As discussed in Sect. 2.4, numerous factors govern the site-specific isotope distribution, and discrimination against heavy or light isotopes is likely to occur all along the elaboration of the product. Decisive differences are therefore expected between samples of a given molecular species resulting either from chemical synthesis or from biosynthesis in natural conditions.

Considering the complexity of the biosynthetical pathways and the diversity of the physical, physiological, and technological effects which may affect isotope fractionation (Fig. 5), it could be suspected that the site-specific isotope ratios of the final products do not really characterize the mode of synthesis. However, we have now demonstrated, using a large number of typical examples, that, under definite experimental conditions, isotope fractionation is remarkably reproducible. Moreover, the environmental effects affecting biosyntheses in field situations are usually of a sufficiently moderate or predictable magnitude not to mask differences which are characteristic of the reaction mechanism. Thus, the differences between the site-specific isotope ratios of synthetic and fennel anetholes (Fig. 13) ranges from 3 ppm to 68 ppm for the whole set of molecular sites, whereas the observed variations due to environmental effects occurring in the biosynthesis of fennel anethole do not exceed 15 ppm [45, 149]. Similarly, characteristic differences between the isotopic pattern of synthetic and naturally occurring *d*-camphor were recognized by Grant et al. [150].

It should be emphasized that the isotopic distribution in molecules from chemical synthesis may be strongly dependent on the reaction scheme and on the origin of the raw materials utilized. Consequently, the sets of deuterium content at the natural abundance level constitute useful criteria for inferring the reaction pathway. Thus anetholes resulting from chemical synthesis using anisole, on one hand, and from chemical isomerization of estragole, on the other hand, are unequivocally distinguished on the basis of their specific isotope ratios [149] (Fig. 13).

Fig. 13a–d. ^{2}H-NMR spectra of four samples of anethole obtained from hemi-synthesis from estragole (**a**), chemical synthesis from anisole (**b**), extraction from fennel (**c**), and from star anise (**d**).

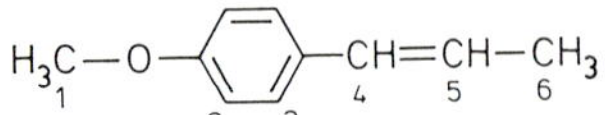

It is observed in particular that the ethylenic site 4 exhibits a significant relative depletion in the sample from chemical synthesis and an enrichment in the sample from hemi-synthesis

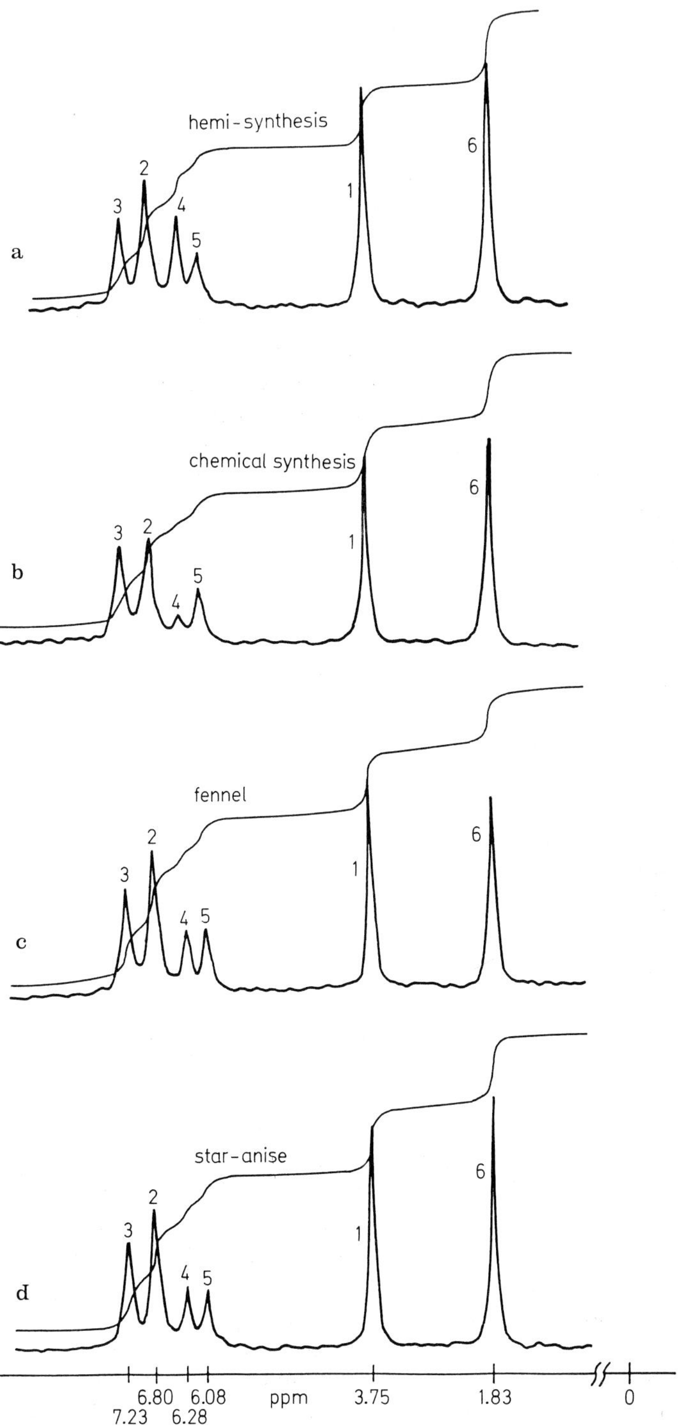
hemi-synthesis
a
chemical synthesis
b
fennel
c
star-anise
d
1
2
3
4
5
6
7.23
6.80
6.28
6.08
ppm
3.75
1.83
0

6.2 Isotopomeric Distribution and Metabolic Pathway

Much effort has been devoted during the last twenty years to the interpretation of the differences, observed by mass spectrometry, in the overall ^{13}C content of species elaborated in plants with a C_3 (Calvin) or C_4 (Hatch-Slack) photosynthetic cycle (Sect. 1). The noticeable ^{13}C depletion generally observed in C_3 plants as compared to C_4 plants is mainly attributed to differences in isotope effects intervening in the carboxylation step of the photosynthesis [20–26, 18, 151]. Other steps of biochemical fractionation of carbon have been invoked to explain the variability in the $^{13}C/^{12}C$ ratios of certain metabolites [152–155]. The information was then made more selective by carrying out appropriate chemical transformations of the investigated metabolites. Thus fractionation occurring in the formation of Acetyl CoA by pyruvate dehydrogenase results in a relative ^{13}C depletion in the carbonyl site of lipids [153–155]. In the same respect, a contribution of neoglucogenesis processes has been assumed to be responsible for small deviations detected in the carbon isotope contents of the C-1 and C-6 sites of starch glucose [251].

The significant difference in isotope discrimination exhibited by C_3 and C_4 plants has interesting applications in the recognition of the type of diets absorbed by humans or animals. Thus, Europeans are expected to possess a smaller ^{13}C level than North Americans since in Europe the nutrients are mainly derived from C_3 plants (wheat, potatoes, beet-sugar,...), whereas C_4 plants, which include maize, sorghum, and cane sugar are more common in the southern and western U.S. [156]. Evolutions in the food chain and therefore in the habits of prehistoric populations can also be investigated on this basis [157, 158]. Similarly, differences existing within the $\delta^{13}C$ values of land plants or marine phytoplankton can be used to trace—food chain relationships among organisms from nearshore waters [159]—animal diet in seagrass meadows [160]—or carbon flow in marine food webs [161].

Although significant variations in overall deuterium/protium ratios have been measured by mass spectrometry, no general behavior could be recognized with regard to the distinction between the main C_3 and C_4 photosynthetic cycles [162]. However, the total deuterium contents of given metabolites can provide important information on the metabolic scheme, on the eventual existence in the plant of separate pools of compounds and on possible steps of biochemical fractionation [33–41, 162–171]. Compounds, such as terpenes, which are biosynthetized through the acetylcoenzyme A, are more depleted than metabolites, such as vanillin or estragole, which derive more directly from the degradation of glucose [164]. Joint investigations of the hydrogen and oxygen isotope ratios have been performed in the course of assimilation of water and elaboration of metabolites by the plant. The relationships between δD and $\delta^{18}O$ provide information on the role of evaporative transpiration and on the incorporation of CO_2 and H_2O into organic matter of terrestrial or aquatic plants [172–175]. From the study of cellulose nitrate of plants growing in the same environment, it is concluded that CAM plants are enriched in deuterium with respect to C_3 and C_4 plants [162–167]. The variability observed in the δD values of cellulose nitrate derived from C_3, C_4, C_3/C_4 intermediates and hybrid plants is attributed to biochemical fractionation occurring in the course of carbohydrate metabolism that leads to cellulose synthesis [166, 169].

Considering the diversity of the possible sources of fractionation effects (Sect. 2.4) the specific information brought about by the NMR parameters is expected to provide more detailed mechanistic information than the mean ratios obtained by mass spectrometry. Such selective information is specially useful in the study of metabolites situated far from the primary photosynthetic products in the biosynthetic pathway. Thus, whereas mass spectrometry provides a single averaged difference of only a few ppm between the overall hydrogen isotope parameters of *R* and *S* enantiomers of α-pinene, up to 10 parameters are accessible by NMR [141] (Fig. 14). The site-specific ratios of the *R* enantiomer, for example, vary in a range of more than 100 ppm and the difference in the isotope content of a given molecular site in the *R* and *S* configurations may be as large as 35 ppm. Such fractionation effects certainly involve strong kinetic isotope effects intervening at different enzymatic steps of branched metabolic pathways (Sect. 4). In this respect it is shown that in fermentation reactions the discrimination against deuterium reaches about 30 ppm for the stereospecific transfers of hydrogen from water to the enantiomeric methylenic sites of ethanol [40]. Changes in biosynthetic mechanisms are therefore, expected to play an important role in site-specific isotopic distribution. However, an influence of physiological effects on the metabolic fluxes is also likely to intervene along with climatic effects capable of modifying the deuterium content of intracellular water. The possibility of distinguishing different metabolic pathways will therefore depend on the position of the considered molecular species in the biosynthetic route. In the case of the key component, glucose, for example, we have shown that deviations exist with respect to a statistical repartition of deuterium and that these deviations are more pronounced in starch glucoses extracted from potato or sunflower (C_3 plants) than in glucoses obtained from maize (C_4 plant) cultivated in comparable climatic conditions [40, 138, 176]. Such distinctions in terms of metabolism or plant physiology is supported by the behavior of the methyl parameter of the corresponding fermentation ethanols which have been shown to partly reflect the deuterium contents in sugars [48, 249]. However, if samples from various plant species grown in a variety of countries are considered, the dispersion is relatively large and results in overlaps of certain botanical families.

6.3 *Climatic Significance of Site-Specific Isotope Fractionation of Hydrogen in Natural Products*

The overall carbon, oxygen, or hydrogen isotope contents of wood or, more specifically, of the cellulose component of tree rings have been shown to contain

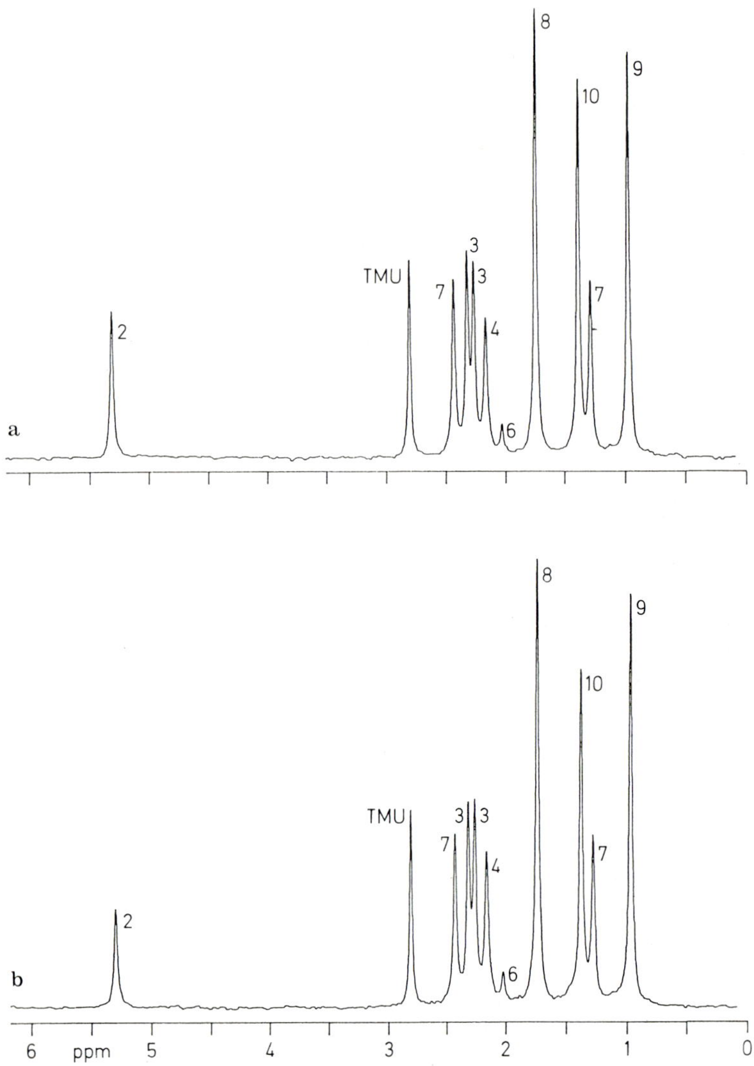

Fig. 14 a, b. Comparison of the ^{2}H-NMR spectra of two samples of (+) α-pinene (**a**) and (−) α-pinene (**b**) from the same commercial origin.

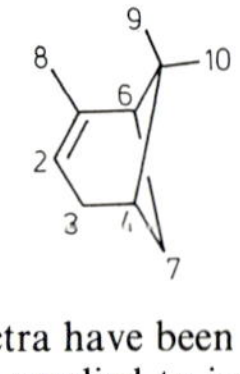

The spectra have been obtained at 61.4 MHz. An exponential multiplication (1 Hz line-broadening) has been applied to improve the signal to noise ratio.

The difference in the site-specific deuterium contents of the enantiomers is particularly large at position 2. Site 6 which is very depleted in both pinenes is formerly the ethylenic site of so-called "active isoprene" moiety which undergoes the cyclization to give the bicyclic pinene skeleton climatic information and they are frequently used for palaeoclimatic reconstruction [167, 177–182]. This field has been somewhat controversial and it has been stressed that a reliable interpretation of the small variations in the δD value requires that appropriate and careful sampling procedures are carried out. In order to avoid uncontrolled contributions due to multiple components in the case of whole wood or due to the exchangeable hydrogens in the case of extracted cellulose, the mean isotope ratio of the only carbon-bound hydrogens is usually measured on cellulose nitrate [183, 184]. In appropriate cases the δD parameter of lipid components has also been claimed to reflect climatic conditions [185]. In this context, inference of evolutions in the mean temperature during ancient geological periods is frequently based on small variations of a single overall parameter δD of fossil peat or tree cellulose.

Compared to such mass spectrometry determinations which average the contributions of various isotopomers, SNIF-NMR provides us with a richer source of mechanistic information. A climatic significance of the site-specific parameters has been demonstrated in the case of three ethanol probes associated with two C_3 plants, *Vitis vinifera* and *Beta vulgaris* and one C_4 plant *Saccharum officinarum* [186, 187]. Since the ethanol and water components resulting from fermentation reactions conducted in standardized conditions provide a reliable representation of the isotopic behavior of their precursors, glucose and water [40], they constitute faithful witnesses of the climatic features which prevailed during the elaboration of sugars. In this case, three partly independent variables are available instead of the single average parameter determined by mass spectrometry. The results are then efficiently analysed in a factorial representation of the isotopic space. Thus, the isotopic data associated with more than four hundred fermented media originating from *Vitis vinifera* have been centered and reduced to the means of the different production area. A two-dimensional representation in the plane of the two main factors, which represent 75.8% and 20.1% of the variance is illustrative of the influence of the environmental conditions (Fig. 15). It is observed that the data associated with Continental and Mediterranean countries occupy opposite sides along the main axis. More generally, points describing countries situated in the same range of northern or southern latitudes and exhibiting similar climatic features are found relatively close to each other. Such is the case for Morocco, Algeria, South Africa (Stellenbosh), Australia (Adelaïde), on one hand, Switzerland (Valais), Yugoslavia (Voivodine), Bulgaria, and Germany (Rhein Pfalz), on the other hand. Even when a more restricted range of variation in meteorological data is considered, a discriminant factor analysis of the isotopic ratios clearly distinguishes two types of climatic features, relatively cold and wet on one hand, relatively hot and dry, on the other hand. Thus in the case of *Beta vulgaris*, sixty samples originating from different districts of the same country, France, can be classified with a 94% power of assignment within two climatic groups characterized by the following mean annual values of the meteorological parameters (temperatures T, precipitation P, and insulation, I): $T = 9.6\,°C$, $P = 63.3$ mm, $I = 143.5$ h (relatively cold and wet) and $T = 10.5\,°C$, $P = 44.3$ mm, $I = 186.8$ h (relatively hot and dry).

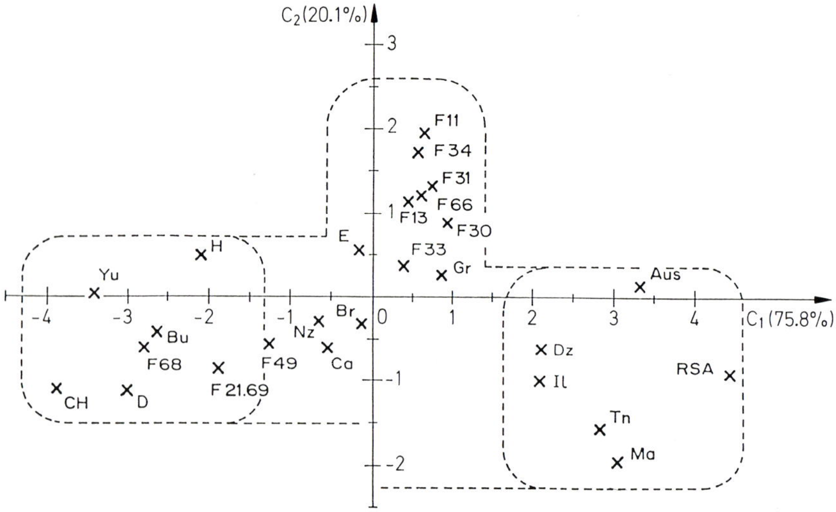

Fig. 15. Factorial representation of the isotopic data typical of *Vitis vinifera*, in the plane of the two principal components. Each point in the graph represents the mean over isotopic values measured for different samples from the same country. The production areas are the following: *Aus*. Australia (Adelaïde). *Br*. Brazil (Porto Alegre), *Bu*. Bulgaria, *Ca*. California, *CH*. Switzerland (Valais), *D*. Germany (Rhein Pfalz), *Dz*. Algeria, *E*. Spain (Rioja), *F*. France, *Gr*. Greece (Thessalonica), *H*. Hungary, *Il*. Israel, *Ma*. Morocco, *Nz*. New-Zealand (Ti Kauwata), *RSA*. Republic of South Africa (Stellenbosh), *Tn*. Tunisia, *Yu*. Yugoslavia (Voivodine). The French regions are characterized by the number of the Department: *11* (Aude), *13* (Bouches du Rhône), *21* (Côte d'Or), *30* (Gard), *31* (Haute Garonne), *33* (Gironde), *34* (Hérault), *49* (Maine et Loire), *66* (Pyrénées Orientales), *68* (Haut-Rhin), *69* (Rhône)

The SNIF-NMR method presents the advantage over mass spectrometry of providing multiple partly independent information for a given sample. Consequently, a more detailed analysis of the results can be realized in terms of both the relative roles of the various meteorological parameters and the time evolution of these parameters. From the latter point of view, a correlation analysis of the results presented in Fig. 15 shows that the first factor C1 is mainly representative of fractionation effects affecting the water medium, whereas the second component, C2, is more related to properties of the sugars. The site-specific isotope ratios therefore contain information on climatic effects prevailing either in the period of photosynthesis of the glucose-fructose mixture or in the later period of must elaboration.

The climatic dependency of the isotope parameters obtained by mass spectrometry has frequently been analysed in terms of possible relationships between a given isotope ratio and either another isotopic variable or a meteorological or environmental parameter. Detailed investigations of leaf isotopic characteristics have cast light on the dependency of the enrichment in the heavy isotope on the isotopic composition of the ambient vapor, on the relative humidity of the air and of the transpiration rate [188–197, 173]. Effects of temperature and precipitation on the overall deuterium content of plant components have also been discussed [198–200, 174, 182, 186]. In particular, the existence of regular increases in δD of cellulose with temperature have been frequently exploited for climate reconstruction [178, 179, 183, 201].

Similar trends are exhibited by the different site-specific parameters associated with *Vitis vinifera* and *Beta vulgaris* metabolites [186, 187]. However, it could be shown that a simple increase in temperature cannot be generally considered. Thus, an opposite effect is exhibited in the case of the C_4 species *Saccharum officinarum* which is cultivated in countries where hot climates may be associated with shorter periods of insulation than less warm climates. Under these circumstances the insulation effect predominates over the temperature effect in the enrichment of sugars in the heavy isotope whereas an increase in the amount of precipitation decreases the deuterium content [186].

The climatic significance of sets of site-specific isotope parameters may also be applied to climate reconstruction using cellulose. For this purpose, a methodology has been developed which involves enzymic hydrolysis of cellulose in order to provide access to the isotopomeric distribution of monomeric units [176]. In this procedure the glucose derivative is first purified, crystallized, and acetylated before being subjected to the NMR analysis.

7 SNIF-NMR Applied to Quality Assessment

7.1 *Isotopic Methods in Quality Assessment*

Characterization and recognition of the origin of raw materials or elaborated products is a considerable problem in the food industry. Once sucroses or starches, for example, have been completely fermented into ethanol, it is no longer easy to determine whether the ethanol comes from maize, malt, potato, or grape. As expected from the general properties recalled in the previous section, mass spectrometry determinations of the overall $^{13}C/^{12}C$ ratios of a given molecular species are frequently very efficient for identifying a synthetic or natural origin and even for distinguishing products from plants with either a C_3 or a C_4 metabolism [202–205]. On this basis it is possible to distinguish sucroses extracted from beet-root or fruits (C_3-plants), on one hand, and from sugar-cane (C_4-plant), on the other hand [206–210]. This distinction is transferable to fermentation ethanols in wines [41, 211–213], whiskies [214], and rums [215, 216], for example, and to acetic acid in vinegars [217–221]. It can also be used to detect added sugars such as cane sugar or corn syrup to maple syrup [210, 204, 222], to citrus juices [209, 203, 223], to apple juice [224], to honey [225, 226], etc. From a related point of view, the $^{13}C/^{12}C$ ratio has been used, in sparkling wines, to distinguish between CO_2 originating from C_3 or C_4 plants and from an industrial source [227].

A number of identifications based on $^{13}C/^{12}C$ determinations have been corroborated by mass spectrometry measurements of overall D/H ratios and eventually of $^{18}O/^{16}O$ ratios [202, 205, 206, 208, 209, 228]. Moreover, these partly independent parameters significantly improve the discriminating power of the carbon ratios. They are particularly helpful in detecting the addition of exogeneous compounds, including beet sucrose or water, to fruit juices [202, 208, 209, 229, 230] or to alcoholic beverages [212, 213, 231, 232] and for characterizing the origin of flavors and spices [37, 205].

However, in a number of cases, the average isotope ratios measured by mass spectrometry do not really provide unambiguous results for the distinction of species pertaining to the same photosynthetic family. On the basis of the properties described in the previous section, it is obvious that the SNIF-NMR method offers a very powerful tool for origin recognition. Its potential can eventually be reinforced by resorting, in conjunction, to isotope mass spectrometry determinations which benefit from high precision and the need of smaller samples. Mass spectrometry is also unique for determining $^{18}O/^{16}O$ isotope ratios since the ^{18}O nucleus is devoid of a nuclear magnetic moment.

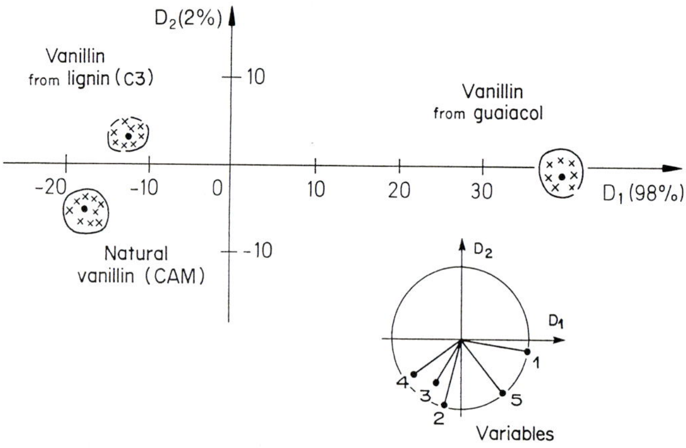

Fig. 16. Discriminant analysis performed on 5 site-specific hydrogen isotope ratios of vanillin.

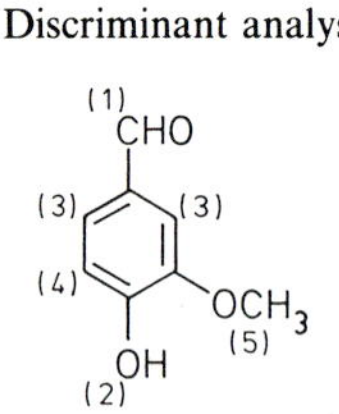

Samples prepared by chemical synthesis from guaiacol are perfectly separated and there is also a 100% discriminating power between vanillin samples extracted from *Vanilla planifolia* or prepared from lignin. The correlation circle shows that site 1 and site 2 highly correlate with the main axes 1 and 2, respectively

7.2 "Natural" Origin Assessment

The NMR method in very efficient in distinguishing natural and synthetic origins and it can even be used in recognizing different routes of chemical synthesis (Sect. 6.1). The story of vanillin authentification is particularly illustrative. The great difference in the prices of vanillin extracted from *Vanilia planifolia*, on one hand, and of synthetic or hemisynthetic vanillin, on the other hand, has stimulated the development of procedures to get round the methods successively proposed for origin recognition. Thus, it was first demonstrated that natural and synthetic vanillin are distinguishable on the basis of the average $^{13}C/^{12}C$ ratio [37, 233, 234]. However, slightly enriched methoxyl groups were subsequently introduced into the chemical product in order to mimic the overall ratio of natural vanillin. Demethylation then became necessary to gain access to the ^{13}C contents of the fragments [235, 236, 211]. Differences were also shown to exist between the overall deuterium ratios measured by mass spectrometry [37, 234]. However, following this observation, appropriate enrichments were performed in order to artificially reproduce the natural mean $\overline{D/H}$ ratio. The SNIF-NMR pattern therefore constitutes at present the most attractive criterium of authentification [237, 238]. A discriminant analysis of the isotopic data shows that vanillin samples extracted from *Vanilia planifolia* are unambiguously distinguished, whatever the region of production, from vanillin samples synthetized from guaiacol or derived from lignin [238] (Fig. 16).

7.3 Origin of the Vegetal Precursor

In Sect. 6.2 we emphasized the discriminating role of the biochemical and physiological effects which intervene either in plants with different photosynthetic cycles or in plants with the same metabolic type but different physiologies. The SNIF-NMR method is therefore expected to be capable of recognizing the botanical nature of the vegetal precursor of certain food components [45, 239, 240]. An analytical procedure has been described to identify the origin of anethole aromas [149, 241] (Fig. 13). A discriminant analysis of the SNIF ratios shows that fennel anethole is clearly distinguished, not only from synthetic anethole, but also from star-anise and green-anise anetholes. Similarly, it is frequently possible to identify alcoholic beverages prepared from various cereals or fruits [46]. Pure malt whiskies and bourbon whiskies partly issued from maize, or rye cereals are easily recognized [242]. In the same line, malt beers and beers prepared from maize grain can be classified with a high discriminating power [243, 249]. More generally, the method is applicable to the characterization of various products such as vinegars [139], vegetal oils, etc. Since a simple weighted averaging over the isotope ratios of the components is approximately satisfied, the percentages of the raw materials used in blended whiskies or in adjunct beers, for example, can also be estimated.

7.4 *Inference of the Geographical Origin and of the Year of Production*

Another type of application is based on the climatic significance exhibited by the SNIF-ratios (Sect. 6.3), a property which makes it possible to infer the geographical origin of the raw material used in the elaboration of the considered product [187]. A discriminant analysis of the isotopic data of ethanols in terms of the region of cultivation of vine demonstrates that wines produced in a country such as Tunisia are easily recognized among wines from New-Zealand or from Bulgaria, for example (Fig. 17). Such geographical characterizations are even possible within a given country. Thus, in France, wines produced in the Loire region or in the Bordeaux area can be differentiated from wines originating from Alsace [187, 244].

The climatic dependency can also be used to determine the temporal origin of food components. The method then provides an attractive possibility of quality assessment of alcoholic beverages in terms of certification of the year of production.

7.5 *Detection of Mixing, Enrichment, and Adulteration*

From a related point of view, the SNIF-NMR method is well suited for detecting whether a given main component of a food or beverage has been diluted by the addition of a substance with the same molecular formula but from a different

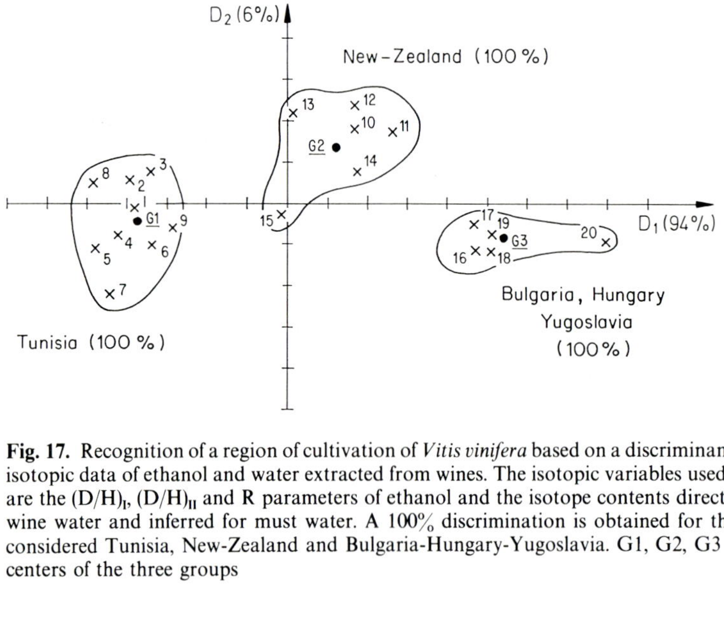

Fig. 17. Recognition of a region of cultivation of *Vitis vinifera* based on a discriminant analysis of the isotopic data of ethanol and water extracted from wines. The isotopic variables used in this analysis are the $(D/H)_I$, $(D/H)_{II}$ and R parameters of ethanol and the isotope contents directly measured on wine water and inferred for must water. A 100% discrimination is obtained for the three regions considered Tunisia, New-Zealand and Bulgaria-Hungary-Yugoslavia. G1, G2, G3 are the gravity centers of the three groups

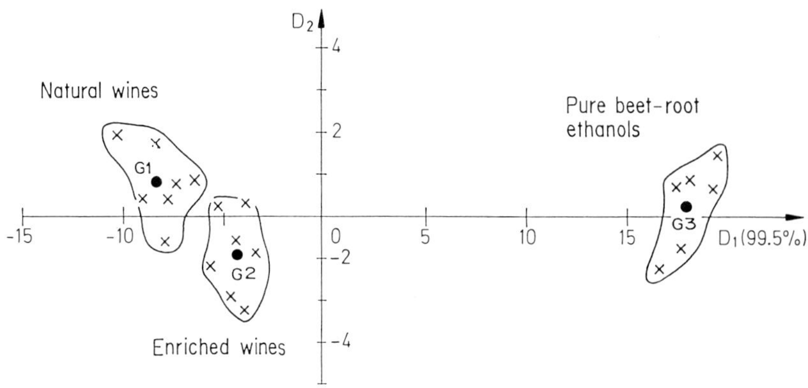

Fig. 18. Detection of wine enrichment with beet-sugar. The discriminant analysis is performed on the hydrogen isotopic data of ethanol and water samples extracted from natural wines, from pure beet-root alcohol, and from wines enriched at 2 degrees in volume. The wines considered in this analysis are from different regions, different years of production, and different varieties. Each point corresponds to a mean over the values for several samples from the same region

origin. Some typical examples are the adjunction of exogeneous sugars to pure orange or apple juice and the addition of water to pure fruit juices or to wines either before or after fermentation [244].

It can also be mentioned that the method is capable of characterizing certain technological procedures used in the preparation of the product. For example, top and bottom fermentation of beers can be distinguished [243]. In the same respect, the fractionation phenomena occurring in the course of distillation or evaporation procedures have been precisely analysed.

Finally, particular emphasis should be laid on the development of standardized protocols for detecting and quantifying the chaptalization practice, i.e., the enrichment of wines by addition of sugar to the must before fermentation [245, 246, 244]. Figure 18 illustrates the discrimination power of the method. A discriminant analysis of isotopic data shows that, whatever the origin of the wine, a two-degree enrichment with beet-sugar is unambiguously detected.

8 Conclusion

In this review our aim was to describe the characteristics and the potential of Site-specific Natural Isotope Fractionation as studied by NMR (SNIF-NMR) in the field of the isotopic methods essentially based until now on mass spectrometry (MS) techniques. Although an exhaustive review of the isotopic mass spectrometry results was not in the scope of this presentation, the information provided by the SNIF-NMR method has always been discussed in the light of the knowledge already acquired from the interpretation of mean MS isotope parameters.

Evidence for very large deviations with respect to a random distribution of deuterium within molecules in natural conditions has opened new research possibilities of both fundamental and applied interest. In appropriate situations, the SNIF parameters provide us with the advantages of a labelling technique exempt from the need for selective isotopic enrichments. The NMR method is therefore very attractive for investigating the biogenesis of natural products. From a quantitative point of view, it enables primary and secondary isotope effects of chemical and biochemical reactions to be determined in a one-pot experiment. A new approach has now been defined for inferring the isotope contents of disappeared precursors from the SNIF parameters of natural products. The site-specific isotopic patterns constitute powerful sources of information about the filiation and the history of the hydrogen atoms. They are faithful witnesses of the chemical, biochemical, physiological, or technological phenomena which have presided over the elaboration of the product. In particular, they provide new probes of climatic significance.

From the methodological point of view, the spectroscopic and sampling conditions have been optimized and a computerized environment has been defined to facilitate data treatment and automation of the analytical procedures.

The SNIF fingerprint has been shown to be of great interest in quality assessment of food products in terms of natural or synthetic origin, botanical species of the precursor, geographical area of plant growing, year of production, and the addition of exogeneous components from a different origin.

9 References

1. Craig H (1961) Science 133: 1833
2. Hagemann R, Nief G, Roth E (1970) Tellus 22: 712
3. Craig H (1957) Geochim Cosmochim Acta 12: 133
4. Junc GA, Svec HJ (1958) Geochim Cosmochim Acta 14: 234
5. De Wit JC, Van Der Straaten CM, Mook WG (1980) Geostandards Newslett 4: 33
6. Gonfiantini R (1978) Nature 271: 534
7. Baertschi P (1976) Earth Planet Sci Lett 31: 341
8. Mc Crea JM (1950) J Chem Phys 18: 849
9. Craig H (1953) Geochim Cosmochim Acta 3: 53
10. Owen T, Maillard JP, De Bergh C, Lutz BL (1988) Science 240: 1767
11. Donahue TM, Hoffman JH, Hodges RR, Watson AJ (1982) Science 216: 630
12. Craig H, Gordon LI, Horibe Y (1963) J Geophys Res 68: 5079
13. Dansgaard W (1964) Tellus 16: 436
14. Fontes JC, Gonfiantini R (1970) Earth Planet Sci Lett 7: 325
15. Merlivat L, Coantic M (1975) J Geophys Res 80: 3455
16. Yurtsever Y, Gat JR (1981) In: Stable isotope hydrology, IAEA Technical Reports Series n° 210, p 103
17. Niewodniczanski J, Grabczak J, Baranski L, Rzepka J (1981) J Glaciology 27: 99
18. O'leary MH (1981) Phytochemistry 20: 553
19. O'leary MH (1988) Bioscience 38: 328
20. Craig H (1954) J Geol 62: 115
21. Baertschi P (1953) Helv Chim Acta 36: 773

22. Park R, Epstein S (1960) Geochim Cosmochim Acta 21: 110
23. Lerman JC, Raynal J (1972) CR Acad Sci D 275: 1391
24. Bender MM (1968) Radiocarbon 10: 468
25. Bender MM (1971) Phytochemistry 10: 1239
26. Smith BN, Epstein S (1971) Plant Physiol 47: 380
27. Bender MM, Rouhani I, Vines HM, Black CC (1973) Plant Physiol 52: 427
28. Nalborczyk I, Lacroix LJ, Hill RD (1975) Can J Bot 53: 1132
29. Lerman JC, Queiroz O (1974) Science 183: 1207
30. Lerman JC, Deleens E, Nato A, Moyse A (1974) Plant Physiol 53: 581
31. Schidlowski M (1988) Nature 333: 313
32. Schoell M (1982) Spectra 8: 32
33. Ziegler H (1979) Ber Deutsch Bot Ges 92S, 169
34. Ziegler H, Osmond CB, Stichler W, Trimborn (1976) Planta 128: 85
35. Sternberg L, Deniro MJ, Ajie H (1984) Phytochemistry 23: 2475
36. Smith BN, Jacobson BS (1976) Plant Cell Physiol 17: 1089
37. Bricout J (1982) In: Stable isotopes, (eds) Schmidt HL, Förstel H, Heinzinger K, p 483, Elsevier 11
38. Yeh HW, Epstein S (1981) Geochim Cosmochim Acta 45: 753
39. Sternberg L, Deniro MJ, Ting IP (1984) Plant Physiology 74: 104
40. Martin GJ, Zhang BL, Naulet N, Martin ML (1986) J Am Chem Soc 108: 5116
41. Gonfiantini R (1981) In: Stable isotope hydrology, IAEA Technical Reports Series n° 210, p 35
42. Martin GJ, Martin ML (1981) CR Acad Sci II, 293: 31
43. Martin GJ, Martin ML (1981) Tetrahedron Lett 22: 3525
44. Martin GJ, Martin ML, Mabon F, Michon MJ (1982) JCS Chem Comm 616
45. Martin GJ, Martin ML, Mabon F, Bricout J (1982) J Am Chem Soc 104: 2658
46. Martin GJ, Martin ML, Mabon F, Michon MJ (1982) Anal Chem 54: 2380
47. Gonfiantini R (1984) Int J Appl Radiat Isot 35: 426
48. Martin GJ, Zhang BL, Martin ML, Dupuy P (1983) Biochem Biophys Res Comm 111: 890
49. Diehl P, Leipert T (1964) Helv Chim Acta 47: 545
50. Diehl P (1974) In: Nuclear magnetic resonance spectroscopy of nuclei other than protons, (eds) T Axenrod, GA Webb, p 275, J Wiley
51. Briggs JM, Farnell LF, Randall EW (1973) JCS Chem Comm 70
52. (a) Mantsch HH, Saito H, Smith ICP (1982) Progr NMR Spectrosc 11: 211 (1977); (b) Smith ICP, Mantsch HH (1982) In: NMR spectroscopy: new methods and applications, (ed) GC Levy, ACS symposium Series 191, Am Chem Soc Washington
53. Brevard D, Kintzinger JP (1978) in NMR and the Periodic Table. (eds) RK Harris, BE Mann, p 107, Academic Press, NY
54. Grivet JP (1980) Spectra 2000, 8: 60
55. Evans DF (1982) JCS Chem Comm 1226
56. Altman LJ, Laungani D, Gunnarsson G, Wennerstrom H, Forsen S (1978) J Am Chem Soc 100: 8264
57. Pitner TP, Edwards WB, Bassfield RL, Whidby JF (1978) J Am Chem Soc 100: 246
58. Casadevall E, Metzger P (1970) Tetrahedron Lett 48: 4199
59. Martin ML, Martin GJ, Couffignal R (1968) J Chem Phys 49: 1985
60. Garson MJ, Staunton J (1979) Chem Soc Rev 8: 539
61. Akitt JM (1987) in Multinuclear NMR, (ed) J Mason, Plenum Press, NY, p 171
62. Jarrel HC, Smith ICP (1983) In: The multinuclear approach to NMR spectroscopy, (eds) JB Lambert and FC Riddell, p 133 and p 151, Nato ASI series D Reidel
63. Seelig J (1977) Quaterly Rev Biophys 10: 353
64. Oldfield E, Chapman D, Derbyshire W (1971) Febs Letters 16: 102
65. Saito H, Schreier-Muccillo S, Smith ICP (1973) Febs Letters 33: 281
66. Fujiwara F, Reeves AS, Wilson LA (1974) J Am Chem Soc 96: 5249
67. Stockton GW, Johnson KG, Butler KW, Polnaszek CF, Cyr R, Smith ICP (1975) Biochim Biophys Acta 401: 535
68. (a) Hansen PE (1984) Ann reports NMR Spectrosc 15: 106; (b) Hansen PE (1988) Progr NMR Spectrosc 20: 207
69. Hertz HG (1967) Progr NMR Spectrosc 3: 159
70. Boere RT, Kidd RG (1982) In: Ann Reports on NMR Spectrosc, (ed) GA Webb 13: 319, Academic Press

71. Martin GJ, Sun XY, Guillou C, Martin ML (1985) Tetrahedron 41: 3285
72. Martin ML, Delpuech JJ, Martin GJ (1980) Practical NMR spectroscopy (a) p 350 (b) 392, Heyden–Wiley, London
73. Martin GG, Pelissolo FJC, Martin GJ (1986) Computer Enhanced Spectroscopy 3: 147
74. Green PE, Caroll JD (1976) Mathematical tools for applied multivariate analysis, p 278, Academic Press, New York
75. Guillou C, Trierweiler M, Martin GJ (1988) Magn Res Chem 26: 491
76. Melander L. (ed) (1960) Isotope Effects on Reaction Rates, The Ronald Press Company, New York
77. Melander L, Saunders WH (eds) (1980) Reaction rates of isotopic molecules, J Wiley, New York
78. Wiberg KB (1955) Chem Rev 55: 713
79. Laszlo P, Welvart (1966) Bull Soc Chim Fr 2412
80. Hanzlik RP, Schaefer AR, Moon JB, Judson CM (1987) J Am Chem Soc 109: 4926
81. Pascal RA, Baum MW, Wagner CK, Rodgers LR (1984) J Am Chem Soc 106: 5377
82. Pascal RA, Baum MW, Wagner CK, Rodgers LR, Huang DS (1986) J Am Chem Soc 108: 6477
83. Saunders WH, Edison DH (1960) J Am Chem Soc 82: 138
84. Zhang BL, Li GG, Gao ZH, Wu JZ (1988) Chem J Chin Univ 9: 808
85. Leopold MF, Epstein WW, Grant DM (1988) J Am Chem Soc 110: 616
86. Zhang BL, Wu WX, Gao ZH, Sun XY (1986) Acta Chim Sin 44: 437
87. Zhang BL (1988) Magn Res Chem 26: 955
88. Kooner ZS, Puthela RC, Fenby DV (1980) Aust J Chem 33: 9
89. Rolston JH, Gale KH (1984) J Phys Chem 88, 163 and 88: 4394
90. (a) Kooner ZS, Fenby DV (1980) Aust J Chem 33 1943; (b) Phutela RC, Kooner ZS, Fenby DV (1979) Aust J Chem 32: 2353
91. Bertrand GL, Burchfield TE (1975) J Phys Chem 79: 1547
92. Halford JO, Pecherer B (1938) J Chem Phys 6: 571
93. Kwart H, Huhn LP Bannister EL (1954) J Am Chem Soc 76: 5998
94. Khurma JR, Fenby DV (1979) Aust J Chem 32: 465
95. Siehl U (1987) Adv Phys Org Chem 23: 63
96. Kresge AJ, Allred AL (1963) J Am Chem Soc 85: 1541
97. Gold V (1963) Proc Chem Soc 141
98. Gold V (1968) Trans Faraday Soc 64: 2770
99. Baltzer L, Bergman NA (1982) JCS Perkin II: 313
100. Schowen RL (1972) Prog Phys Org Chem 9: 275
101. Bone R, Wolfenden R (1985) J Am Chem Soc 107: 4772
102. Saunders M, Saunders S, Johnson CA (1984) J Am Chem Soc 106: 3098
103. Jarret RM, Saunders M (1985) J Am Chem Soc 107: 2648
104. Paquette LA, Kitching W, Heyd WE, Meisinger RH (1974) J Am Chem Soc 96: 7371
105. Goldstein MJ, Pressman EJ (1981) J Am Chem Soc 103: 6533
106. Reuben J (1986) J Am Chem Soc 108: 1082
107. Martin ML, Zhang BL, Trierweiler M, Martin GJ, unpublished results
108. Günther H, Alizade MA, Kellner M, Biller F, Simon H (1973) Z Naturforsch. 28c, 241
109. Hagele G, Boisniere B, Helie I, Rabiller C, Martin ML (1990) to be published
110. Rabiller C, Mesbahi M, Martin ML (1990) Chirality 2: 85
111. Rambeck W, Simon H (1971) Hoppe Seyler's Z Physiol Chem 352: 65 and (1972) 353: 1107
112. Hartshorn SR, Shiner VJ (1972) J Am Chem Soc 94: 9002
113. Cleland WW (1980) Methods in Enzymology 64: 104
114. Hansen EW, Ruoff P (1988) J Phys Chem 92: 2641
115. Morris GA (1986) Magn Res Chem 24: 371
116. Ernst RR, Bodenhausen G, Wokaun A (1987) Principles of nuclear magnetic resonance in one and two dimensions, Clarendon Press, Oxford
117. Kessler H, Gehrke M, Griesinger (1988) Angew Chem Int Ed 27: 490
118. Schmitt P, Wesener JR, Günther H (1983) J Magn Res 52: 511
119. Wesener JR, Schmitt P, Günther H (1984) J Am Chem Soc 106: 10
120. (a) Rinaldi PL, Baldwin NJ (1982) J Am Chem Soc 104: 5791
(b) Rinaldi PL, Baldwin NJ (1983) J Am Chem Soc 105: 7523
121. Bendall MR, Pegg DT, Wesener JR, Günther H (1984) J Magn Reson 59: 223
122. Nakashima TT, Mc Clung RED, John BK (1984) J Magn Res 58: 27

123. Doddrell DM, Staunton J, Laue ED (1983) J Magn Res 52: 523
124. Doddrell DM, Staunton J, Laue ED (1983) J Chem Soc Chem Comm 602
125. Abell C, Doddrell DM, Garson MJ, Laue ED, Staunton J (1983) JCS Chem Comm 694
126. (a) Bleich H, Gould S, Pitner P, Wilde J (1984) J Magn Res 56: 515
(b) Gould SJ, Palaniswamy VA, Bleich H, Wilde J (1984) JCS Chem Comm 1075
127. (a) Wesener JR, Schmitt P, Günther H (1984) Org Magn Reson 22: 468; (b) Wesener JR, Günther H (1985) J Am Chem Soc 107: 1537
128. Trimble LA, Reese PB, Vederas JC (1985) J Am Chem Soc 107: 2175
129. Wilkins CL, Marianelli RS (1968) Tetrahedron 26: 4131
130. Stothers JB, Tan CT (1974) JCS Chem Comm 738
131. Al-Rawi JMA, Behnam GQ, Taha NI (1981) Org Magn Res 17: 204
132. Hanson JR, Reese PB, Sadler IH (1984) JCS Perkin Trans I, 2937
133. Sato Y, Oda T, Saito H (1976) Tetrahedron Letters 2695; (1977) JCS Chem Comm 415
134. Cane DE, Murthy PPN (1977) J Am Chem Soc 99: 8327
135. Aberhart DJ, Caspi E (1979) J Am Chem Soc 101: 1013
136. Cane DE, Buchwald SL (1977) J Am Chem Soc 99: 6132
137. Lauffer RB, Que L (1982) J Am Chem Soc 104: 7324
138. Martin GJ, Zhang BL, Saulnier L, Colonna P (1986) Carbohydr Res 148: 132
139. Vallet C, Arendt M, Martin GJ (1988) Biotechnol Letters 2: 83
140. Sancho A, Combaut G, Jupin H, Giraud G, Quemerais B, Naulet N, Martin GJ (1989) Plant Physio and Biochem 27: 3
141. Martin GJ, Janvier P, Akoka S, Mabon F, Jurczak J (1986) Tetrahedron Lett 27: 2855
142. Martin ML, Zhang BL, Martin GJ (1983) FEBS Letters 158: 131
143. Fraser RR, Petit MA, Miskow M (1972) J Am Chem Soc 94: 3253
144. Harrison DM, Quinn P (1983) Tetrahedron Lett 24: 831
145. Gerlach H, Zagalak B (1973) JCS Chem Comm 274
146. Brown JM, Parker D (1981) Tetrahedron Lett 22: 2815
147. Parker D (1983) JCS Perkin Trans II: 83
148. Schwab JM (1981) J Am Chem Soc 103: 1876
149. Martin GJ, Janvier P, Mabon F (1985) Analusis 13: 267
150. Grant DM, Curtis J, Croasmun WR, Dalling DK, Wehrli FW, Wehrli S (1982) J Am Chem Soc 104: 4492
151. Deleens E (1976) Physiol Veget 14: 641
152. Abelson PH, Hoering TC (1961) Proc Natl Acad Sci 47: 623
153. Monson KD, Hayes JM (1982) Geochim Cosmochim Acta 46: 139 and (1980) J Biol Chem 225: 11435
154. Bengsch E, Grivet JP, Schulten HR (1981) Z Naturforsch 36b: 1289
155. De Niro MJ, Epstein S (1977) Science 197: 261
156. (a) Lacroix M, Mosora F, Pontus M, Lefebvre P, Luyckx A, Lopez-Habib G (1973) Science 181: 445; (b) Nakamura K, Schoeller DA, Winkler FJ, Schmidt HL (1982) Biomed Mass Spectrometry 9: 390; (c) Schoeller DA, Klein PD, Watkins JB, Heim T, Mac Lean WC (1980) Am J Clin Nutr 33: 2375
157. Schoeninger MJ, Deniro MJ (1983) Science 220: 1381
158. Mosora F, Lefebvre P, Pirnay F, Lacroix M, Luyckx A, Duchesne J (1976) Metabolism 25: 1575
159. (a) Rodelli MR, Gearing JN, Gearing PJ, Marshall N, Sasekumar A (1984) Oecologia 61: 326
(b) Gearing JN (1988) in Lecture notes on coastal and estuarines studies, vol. 22, (ed) BO Jansson, p. 70, Springer-Verlag, Berlin Heidelberg
160. Fry B, Parker L (1979) Estuarine and Coastal Marine Science 8: 499
161. Fry B, Lutes R, Northam M, Parker PL, Ogden J (1982) Aquatic Botany 14: 389
162. Sternberg L, Deniro NJ, Johnson HB (1984) Plant Physiol 74: 557
163. Smith BN, Epstein S (1970) Plant Physiol 46: 738
164. Bricout J, Merlivat L, Koziet J (1973) CR Acad Sci 277: 885
165. Estep MF, Hoering TC (1981) Plant Physiol 57: 474
166. Sternberg L, De Niro MJ (1983) Science 220: 947
167. Leaney FW, Osmond CB, Allison GB, Ziegler H (1985) Planta 164: 215
168. Sternberg L, De Niro MJ, Keeley JE (1984) Plant Physiol 76: 68
169. Sternberg L, De Niro MJ, Sloan ME, Black CC (1986) Plant Physiol 80: 242
170. Sternberg L, De Niro MJ, Ajie HO (1986) Planta 169: 320
171. Estep MLF (1984) Geochim Cosmochim Acta 48: 591

172. Merlivat L, Jouzel J (1979) J Geophys Res 84: 5029
173. Lesaint C, Merlivat L, Bricout J, Fontes JC, Gautheret R (1974) CR Acad Sci (D) 278: 2925
174. Allison GB, Gat JR, Leaney FWJ (1985) Chemical Geology 58: 145
175. Epstein S, Thompson P, Yapp CJ (1977) Science 198: 1209
176. Deleens E, Caer V, Schwebel-Dugue N, Naulet N, Martin ML, Carbohydr Res (1990) in print
177. Schiegl WE, Vogel JC (1970) Earth Planet Sci Lett 7: 307
178. Schiegl WE (1974) Nature 251: 582
179. Epstein S, Yapp CJ (1977) Nature 266: 477
180. Francey RJ, Farquhar GD (1982) Nature 297: 28
181. Lawrence JR, White JWC (1984) Nature 311: 558
182. Dupont LM, Mook WG (1987) Chemical Geology 66: 323
183. Epstein S, Yapp CJ, Hall JH (1976) Earth Planet Sci Lett 30: 241
184. De Niro MJ (1981) Earth Planet Sci Lett 54: 177
185. Northfelt DW, De Niro MJ, Epstein S (1981) Geochim Cosmochim Acta 45: 1895
186. Martin ML, Filleux ML, Martin GJ, unpublished results
187. Martin GJ, Guillou C, Martin ML, Cabanis MT, Tep Y, Aerny J (1988) J Agric Food Chem 36: 316
188. Gonfiantini R, Gratziu S, Tongiorgi E (1965) IAEA Vienna 5: 405
189. Ferhi A, Letolle R (1977) Physiol Veget 15: 363
190. Ferhi A, Bariac T, Letolle R (1983) In: Palaeoclimates and palaeowaters: A collection of environmental isotope studies. International Atomic Energy Agency, Vienna, p. 85
191. Farris F, Strain BR (1978) Rad and Environm Biophys 15: 167
192. Förstel H (1978) Rad and Environm Biophys 15: 323
193. Förstel H, Hutzen H (1982) In: Stable isotopes, (eds) HL Schmidt, H Förstel, K Heinzinger, Elsevier, p 511
194. Farquhar GD, Richards RA (1984) Aust J Plant Physiol 11: 539
195. Deleens E, Gregory N, Bourdu R (1984) Plant Science Letters 37: 35
196. Martin GJ, Odiot D, Godineau V, Naulet N, Morlat R, Asselin C, Robichet J (1989) Applied Geochemistry 4: 1
197. Dongmann G, Nürnberg HW, Förstel H, Wagener K (1974) Rad and Environm Biophys 11: 41
198. Wilson AT, Grinsted MJ (1975) Nature 257: 387
199. De Niro MJ, Epstein S (1981) Geochim Cosmochim Acta 45: 1885
200. Brenninkmeijer CAM, Van Geel B, Mook WG (1982) Earth Planet Sci Lett 61: 283
201. Yapp CJ, Epstein S (1977) Earth Planet Sci Lett 34: 333
202. Bricout J (1978) Rev Cytol Biol Veget Bot 1: 133
203. Winkler FJ, Schmidt HL (1980) Z-Lebensm Unters Forsch 171: 85
204. Krueger HW, Reesman RH (1982) Mass Spectrometry Reviews 1: 205
205. Schmidt HL (1986) Fresenius Z Anal Chem 324: 760
206. Smith BN (1975) Naturwissenschaften 62: 390
207. Bricout J, Fontes JC (1974) Ann Fals Exp Chim 716: 211
208. Bricout J, Koziet J (1985) Science des Aliments 5: 197
209. Nissenbaum A, Lifshitz A, Stepek Y (1974) Lebensm Wiss Technol 7: 152
210. Hillaire-Marcel C, Carro-Jost O, Jacob C (1977) J Inst Can Sci Technol Aliment 10: 333
211. Martin GE, Alfonso FC, Figert DM, Burggraff JM (1981) J Ass Off Anal Chem 64: 1149
212. Dunbar J (1982) In: Stable isotopes, (eds) HL Schmidt, H Förstel, Heinzinger K, Elsevier p 495
213. Dunbar J, Schmidt HL, Woller R (1983) Vitis 22: 375
214. Koziet J, Bricout J (1978) Ann Nutr Alim 32: 941
215. Bricout J, Menoret Y (1975) Ann Technol Agric 24: 247
216. Rauschenbach P, Simon H (1979) Z Naturforsch 34c: 1
217. Yankwich PE, Promislow AL (1953) J Am Chem Soc 75: 4881
218. Rinaldi G, Meinschen WG, Hayes JM (1974) Biomed Mass Spectrom 1: 412
219. Meinschen WG, Rinaldi GGL, Hayes JM, Schoeller DA (1974) Biomed Mass Spectrom 1: 172
220. Schmid ER, Fogy I, Schwarz P (1979) Z Lebensm Unters Forsch 166: 89
221. Schmid ER, Grundmann H, Fogy I, Papesch W, Rank D (1981) Biomed Mass Spectrom 8: 496
222. Morselli MF, Baggett KL (1984) J Ass Off Anal Chem 67: 22
223. Doner LW (1985) J Agric Food Chem 33: 770
224. Doner LW, Krueger HW, Reesman RH (1980) J Agric Food Chem 28: 362
225. Doner LW, White JW (1977) Science 197: 891

226. White JW, Doner LW (1978) J Assoc Off Anal Chem 61: 746
227. Dunbar J (1982) Fresenius Z Anal Chem 311: 578
228. Sudraud P, Koziet J (1978) Ann Nutr Alim 32: 1063
229. Doner LW, Ajie HO, Sternberg LSL, Milburn JM, De Niro MJ, Hicks KB (1987) J Agric Food Chem 35: 610
230. Bricout J, Mouaze Y (1971) Fruits 26: 77
231. Misselhorn K, Brückner H, Müssig-Zufika M, Grafahrend W (1983) Die Branntweinwirschaft 9: 162
232. Forstel H, Hützen H (1984) Weinwirtschaft-Technik 3: 71
233. Bricout J, Fontes JC, Merlivat L (1974) J Ass Off Anal Chem 57: 713
234. Hoffman PG, Salb M (1979) J Agric Food Chem 27: 352
235. Krueger DA, Krueger HW (1983) J Agric Food Chem 31: 1265
236. Krueger DA, Krueger HW (1985) J Agric Food Chem 33: 324
237. Toulemonde B, Horman I, Egli H, Derbesy M (1983) Helv Chim Acta 66: 2342
238. Maubert C, Guerin C, Mabon F, Martin GJ (1988) Analusis 16: 434
239. Martin GJ, Martin ML (1984) L'Actualité Chimique 31
240. Kalinowski HO (1988) Chemie in unserer Zeit 22: 162
241. Martin GJ, Martin ML, Mabon F, Bricout J (1983) Science des Aliments 3: 147
242. Martin GJ, Martin ML, Mabon F, Michon MJ (1983) J Agric Food Chem 31: 311
243. Martin GJ, Benbernou M, Lantier F (1985) J Inst Brew 91: 242
244. Martin GJ, Martin ML (1988) In: Modern methods of plant analysis, (eds) HF Linskens, JF Jackson 6: 258, Springer-Verlag, Berlin Heidelberg
245. Martin GJ, Martin ML (1983) J Chim Phys 80: 293
246. Martin GJ, Guillou C, Naulet N, Brun S, Tep Y, Cabanis JC, Cabanis MT, Sudraud P (1986) Science des Aliments 6: 385
247. Shaka AJ, Keeler J, Freeman R (1983) J Magn Res 53: 313
248. Haasnoot CAG (1983) J Magn Res 52: 153
249. Franconi JM, Naulet N, Martin GJ (1989) J Cereal Science 10: 69
250. Mariotti A, Germon JC, Hubert P, Kaiser P, Letolle R, Tardieux A, Tardieux P (1981) Plant and Soil 62: 413
251. Irlev AA, Lapin AV, Brizanova LY (1987) Fiziologiya Rastenii 34: 493

Dynamic NMR Spectroscopy in the Presence of Kinetic Hydrogen/Deuterium Isotope Effects

Hans-Heinrich Limbach*

Institut für Physikalische Chemie der Universität Freiburg, Albertstr. 21, D-7800 Freiburg i. Br., FRG

Table of Contents

---

* Present address: Fachbereich Chemie, Takustr. 3, D-1000 Berlin 33

NMR Basic Principles and Progress, Vol. 23

The determination and interpretation of kinetic hydrogen/deuterium isotope effects of chemical and biochemical reactions constitutes a challenging problem to physicists and chemists. So far, kinetic isotope effects have mainly been measured using conventional kinetic methods. In view of the great potential of dynamic multinuclear NMR spectroscopy—which includes the techniques of lineshape analysis, polarization transfer and relaxation time measurements—the question arises whether this method can also be a useful kinetic tool in the chemistry of hydrogen isotopes. The scope of this review is to show that dynamic NMR spectroscopy can indeed contribute to the development of the field of isotopic reaction kinetics. Certain kinetic isotope effects can even only be observed using this method.

The first sections provide a theoretical background of dynamic NMR spectroscopy in the presence of kinetic isotope effects. Then, vacuum methods for the preparation of sealed NMR samples of high purity with defined reactant concentrations and deuterium fractions in the mobile proton sites are described. In the subsequent sections examples for the determination of multiple kinetic hydrogen/deuterium isotope effects on intramolecular and intermolecular double and triple proton transfer reactions in the liquid and the solid state by NMR are given. Finally, the results obtained so far, as well as possible areas of isotope chemistry where the NMR technique could be applied in the future are briefly discussed.

Explanation of Symbols

c_{A_i} concentration of A in state i
dC_{rs}/dt number of molecules per volume leaving from environment r to s during the time dt
g_{κ_r} statistical weight of the spin function k in the environment r
$\hbar$ Planck's constant devided by 2π
e ± 1
i $= \sqrt{-1}$
k exchange rate constant of a degenerate process
k_{rs} pseudo first order exchange rate between environment r and s
k_{rls} unimolecular rate constant of the rate limiting step of a reaction
$k_{rs}(A)$ pseudo first order rate constant of exchange of group A in environment r to environment s, same as τ_{rs}^{-1}
k_+, k_- forward and backward rate constants
k_i rate constant of reaction i
n_{Ar} number of spin states in group A of environment r
p_r population or mole fraction of an environment r
r, s molecular environments, species, states etc.
t time, mixing time in polarization transfer experiments
t_1 evolution time in two-dimensional NMR experiments
t_2 detection time in two-dimensional NMR experiments
A^rB^r molecule or molecular environment r consisting of two parts A and B
B_0, B_1 external magnetic fields in the laboratory and the rotating frame
C^l, C^r "left" and "right" matrix diagonalizing $\mathcal{M}$
C_r total concentration of environment r
C_A total concentration of species AH + AD
C_{AL} total concentration of species AH or AD
D overall deuterium fraction
D_A deuterium fraction in environment A
$\mathcal{H}$ Hamilton operator
I_i, I_{zi} angular momentum operators of spin i in the subsystem r.
I^- lowering operator

I^+	raising operator
J_{ij}, $J_{A^rB^r}$	coupling constant in Hz between spins i and j or between A and B in environment r
K_{ij}	reduced coupling constant between spin i and j
K_{rs}	equilibrium constant of the interconversion of environment r and s
K_i	equilibrium constant of reaction i
L	hydrogen isotope
$\mathscr{L}$	Liouville superoperator
$\mathscr{L}_r$	Liouville superoperator of environment r
$\mathscr{L}_{A^r}$, $\mathscr{L}_{B^r}$	Liouville superoperator of group A and B in environment r
$\mathscr{L}_{A^rB^r}$	part of the Liouville superoperator describing the scalar spin–spin interaction between groups A and B in environment r
$M_r(t)$	magnetization of environment r as a function of time
$\mathscr{M}$	superoperator in Liouville space describing the time dependence of the density matrix
N_r	number of spin states in the environment r
O_{rs}	permutation operator which changes a spin function of environment r into the corresponding spin function of environment s
P	primary kinetic isotope effect
P_{rs}	probability of interconversion of environment r to environment s
Q_m	complex intensity of transition m
Q_m^{re}	real part of Q_m
Q_m^{im}	imaginary part of Q_m
$\mathscr{R}$	Redfield relaxation superoperator
$\mathscr{R}^{(1)}$, $\mathscr{R}^{(2)}$	longitudinal and transverse part of the Redfield relaxation superoperator
S	secondary kinetic isotope effect
T_{1r}	longitudinal relaxation time in the environment r
T_{2r}	transverse relaxation time in the environment r
T_{2r}^*	effective transverse relaxation time in the environment r which includes the inhomogeneity of the magnetic field B_0
$T_{1\rho}$	longitudinal relaxation time in the rotating frame
$Y(\nu)$	lineshape function
W_{0r}	effective NMR-line width of environment r in the absence of exchange
α	spin state of a spin 1/2 nucleus
β	spin state of a spin 1/2 nucleus
γ_i	gyromagnetic ratio of spin i
δ_{rs}	Kronecker symbol, $=1$ for $r=s$ and $=0$ for $r \neq s$
μ_r, ν_r	spin functions of environments r
ν	frequency in Hz
ν_i	Larmor frequency of spin i in Hz

π	3.14159
$\rho, \rho(t)$	density matrix at time t
σ_i	the screening constant of the magnetic field of nucleus i
τ_r	average lifetime of environment r
τ_{rs}^{-1}	average inverse lifetime of environment r before the conversion to environment s, same as k_{rs}
$\tau_{rx}^{-1}(A)$	average inverse lifetime of group A in environment r before reaction to environment x, same as $k_{rx}(A)$
τ_{AL}^{-1}	inverse lifetime of group L in A
τ_{AHXH}^{-1}	average lifetime of H on A before the jump to X
$\tau_{AHXD}^{-1}, \tau_{AHXH*}^{-1}$	average inverse lifetime of A before H it is replaced by D or another H and before A rearranges to X
τ_{ALXL}^{-1}	average inverse lifetime of L in AL before it is incorporated into XL
τ_{ALA}^{-1}	inverse lifetime of group A in AL before L leaves A
Φ	inverse equilibrium constant of fractionation of isotopes between two different sites (fractionation factor)
$\Phi^{\ddagger}$	inverse kinetic isotope effect or fractionation factor between ground and transition state
$\Delta\nu$	frequency interval in Hz
$\Delta\Omega$	frequency interval in rad/sec
Λ_m	mth eigenvalue of $\mathscr{M}$
Λ_m^{re}	real part of Λ_m
Λ_m^{im}	imaginary part of Λ_m
ω_1	$=\gamma B_1$
Ω_r	Larmor frequency of environment r in radian/sec
Ξ	exchange superoperator

1 Introduction

In 1931 Urey, Brickwedde, and Murphy discovered the hydrogen isotope deuterium. The isotopic enrichment was found to arise from the fact that the electrolysis of light water is faster than of heavy water [1, 2]. This success showed that although different isotopes of an element behave identically from a chemical standpoint the different isotopic masses nevertheless lead to both isotope effects on equilibrium as well as on rate constants of chemical reactions. Soon, ratios of equilibrium constants of isotopic reactions were called "equilibrium isotope effects" (EIE), ratios of isotopic rate constants "kinetic isotope effects" (KIE). Isotope effects have been found to be especially large for those elements which are directly involved in bond breaking and bond formation during the reaction studied [3]. Such effects are, therefore, referred to as "primary". Isotopic substitution in atomic sites which maintain all chemical bonds with their neighbors during the reaction of interest leads then only to smaller "secondary" isotope effects. Because of the unique mass relation between the different hydrogen isotopes hydrogen/deuterium isotope effects are particularly large and have attracted most attention. The largest contributions to these effects arise from changes in the vibrational frequencies of the reactants. The theory of equilibrium isotope effects has been founded by Urey [4] and Bigeleisen [5, 6] and has widely been accepted [3]. Generally, kinetic isotope effects can be much larger than equilibrium isotope effects; this is because of the particular vibrational structure of transition states of chemical reactions which involve "non bonded" atoms. However, since transition states cannot directly be observed, the interpretation of kinetic isotope effects is not always straightforward. Nevertheless, KIE are often the only source of information on molecular events in a variety of chemical and biochemical reactions. By combining the latter with Eyrings's transition state theory [7] Bigeleisen [8] derived a general theory of kinetic isotope effects for over barrier reactions [3]. This theory was modified by Bell in order to include tunneling in the case of hydrogen transfer reactions, a phenomenon Bell had already discussed in the thirties [9, 10]. In the last decades efforts have been made to develop models of kinetic hydrogen/deuterium isotope effects [11–16] and of proton transfer in general [9, 10, 17–27] based on more rigorous quantum mechanical treatments.

Hydrogen transfer reactions in which either protons, hydrogen atoms or hydride ions are transferred are not only interesting from a theoretical standpoint but play a major role in different areas of chemistry as a quick look in textbooks of organic, inorganic or biological chemistry will show. One finds complicated nucleophilic substitution and reduction/oxidation reactions as well as simple proton transfer processes [27]. Multiple proton transfer reactions are related to bifunctional and enzymatic catalysis [28–54]. When hydrogen transfer systems are incorporated in dyes they can also be induced by light, even at cryogenic temperatures if certain requirements are met [55–60].

In the past decades kinetic isotope effects in chemical reactions have mainly been studied using conventional kinetic [3, 27] or fast reaction methods [27, 61–63] where concentrations of reactants or products are monitored as a function of time. Although various spectroscopic techniques have been used for this purpose such as UV/VIS-, IR- and mass spectroscopy etc., dynamic Nuclear Magnetic Resonance (DNMR) techniques [64–81] have been employed only very recently in this research area. In view of the great potential of multinuclear pulse Fourier Transform NMR spectroscopy [72, 73] to selectively monitor different isotopes of a given nucleus as spin probes for molecular structure and dynamics, this situation is surprising. The main purpose of this article is, therefore, to show that and how the so far separate fields of isotope chemistry and dynamic NMR spectroscopy can be linked together.

NMR can be used in different ways to study kinetic isotope effects. First, as in conventional techniques the concentrations of reacting isotopic species can be measured as a function of time if the reactions studied are slow enough and if non-equilibrium initial states can be prepared. Secondly, when the reactions become too fast it is sometimes still possible to obtain kinetic isotope effects from an NMR isotope analysis of the products. Especially ^{2}H NMR spectroscopy has been applied for this purpose [82–84]. The third use of NMR spectroscopy in isotope chemistry, the topic of this article takes full advantage of NMR as a fast reaction method, which includes the techniques of line shape analysis [64–72], and polarization transfer [73–81], in one or two dimensions.

NMR is a special method covering reactions which are not easily studied by other techniques. Therefore, one has to discuss for which kind of reactions kinetic isotope effects can be obtained by this approach. In principle, any reversible reaction can be studied. In practice, however, NMR performs best when the concentrations of reactants and products are of equal magnitude, i.e. when the equilibrium constant of the reaction studied is unity. This is just the type of reaction which cannot easily be followed by other techniques, which require that $K \neq 1$. It is, therefore, understandable, that the literature on kinetic isotope effects mainly refers to asymmetric reactions. From a theoretical standpoint, however, the knowledge of kinetic isotope effects on symmetric or quasisymmetric reactions is especially important. Therefore, NMR constitutes a valuable complement to conventional kinetic methods.

Since the largest kinetic isotope effects are found in reactions where hydrogen isotopes are transferred, NMR will be applied in the near future mostly to this type of reactions. Proton transfer reactions have been studied by dynamic NMR spectroscopy since its early days [85–88]. Mostly proton exchange in water and other protic solvents were studied where H_3O^+, OH^- or other acids and bases act as catalysts of the exchange. These reactions are, generally, very complex and it is, therefore, not surprising that only very few proton and deuteron exchange rates have been studied by NMR [88–90]. Kinetic hydrogen/deuterium isotope effects of reactions in protic media have, therefore, mainly been measured using other kinetic methods [9, 10, 27, 43].

Because of the high complexity of proton exchange in protic media there has been a trend in the past decade to study proton exchange reactions in aprotic solvents [3, 27, 39, 91–129]. However, in these media the problem of hydrogen/deuterium exchange with residual mobile protons on glass walls and traces of water arises. This problem is smaller in the case of hydrogen transfers to strong bases. Therefore, mostly the latter reactions were studied using conventional kinetic techniques. In early dynamic NMR studies of kinetic isotope effects residual mobile protons in the NMR samples were removed by adding deuterated alcohols to the NMR samples [109, 111], a procedure which is not practical. The problem of preparing pure NMR samples of proton donors in organic solvents with defined deuterium fractions in the proton sites was, however, solved [87, 104–106]. Thus, it has been possible to measure KIE of symmetric proton transfer reactions between nitrogen atoms and oxygen atoms by DNMR. Both intermolecular [86, 87, 100–108] and intramolecular proton exchange reactions [105, 109, 111, 115, 120, 122, 124–129] in aprotic solvents were studied. These reactions have attracted considerable theoretical interest [15, 16, 130–144].

One problem in the study of proton transfer reactions is the determination of the number of protons which are transferred in the rate limiting step. E.g., when proton donors like H_2O or more complicated bifunctional catalysts are present in is often difficult to distinguish between the processes

$$AL \cdots B \rightleftharpoons A \cdots LB, \quad L = H, D \tag{1.1}$$

and

$$AL \cdots XL \cdots B \rightleftharpoons A \cdots LX \cdots LB, \quad L = H, D \tag{1.2}$$

etc. and to measure their kinetic hydrogen/deuterium isotope effects. Whereas the process described by Eq. (1.1) depends only on the two isotopic rate constants k^H and k^D the process described by Eq. (1.2) is characterized by four different forward rate constants k^{HH}, k^{HD}, k^{DH}, and k^{DD} which are difficult to determine but which are interesting to know from a theoretical standpoint. In order to determine the number of protons transferred, generally, kinetic measurements have to be performed as a function of the deuterium fraction in the mobile proton sites [42–48]. This method has been called "proton inventory technique" [43]. However, the extraction of reliable numbers requires knowledge of the multiple kinetic isotope effects which is difficult to obtain when the exact reaction mechanism is unknown. Therefore, most workers have assumed the validity of the so called "rule of the geometric mean (RGM)", which states for a double proton transfer that [43]

$$k^{HD} = (k^{HH} \cdot k^{DD})^{1/2}, \quad \text{i.e.} \quad k^{HH}/k^{HD} = k^{HD}/k^{DD}. \tag{1.3}$$

This rule was originally derived by combination of equilibrium isotope effect [5] and transition state theory [7] for concerted double proton transfers.

Since NMR has the advantage of being able to perform experiments selectively on different nuclei it can be helpful in the study of multiple kinetic hydrogen/deuterium isotope effects of proton transfer reactions. Thus, it has become possible to determine such effects in the case of some double proton transfer reactions without using the RGM. E.g., full kinetic HD/HD/DD isotope effects were reported in 1982 for proton exchange between methanol and acetic acid in tetrahydrofuran [105, 106] and for the porphyrin tautomerism [105]. Subsequently, such effects have been determined by NMR for several other inter- and intramolecular double proton transfer systems [107, 108, 120, 124–129]. In these studies major unexpected deviations from the RGM have been found, from which interesting insights into the reaction mechanisms could be obtained. Thus, questions such as whether tunneling plays a role and whether the double proton motion is concerted or stepwise were studied. Note that kinetic H/D isotope effects of intramolecular proton transfer reactions have also been obtained by dynamic EPR spectroscopy [145–151] which is, however, restricted to radicals. Recently, the development of high resolution solid state NMR techniques [152–165] has paved the way for studies of fast reactions in the solid state [162, 166–191]. Especially variable temperature NMR studies [162] under the conditions of cross polarization (CP) [152–154], magic angle spinning (MAS) [155–157], and proton decoupling [154] have been used in the past in order to determine some kinetic hydrogen/deuterium isotope effects [164, 167, 179]. The whole potential of this technique for the study of KIE has, however, not yet been fully exploited.

The purpose of this review is to present a framework for the study of kinetic hydrogen/deuterium isotope effects by dynamic NMR spectroscopy and to discuss stratagems, possibilities and drawbacks of this method. For the reasons stated above experimental examples will be presented from the area of multiple hydrogen transfer reactions, although strategies arising from these studies are general and could equally well be applied to other atom transfer processes.

The article is organized as follows: Section 2 contains a short description of general techniques of dynamic NMR spectroscopy based on the quantum mechanical density matrix formalism. Since exchange reactions can be followed only if they are associated to a change in the nuclear spin hamiltonian, we discuss in Sect. 3 the aptitude of the different nuclear magnetic interactions for use in the study of kinetic isotope effects. So far, NMR techniques have mostly be applied to single reactions systems. Therefore, the concept of group exchange is introduced in Sect. 4, which allows us to give a description of dynamic NMR lineshapes in the presence of different superposed complex bond breaking/bond formation processes. In Sect. 5 a theoretical background of the theory of the NMR proton inventory technique is provided where kinetic measurements are performed as a function of the deuterium fraction in the mobile proton sites. Especially, relations between the kinetic parameters obtained by NMR and rate constants of isotopic reactions are derived. These relations depend on the type of processes studied and can be quite complex especially in intermolecular multiple proton transfer reactions. In Sect. 6 typical NMR lineshape equations encountered in the study of intra- and intermolecular proton exchange reactions are presented. In Sects. 7 and 8 computational aspects of NMR lineshape analysis and NMR sample preparation techniques are described, adapted for the study of kinetic hydrogen/deuterium isotope effects in aprotic solvents.

The following sections are then devoted to practical examples of the determination of kinetic isotope effects in intramolecular (Sect. 9) and intermolecular (Sect. 10) multiple proton transfer reactions in liquids as well as in solids (Sect. 11). The review ends with a short discussion (Sect. 12) and a summary (Sect. 13).

2 General Dynamic NMR Spectroscopy

In this section we briefly review the theory of dynamic pulse Fourier transform NMR spectroscopy [64–73, 159, 160]. First, we will treat the theory of lineshapes in the presence of chemical exchange; for this purpose, the notation of Binsch [71] will be employed. Then, polarization transfer techniques in the slow exchange range, and, finally, relaxation methods will be discussed. For all techniques introductory examples will be given.

2.1 NMR Lineshape Theory

We consider ensembles of nuclear spins in different environments r, s, ... which are subject to chemical exchange. Such superposed spin systems can be characterized in Hilbert space by the density matrix ρ with the elements $\rho_{\mu_r \nu_r}$, where μ_r and ν_r are the spin states of the spin system in the environment r. The diagonal elements $\rho_{\mu_r \mu_r}$ correspond to polarizations, the off-diagonal elements $\rho_{\mu_r \nu_r}$ to quantum coherencies whose order is given by the difference in magnetic quantum numbers of the associated spin states. In Liouville space ρ corresponds to a vector whose time dependence is governed by the master equation in the rotating frame [71, 73]

$$\frac{\mathrm{d}\rho}{\mathrm{dt}} = \mathcal{M}(\rho - \rho(\infty)), \tag{2.1}$$

where $\rho(\infty)$ is the density matrix for $\mathrm{t} \to \infty$. The matrix $\mathcal{M}$ is given by [71]

$$\mathcal{M} = \mathrm{i}\mathcal{L} + \mathcal{R} + \Xi, \tag{2.2}$$

where $\mathrm{i} = \sqrt{-1}$, $\mathcal{L}$ the Liouville operator, $\mathcal{R}$ the Redfield relaxation operator and Ξ the operator describing the chemical exchange. The elements of $\mathcal{L}$ are given by

$$\mathcal{L}_{\mu_r \nu_r \kappa_s \lambda_s} = -2\pi(\mathcal{H}_{\mu_r \kappa_s} \delta_{\lambda_s \nu_r} - \mathcal{H}_{\lambda_s \nu_r} \delta_{\mu_r \kappa_s}), \tag{2.3}$$

where δ is the Kronecker symbol. $\mathcal{H}$ is the nuclear spin Hamiltonian whose elements will be discussed in the following section. In the absence of radio frequency pulses as typical for free induction decays the above master Eq. (2.1) splits up in a longitudinal and a transverse part:

$$\frac{d}{dt}\begin{bmatrix}\rho_{\mu_r\mu_r}\\ \rho_{\mu_s\mu_s}\\ \rho_{\mu_r\mu_r}\\ \rho_{\mu_s\mu_s}\end{bmatrix} = \begin{bmatrix}\mathscr{R}^{(1)}_r, \Xi & \Xi & & \\ \Xi & \mathscr{R}^{(1)}_s, \Xi & & \\ & & i\mathscr{L}_r, \mathscr{R}^{(2)}_r, \Xi & \Xi \\ & & \Xi & i\mathscr{L}_s, \mathscr{R}^{(2)}_s, \Xi\end{bmatrix} \begin{bmatrix}\rho_{\mu_r\mu_r} - \rho_{\mu_r\mu_r}(\infty)\\ \rho_{\mu_s\mu_s} - \rho_{\mu_s\mu_s}(\infty)\\ \rho_{\mu_r\mu_r} - \rho_{\mu_r\nu_r}(\infty)\\ \rho_{\mu_s\nu_s} - \rho_{\mu_s\nu_s}(\infty)\end{bmatrix} \tag{2.4}$$

$\mathscr{R}^{(1)}_r$ describes the longitudinal relaxation of the spin system in r, $\mathscr{R}^{(2)}_r$ the transverse relaxation. In the absence of spin–spin coupling $\mathscr{R}^{(1)}_r$ and $\mathscr{R}^{(2)}_r$ are diagonal with the elements given by

$$\mathscr{R}^{(1)}_{\mu_r\mu_r\mu_r\mu_r} = -1/T_{1r} \quad \text{and} \quad \mathscr{R}^{(2)}_{\mu_r\nu_r\mu_r\nu_r} = -1/T_{2r}, \tag{2.5}$$

where T_{1r} and T_{2r} are the longitudinal and transverse relaxation times in the environment r. In high resolutions studies of liquids, generally, artificial and apparative line broadening is included in $\mathscr{R}^{(2)}$, i.e.

$$\mathscr{R}^{(2)}_{\mu_r\nu_r\mu_r\nu_r} = -1/T^*_{2r} = -\pi W_{0r}, \tag{2.6}$$

where W_{0r} is the effective NMR line width of environment r in the absence of exchange. The approximation Eq. (2.5) holds also in the presence of spin–spin coupling as long as $|\Xi| \gg |\mathscr{R}|$.

The exchange operator Ξ depends on the particular exchange problem studied. The simplest exchange problem can be written as

$$r \underset{k_{sr}}{\overset{k_{rs}}{\rightleftharpoons}} s. \tag{2.7}$$

k_{rs} is the pseudo-first order reaction rate constant of the forward reaction defined by

$$k_{rs} = -\frac{1}{C_r}\frac{dC_{rs}}{dt}. \tag{2.8}$$

C_r is the concentration of environment r and dC_{rs} the number of moles per volume leaving from environment r to s during the time dt. k_{rs} is equal to the average inverse lifetime τ_{rs}^{-1} of environment r before the conversion to environment s occurs. Defining τ_r as total lifetime of r and P_{rs} as probability of interconversion to environment s it follows that

$$\tau_{rs}^{-1} = k_{rs} = P_{rs}\tau_r^{-1}. \tag{2.9}$$

The equilibrium constant of the above reaction is given by

$$K_{rs} = k_{rs}/k_{sr} = C_s/C_r = p_s/p_r, \tag{2.10}$$

where p_r is the mole fraction of environment r. If K_{rs} is known, only one pseudo first order rate constant, either k_{rs} or k_{sr} needs to be determined in order to characterize the kinetics of the reversible reaction in Eq. (2.8). In other words, the backward reaction step does not need to be treated explicitly.

In the simple case of intramolecular exchange or in the case of intermolecular exchange in the absence of spin–spin coupling the effect of the exchange $r \rightarrow s$ on the spin functions can be described by the operator equation

$$O_{rs}\mu_r = \kappa_s \quad \text{and} \quad O_{rs}\nu_r = \lambda_s \tag{2.11}$$

where O_{rs} is a permutation operator. The elements of the exchange operator are then given by [71]

$$\Xi_{\mu_r\nu_r\kappa_s\lambda_s} = -\delta_{rs}\delta_{\mu_r\kappa_s}\delta_{\nu_r\lambda_s}\left(\sum_{r \neq t} k_{rt}\right) + (1-\delta_{rs})\delta_{\mu_r\kappa_s}\delta_{\nu_r\lambda_s}k_{rs}. \tag{2.12}$$

As indicated in Eq. (2.4) the master equation splits up in two separate sets of equations for the polarizations and the coherences. The first describes polarization transfer due to chemical exchange which can be detected in the slow exchange regime by appropriate pulse experiments (see Sect. 2.3). Thus, from such experiments rate constants of the exchange can be obtained. In order to extract kinetic data from exchange broadened NMR lineshapes only the second set of equations describing the time evolution of the coherences—for which $\rho(\infty) = 0$—has to be solved. Since only single quantum coherences are observable density matrix elements corresponding to multiple quantum coherences are omitted in conventional lineshape calculations of spectra obtained by Fourier transformation of simple free induction decay signals. The dimension of the complex matrix $\mathcal{M}$ is equal to the number of observable transitions in the NMR spectrum of interest. Each transition m is characterized by the width Λ_m^{re} and the position Λ_m^{im}, where $\Lambda_m = \Lambda_m^{re} + i\Lambda_m^{im}$ is the mth complex eigenvalue of the matrix $\mathcal{M}$, calculated by diagonalization of $\mathcal{M}$ according to the transformation

$$\Lambda = C^{l}\mathcal{M}C^{r}. \tag{2.13}$$

The lineshape function is then conveniently written in the form

$$Y(\nu) \sim \sum_m \{[Q_m^{re}\Lambda_m^{re} - Q_m^{re}(\Lambda_m^{im} - 2\pi\nu)]/[(\Lambda_m^{re})^2 + (\Lambda_m^{im} - 2\pi\nu)^2]\}, \tag{2.14}$$

where each transition m is further characterized by the complex intensity

$$Q_m = Q_m^{re} + iQ_m^{im} = Q_m = \left(\sum_k I_k^- C_{km}^r\right)\left(\sum_l \rho_l(0)C_{lm}^l\right). \tag{2.15}$$

In Eq. (2.15) I_k^- are the elements of the lowering operator

$$I_k^- = I_{\mu_r\nu_r}^- = N_r^{-1}\langle\mu_r/I_r^-/\nu_r\rangle, \tag{2.16}$$

and N_r the number of spin states in the environment r. $\rho_l(0)$ represents the elements of the density matrix at the beginning of the acquisition at $t = 0$. For one-pulse experiments these elements are given by

$$\rho_1(0) = \rho_{\mu_r \nu_r}(0) = p_r \langle \mu_r / I_r^- / \nu_r \rangle . \tag{2.17}$$

p_r is the mole fraction of the environment r to which the spin states μ_r and ν_r are associated. In actual calculations it is only necessary to set up the matrix $\mathcal{M}$ and the vector $\rho(0)$. $\rho(0)$ is equal to the population vector p in the absence of scalar coupling.

2.2 Example of a Simple NMR Lineshape Analysis

As an example let us consider the simplest lineshape problem, i.e. exchange between two states r = 1 and s = 2, each containing only one spin. We assume that the Hamiltonian is such that both environments can be characterized by the absorption frequencies $\Omega_1 = 2\pi\nu_1$ and $\Omega_2 = 2\pi\nu_2$ in the rotating frame. The matrix $\mathcal{M}$ and the vector $\rho(0)$ are then given by

$$\mathcal{M} = \begin{bmatrix} -k_{12} - \pi W_{01} + i\Omega_1 & k_{21} \\ k_{12} & -k_{21} - \pi W_{02} + i\Omega_2 \end{bmatrix}, \quad \rho(0) = p = \begin{bmatrix} p_1 \\ p_2 \end{bmatrix}, \tag{2.18}$$

because the elements of the vector I^- are equal to 1. The lineshape can now easily be calculated using the equations of Sect. 2.1. An example is shown in Fig. 2.1. In the slow exchange region two lines of unequal intensity appear at

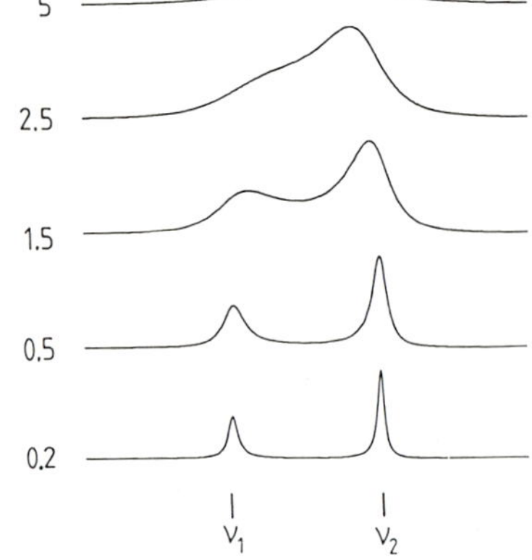

Fig. 2.1. NMR line shapes in the presence of exchange between two states, each containing one uncoupled spin, calculated according to Eq. (2.18)

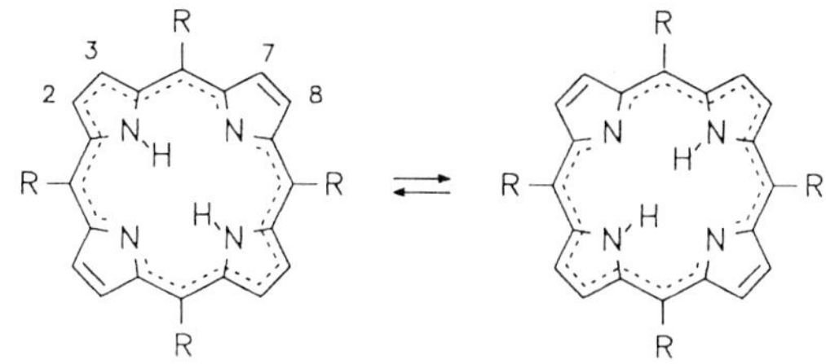

Fig. 2.2. The tautomerism of porphyrin (R = H) and of *meso*-tetraphenylporphyrin (TPP, R = phenyl)

positions given by the chemical shifts of the two states, i.e. ν_1 and ν_2. The line intensity ratios correspond to the equilibrium constant K_{12}. As k_{12} is increased the lines broaden and coalesce when $k_{12} > |2\pi\Delta\nu|$, where $\Delta\nu = \nu_1 - \nu_2$. The positions of the averaged line is given by

$$\nu = p_1\nu_1 + (1 - p_1)\nu_2 = (\nu_1 + K_{12}\nu_2)/(1 + K_{12}). \tag{2.19}$$

Thus, the equilibrium constant K_{12} can be obtained from the line position in the fast exchange regime if the chemical shifts are known.

As an application of Eq. (2.18) let us consider the tautomerism of meso-tetraphenylporphyrin (TPP, Fig. 2.2). This reaction was discovered by Storm et al. [109] and has been studied by several authors [109–116, 119, 125, 167, 168]. As shown in Fig. 2.3 two separate signals of equal intensity are observed in the low temperature ^{1}H NMR spectra of TPP dissolved in toluene-d_8. The low field line was assigned to the 2,3 protons and the high field line to the 7,8 protons [109]. As the proton transfer in TPP becomes fast both types of protons interconvert, leading to line broadening and line coalescence. In the fast exchange regime only one single line survives. Therefore, $K_{12} = 1$ in Eq. (2.19) and $k_{12} = k_{21} = k$. The spectra in Fig. 2.3 were simulated using Eq. (2.18) varying mainly the rate constant k. In the fast exchange regime the chemical shifts ν_1 and ν_2 had to be extrapolated from the slow exchange range. Examples of more complicated lineshape changes involving more complex spin systems will be discussed in later sections.

2.3 *Polarization Transfer Methods*

A problem with the low temperature experiments of Fig. 2.3 is that the residual line widths W_{01} and W_{02} strongly increase as the temperature is lowered, due to slow rotational diffusion and shortening of T_2 [116, 119]. Therefore, the value of $k = 5.5\,s^{-1}$ at 200 K obtained by lineshape analysis is affected by a large systematic error. Such uncertainties of rate constants in the slow exchange regime are encountered in all dynamic NMR experiments. As a remedy to this problem, polarization transfer experiments in the laboratory frame [75–77, 79] and in the rotating frame [80, 81, 164], in one and two dimensions, have been proposed. Whereas the first method is limited by the longitudinal relaxation time T_1 the latter is limited by $T_{1\rho}$, the longitudinal relaxation time in the rotating

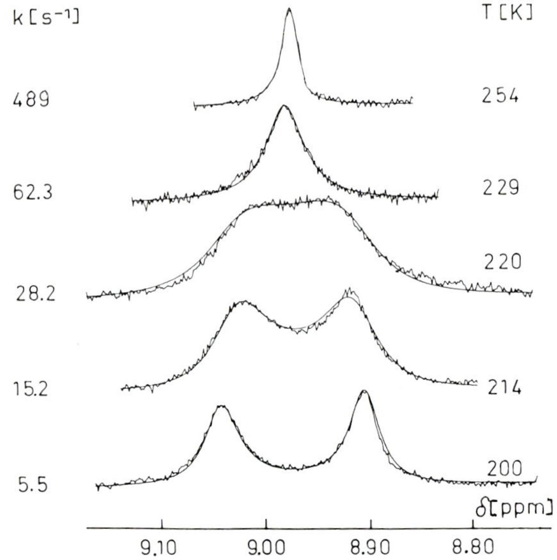

Fig. 2.3. Superposed experimental and calculated CW-100 MHz ^{1}H NMR line shapes of the β-pyrrole protons 2,3 and 7,8 of TPP dissolved in toluene-d_8. At low temperatures, the low field line (protons 2,3) shows an extra broadening due to an unresolved scalar long range coupling to the inner protons [109]. k is the calculated rate constant of the tautomerism in Figure 2.2. Reproduced with permission from Ref. [115]. See also Sect. 9.1.3 and 11.1.1

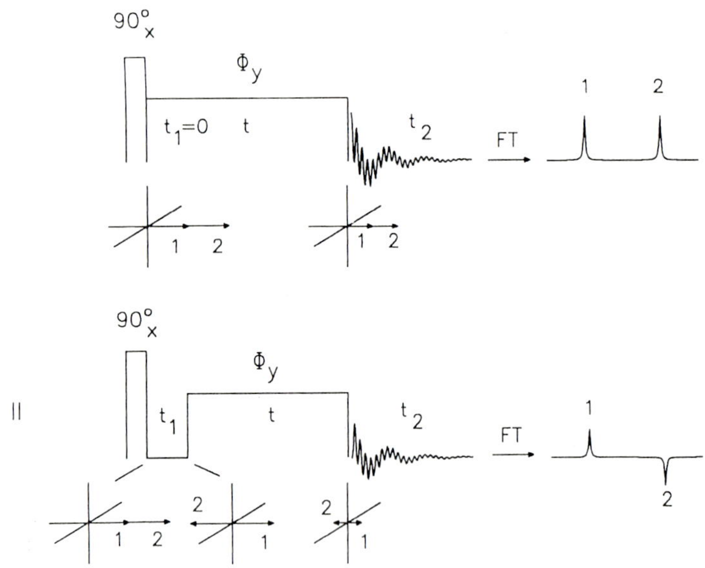

Fig. 2.4 Top: Pulse sequence for the measurement of the longitudinal relaxation time $T_{1\rho}$ in the rotating frame under the influence of a spin-locking pulse ("parallel experiment I"). **Bottom**: Pulse sequence for the detection of polarization transfer in the rotating frame according to Ref. [80] with $t_1 = 1/(2\Delta\nu)$ ("antiparallel experiment II"). The *arrows* indicate the exchanging polarizations

frame. Let us discuss here briefly the rotating frame exchange method for which experimental examples in DNMR of isotope effects will be given in Sects. 8 and 9.

The pulse sequence used is briefly discussed in Fig. 2.4. During the preparation period transverse magnetization is created by a 90^0_x pulse. After an evolution period t_1 a spin locking pulse of the strength $\omega_1 = \gamma B_1$ is applied in the y-direction of the rotating frame for the time t. The magnetizations become then polarizations in the spin locking field. After this mixing period follows the detection period t_2. This method is especially simple if neither scalar nor dipolar coupled spins are involved, and if only two exchanging polarizations M_1 and M_2 of equal probability are present, i.e. if $k_{12} = k_{21} = k$. In a typical one-dimensional experiment two sets of experiments with variable time t are performed. In experiment I, $t_1 = 0$; the exchanging polarizations M_r are parallel and decay exponentially with $T_{1\rho}$

$$M_r = M_r(0)\exp(-t/T_{1\rho}), \quad r = 1, 2. \tag{2.20}$$

This decay is monitored choosing several values for the mixing time t. In Eq. (2.20) it is assumed that $T_{1\rho}$ is equal for both polarizations or that the difference has been averaged out by the exchange. In experiment II the polarizations are prepared in an antiparallel way using a delay $t_1 = 1/(2\Delta\nu)$, where $\Delta\nu = \nu_1 - \nu_2$, between the two pulses. The carrier frequency is set to one of the exchanging lines. The exchange between the polarizations is given by [80]

$$M_r = M_r(0)\exp(-(1/T_{1\rho} + 2k)t), \quad r = 1, 2, \quad M_1(0) = -M_2(0). \tag{2.21}$$

Since $k_{12} = k_{21} = k$, both polarizations have the same decay constant $1/T_{1\rho} + 2k$. Thus, k can easily be obtained from the difference of the polarization decay in

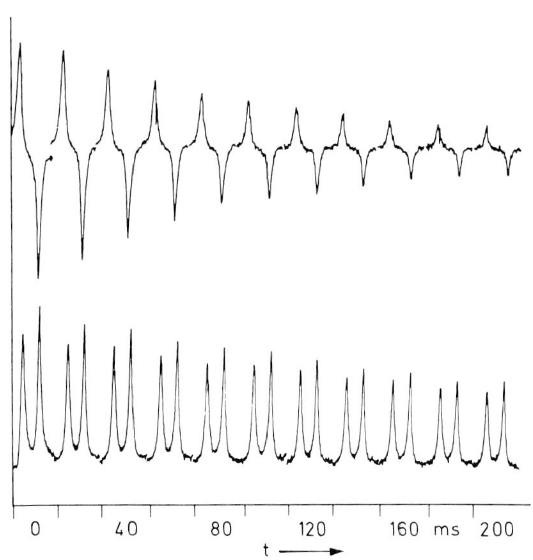

Fig. 2.5. Polarization transfer experiment performed on TPP dissolved in tetrahydrofuran-d_8 at 203 K and 90 MHz. **Bottom**: $T_{1\rho}$-experiment, **top** (i): polarization transfer in the rotating frame, with $t_1 = 1/(2\Delta\nu)$, $\Delta\nu = 25$ Hz. For further description of the experiment see text. Reproduced with permission from Ref. [80]

experiments I and II. The generalization to asymmetric reaction systems is straightforward [164].

As an example we consider again the case of TPP (Fig. 2.2) whose exchange broadened ^{1}H NMR signals were already shown in Fig. 2.3. The results of a ^{1}H NMR polarization transfer experiment in the rotating frame, performed on TPP in tetrahydrofuran at 203 K [80] are shown in Fig. 2.5. In both experiments exponential decays of the polarizations were observed. According to Eq. (2.20) the decay is governed in the "parallel" experiment by the longitudinal relaxation rate $T_{1\rho}^{-1} = 2.6 \pm 0.4\,s^{-1}$. By contrast, the decay in the "antiparallel" experiment is much faster, i.e. given according to Eq. (2.21) by $T_{1\rho}^{-1} + 2k = 8.4 \pm 0.4\,s^{-1}$. Thus, a value of $k = 2.9\,s^{-1}$ was obtained at 203 K which is substantially smaller but also much more precise than the low temperature value obtained by ^{1}H NMR lineshape analysis in Fig. 2.3.

In the presence of many exchanging polarizations a two-dimensional version of the experiment is to be preferred, where, in addition, t_1 is systematically incremented and where Fourier transformation is employed along t_1 [81].

Note that polarization transfer due to nuclear Overhauser effects (NOE) in the rotating frame must be taken into account [80] when the exchanging magnetizations are dipolar coupled. This phenomenon has been employed also to study cross-relaxation dynamics (Camelspin [192], Roesy [193]) in slow tumbling biomolecules. Note also that in the presence of scalar coupling coherent polarization transfer in the rotating frame occurs which can also be used for analytical purposes (Tocsy [194] and Hohaha [195]).

2.4 Relaxation Time Measurements

Kinetic information cannot only be obtained from magnetization transfer and lineshape experiments but also from the measurement of longitudinal relaxation times T_1, longitudinal relaxation times in the rotating frame, $T_{1\rho}$, and transverse relaxation times T_2. Since T_1 is sensitive only to motions in the MHz-range where—in the liquid phase—molecular motions dominate, it can rarely be used for the determination of rate constants of chemical reactions, including their kinetic isotope effects. Exceptions are longitudinal relaxation time measurements in the solid state where T_1 is solely governed by the proton and deuteron transfers of interest [180–187]. By contrast, the measurement of T_2 and $T_{1\rho}$ of coalesced lines provides similar information as lineshape analysis in the fast exchange regime [72], with the difference that the relaxation measurements are more precise and that the difference of chemical shifts, $\Delta\nu$, can be obtained even in the fast exchange regime.

The $T_{1\rho}$ method is especially simple in the case of symmetric exchange processes, the absence of scalar spin–spin coupling, and the case of extreme line narrowing. In this case it has been shown that [67, 73]

$$T_{1\rho}^{-1} - T_1^{-1} = \frac{\pi^2 \Delta\nu^2}{2k}\left(\frac{1}{1 + \omega_1^2/4k^2}\right), \qquad \omega_1 = \gamma B_1. \tag{2.22}$$

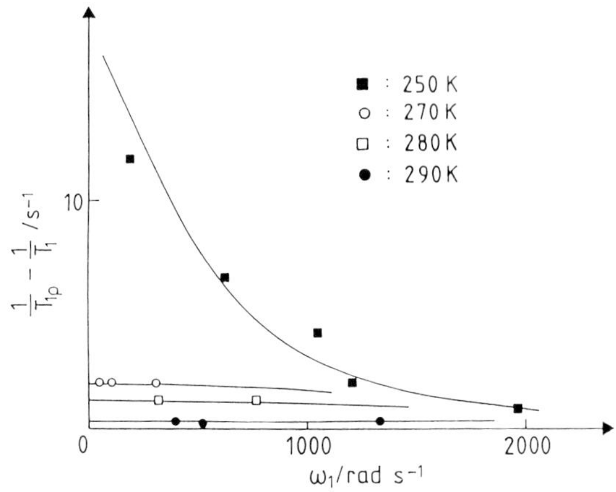

Fig. 2.6. Analysis of a ^{1}H–$T_{1\rho}$-experiment in the fast exchange region on the β-pyrrole protons of TPP dissolved in toluene-d_8. Values of $T_{1\rho}^{-1} - T_1^{-1}$ as a function of the strength of the rotating magnetic field $\omega_1 = \gamma B_1$. The data were adapted to Eq. (2.22) by non-linear least squares fit. ■: 250 K, ○: 270 K; □: 280 K; ●: 290 K. Reproduced with permission from Ref. [115]

By measuring $T_{1\rho}^{-1}$ as a function of ω_1 the rate constant k as well as the chemical shift difference $\Delta\nu$ can be obtained. As an example, consider the experimental $T_{1\rho}^{-1} - T_1^{-1}$ values of the coalesced line of Fig. 2.3 shown in Fig. 2.6 as a function of temperature. At 250 K the $T_{1\rho}^{-1}$ vs. ω_1 dispersion is well pronounced. At higher temperatures the term $\omega_1^2/4k^2$ becomes much smaller than 1, and

$$T_{1\rho}^{-1} - T_1^{-1} \cong \frac{\pi^2 \Delta\nu^2}{2k} \tag{2.23}$$

becomes independent on ω_1. Still, k can be obtained if $\Delta\nu$ is known. At very large k—values the difference $T_{1\rho}^{-1} - T_1^{-1}$ will, however, become so small that rate constants can no more be determined using this method.

Note that when performing ^{13}C $T_{1\rho}$ experiments proton decoupling has to be suppressed during the spin lock period [74]. Note also that the measurement of T_2 in a Carr–Purcell sequence as a function of the pulse distance leads to similar results as the $T_{1\rho}$ method [67, 72].

3 The Modulation of Nuclear Spin Hamiltonians in the Presence of Isotopic Bond Breaking/Bond Formation Processes

One condition for the study of dynamic processes by NMR is that a modulation of the nuclear spin Hamiltonian $\mathscr{H}$ occurs during the process to be studied, i.e. that $\mathscr{H}_r \neq \mathscr{H}_s$ during an interconversion between the environments r and s. This requirement is not always easy to fulfill when studying symmetric exchange reactions, the main domain of dynamic NMR spectroscopy. In this section we discuss the sensitivity of different magnetic interactions with respect to bond breaking and bond formation processes of isotopically labeled molecules of the type r ≡ AL, where L represents the hydrogen isotopes H and D. In the following we will call the L spins also the "mobile" spins since they are transferred during the reactions studied. Spins in the remaining group A will be called "remote" spins.

For most NMR studies of condensed matter it is sufficient to include the following terms in the Hamiltonian $\mathscr{H}_r$ of a nuclear spin system in a given environment r [67]:

$$\mathscr{H}_r = \mathscr{H}_r^{CS} + \mathscr{H}_r^{CSA} + \mathscr{H}_r^{S} + \mathscr{H}_r^{SA} + \mathscr{H}_r^{DD} + \mathscr{H}_r^{Q}. \tag{3.1}$$

$\mathscr{H}_r^{CS}$ is the Hamiltonian of the isotropic chemical shift interaction, $\mathscr{H}_r^{CSA}$ of the chemical shift anisotropy, $\mathscr{H}_r^{S}$ of the isotropic scalar spin–spin interaction. $\mathscr{H}_r^{SA}$ represents the anisotropy of the latter, $\mathscr{H}_r^{DD}$ the Hamiltonian of the dipole–dipole interaction, $\mathscr{H}_r^{Q}$ the Hamiltonian of the quadrupole interaction. Expressions of all spin Hamiltonians can be found in the literature. Since $\mathscr{H}_r^{CSA}$, $\mathscr{H}_r^{SA}$, $\mathscr{H}_r^{DD}$, and $\mathscr{H}_r^{Q}$ are averaged out in the liquid state where most kinetic studies of isotope effects are performed only $\mathscr{H}_r^{CS}$ and $\mathscr{H}_r^{S}$ [67] need to be discussed here:

$$\mathscr{H}_r^{CS} = \sum_i (\nu_i - \nu) I_{zi}, \quad \nu_i = \gamma_i \hbar B_0 (1 - \sigma_i), \tag{3.2}$$

$$\mathscr{H}_r^{S} = \sum_{i<j}\sum J_{ij} I_i I_j, J_{ij} = \gamma_i \gamma_j K_{ij}. \tag{3.3}$$

The sums have to be carried out over all spins, remote and mobile, of the environment r. ν_i is the Larmor frequency of spin i in environment r, γ the gyromagnetic ratio, B_0 the external magnetic field, $\hbar$ Planck's constant, σ_i the screening constant of the magnetic field of nucleus i in the environment r, J_{ij} the coupling constant between i and j in Hz, J'_{ij} the reduced coupling constant, and I_i, I_{zi} the usual angular momentum operators of spin i in the subsystem r.

In the following, the use of the different Hamiltonians in the study of kinetic isotope effects by dynamic NMR spectroscopy will be discussed. Such studies depend on whether one looks at the exchanging hydrogen isotopes L or on the remote spins A.

3.1 *The Determination of Kinetic Isotope Effects via a Modulation of the Isotropic Chemical Shift Interaction*

The magnetic interaction of nuclei that is the most sensitive to different chemical environments is the isotropic chemical shift interaction. Thus, a modulation of this interaction during the course of the reaction of interest makes it possible to obtain rate constants by NMR. As stated above, here bond breaking processes of molecules AL, L = H, D are studied. Depending on the chemical structures involved, these processes may induce chemical shift changes either of the mobile spins L or of the remote spins located in group A. Thus, the information content depends on whether the spectra of group A or of the isotopes L are analyzed. Examples for both types of spectral analyses will be given in Sects. 9 to 11. It is clear that the use of the isotropic chemical shift interaction necessitates the measurement of high resolution NMR spectra. This does not constitute a problem when liquids are studied. In the case of solids, it is necessary to remove all orientation dependent interactions for this purpose.

We will now discuss which nuclei are the most apt for the study of kinetic hydrogen/deuterium isotope effects, making use of the isotopic chemical shift interaction.

3.1.1 NMR Studies of Mobile Hydrogen Isotopes

3.1.1.1 ^{1}H NMR Spectroscopy

The first experiments which are, generally, performed before entering deeper into an isotopic research project by NMR are ^{1}H NMR experiments. This method is the most sensitive NMR technique and no isotopic enrichment is necessary. Also, concentration dependent studies are most easily performed by ^{1}H NMR spectroscopy. One disadvantage is the presence of additional ^{1}H signals, e.g. solvent signals which might interfere with those carrying the kinetic information. Therefore, during the course of the research it may often be necessary to suppress certain proton signals, e.g. by deuteration. Elimination of unwanted signals is especially necessary if experiments have to be performed as a function of the deuterium fraction D in the mobile proton sites because of the limited dynamic range of the NMR receiver system. The use of signal suppression techniques or selective pulse experiments often leads to base line errors which are awkward if exchange broadened lines are to be monitored.

3.1.1.2 ^{2}H NMR Spectroscopy

The problem of the limited dynamic range of the NMR receiver is less severe in ^{2}H NMR spectroscopy because, generally, the signals of the interesting deuterated compounds are stronger than the signals of the natural abundant (0.2%) solvent deuterons. However, the smaller gyromagnetic ratio $\gamma_{^2H} \approx \gamma_{^1H}/6$ of the deuteron constitutes a severe problem. Since the gyromagnetic ratio enters the chemical shift Hamiltonian in Eq. (3.2) the typical frequency differences $\Delta\nu$ between two chemically different environments are reduced by $\gamma_{^2H}/\gamma_{^1H}$, which limits the dynamic range of ^{2}H NMR spectroscopy as a kinetic tool. This effect can be overcome partly by performing experiments at higher magnetic field and partly by the occurrence of kinetic hydrogen/deuterium isotope effects.

A more serious drawback of ^{2}H NMR spectroscopy is the quadrupole moment of ^{2}H which may lead to short longitudinal and transverse relaxation times T_1 and T_2, i.e. to broad lines. Since relaxation becomes faster when the molecular correlation times increase, only small molecules can be studied at not too low temperatures.

3.1.1.3 ^{3}H NMR Spectroscopy

From the standpoint of chemical kinetics the measurement of kinetic hydrogen/deuterium/tritium isotope effects is extremely desirable. In view of the progress made in ^{3}H NMR spectroscopy [196, 197] and the favorable gyromagnetic ratio $\gamma_{^3\mathrm{H}}/\gamma_{^1\mathrm{H}} = 1.1$, it seems surprising that kinetic hydrogen/tritium isotope effects have not yet been reported. The reason is that radioactive tritium is absent in natural samples, which necessitates isotopic enrichment. Unfortunately, because of the radioactivity of this isotope samples of interest can, generally, be tritiated only to a small extent. Such degrees of tritiation are sufficient for structural NMR studies or for kinetic studies of single proton transfer reactions; however, for the elucidation of the kinetics of multiple proton transfer fully tritiated molecules would have to be studied which constitutes a substantial hazard, apart from the fact that high tritium content may lead to sample decomposition as a consequence of radiation.

3.1.2 NMR Spectroscopy of Remote Spins

Let us recall again that remote spins remain bonded to their respective neighbor atom but may experience a modulation of their spin Hamiltonian during the hydrogen transfer process. Remote spins in this sense can be all types of nuclei including non-labile protons such as protons bound to carbon.

3.1.2.1 Liquid State Remote Spin Probes

It is clear that ^{1}H NMR spectroscopy of remote ^{1}H spins, e.g. C–H protons, will always be the first choice because of the sensitivity of this method. The second choice will be naturally abundant ^{13}C NMR spectroscopy. In the presence of exchange broadened lines and low solubilities it might, however, sometimes be necessary to artificially enrich the compounds studied. From the standpoint of NMR sensitivity, an almost ideal alternative is the introduction of a ^{19}F label, which might, however, constitute a major perturbation of the chemical kinetics.

At first sight, the most ideal choice for the study of ^{15}N enriched NH$\cdots$N proton transfer systems is ^{15}N NMR spectroscopy because of the large chemical shift change between protonated and non-protonated nitrogen atoms. A problem of ^{15}N NMR spectroscopy is, however, the fact that most often ^{15}N atoms relax via dipole–dipole coupling to protons. Therefore, ^{15}N atoms in deuterated ^{15}ND$\cdots$^{15}N transfer systems can have very long relaxation times which makes it very difficult to record their exchange broadened NMR lineshapes. The situation is improved when working with high magnetic fields where the anisotropy of the chemical shifts becomes a major mechanism of longitudinal relaxation.

3.1.2.2 Solid State CPMAS Remote Spin Probes

Note that the situation is quite different in solid state ^{15}N NMR studies of ^{15}N enriched molecules under the conditions of cross polarization (CP) and magic angle spinning (MAS) [152–162]. ^{15}N CPMAS NMR spectra of ^{15}ND$\cdots^{15}$N units have successfully been measured [167, 179]. This is because in this technique the bottleneck is not the longitudinal relaxation time of the ^{15}N but of the ^{1}H spins. In addition, there are no solubility problems.

Surely, for HO$\cdots$O protons transfer systems embedded in organic compounds it will, in general, be necessary to use ^{13}C CPMAS NMR.

3.2 The Determination of Kinetic Isotope Effects via a Modulation of the Scalar Spin–Spin Coupling Interaction

By contrast to isotropic chemical shifts coupling constants are much less affected by chemical exchange. This is especially true for coupling constants among remote spins. However, the situation is different when scalar coupling between a mobile hydrogen isotope L and a remote spin A is present. Then, hydrogen exchange will lead to observable changes of the spin–spin splitting pattern of the L signals. As will be shown below, this change will depend on whether the proton is transferred along an intramolecular or an intermolecular pathway. Note that this modulation of the scalar spin–spin interaction is the only way to monitor a proton self exchange reaction between like molecules.

Unfortunately, scalar coupling constants are, generally, not very large and cannot be increased by increasing the magnetic field which limits the dynamic range of this method. Of practical importance is, however, scalar coupling between ^{15}N and ^{1}H spins in ^{15}NH$\cdots$X proton transfer systems because $^{1}J_{^{1}H-^{15}N}$ coupling constants are of the order of 80 to 100 Hz and provide, therefore, a quite large dynamic range.

A problem arises, however, in the study of the related ^{15}ND$\cdots$X deuteron transfer systems. Because of Eq. (3.3) the coupling constant J_{AD} is $\gamma_{^{1}H}/\gamma_{^{2}H}$ times smaller than the corresponding constant J_{AH}. Therefore, the scalar A–D interaction will, generally, be of limited use for the determination of kinetic isotope effects of proton transfer reactions.

There are, however, exceptions to this statement. Thus, as will be shown in Sects. 9 and 10, the scalar spin–spin interaction $J_{^{1}H-^{15}N}$ can be employed to study the partial kinetic HH/HD isotope effects of intra- and intermolecular hydrogen transfer reactions.

3.3 *The Determination of Kinetic Isotope Effects via a Modulation of Orientation Dependent Nuclear Spin Interactions in Solids*

Since proton and deuteron transfers are proton dislocations in space these processes also lead to a modulation of the orientation dependent nuclear spin interactions in solids, i.e. of the anisotropy of the chemical shift, the dipole–dipole interaction and the quadrupole interaction. The determination of kinetic hydrogen/deuterium isotope effects on solid state hydrogen transfer reactions by taking advantage of these interactions is, however, not easy, because of the different gyromagnetic ratios of ^{1}H and ^{2}H, the fact that ^{1}H has no quadrupole moment, and the problem that, generally, more than one interaction is modulated when the transfer occurs. The latter problem may be circumvented by studying single crystals instead of microcrystalline powders. This is because the spectra of single crystals are much simpler since all molecules have the same orientation with respect to the magnetic field. In the case of ^{2}H NMR the situation is furthermore simplified because only the quadrupole interaction needs to be taken into account. Using this method partial kinetic HD/DD isotope effects of double hydrogen transfer reactions can be studied. The determination of the corresponding HH reaction is more complicated. In favorable cases it can be achieved when the jumping protons are dipolar coupled among themselves or coupled to a remote spin. This method will generally require deuteration of all immobile protons in the sample. Finally, it seems to be possible also to look at remote spins which are dipolar coupled to the jumping protons. One also could take advantage of the modulation of the chemical shift anisotropy of the remote spins in order to obtain kinetic H/D isotope effects under conditions of ^{1}H decoupling in order to remove heteronuclear dipolar couplings.

4 Dynamic NMR Spectroscopy in the Presence of Complex Bond Breaking/Bond Formation Processes

4.1 *The Concept of Group Exchange*

In chemical reactions involving bond breaking and bond formation processes the exchange operator in Eq. (2.12) is valid only in the absence of scalar spin–spin coupling. The exchange broadened NMR spectra of general spin systems can, however, conveniently be described using the concept of "group exchange" [94]. This concept is further developed in the following, with special emphasis on the description of isotopic exchange reactions.

Let r, s, x, y be molecular species subject to the general forward reaction

$$r + s \xrightarrow{k_+} x + y, \tag{4.1}$$

characterized by the second order forward rate constant k_+. Note that this reaction is not one-sided but that, as stated above, the backward reaction

$$x + y \xrightarrow{k_-} r + s \tag{4.2}$$

does not need to be treated explicitly as long as the equilibrium constant of the reaction is known.

In order to trace the pathways of atoms or groups of atoms during the reaction described by Eq. (4.1) we formally rewrite each reactant r as a combination of groups of nuclei A^r, B^r... In the simplest case we have $r \equiv A^rB^r$. A^r and B^r are defined in such a way that the nuclei within each group stay together during the reaction of interest. In the case where all molecules contain two groups Eq. (4.1) becomes

$$A^rB^r + C^sD^s \xrightarrow{k_{+1}} A^xC^x + B^yD^y, \tag{4.3}$$

which is characterized by the rate constant k_{+1}. This notation takes into account that a group of nuclei A^r may rearrange or may change its spin Hamiltonian when it is incorporated as group A^x into the molecule $x \equiv A^xC^x$. This notation also allows to trace reaction pathways of the spins in the different groups.

In order to facilitate the line shape calculations we decompose Eq. (4.3) into the following group exchange reactions:

$$A^r \xrightarrow{\tau_{rx}^{-1}(A)} A^x, \quad B^r \xrightarrow{\tau_{ry}^{-1}(B)} B^y, \quad C^s \xrightarrow{\tau_{sx}^{-1}(C)} C^x, \quad D^s \xrightarrow{\tau_{sy}^{-1}(D)} D^y. \tag{4.4}$$

$\tau_{rx}^{-1}(A)$ is the average inverse lifetime of group A in the reactant r before reaction to x occurs. $\tau_{rx}^{-1}(A)$ is equal to the pseudo first order group exchange rate constant $k_{rx}(A)$ of the exchange process in Eq. (4.3). Since the elements of the exchange operator Ξ depend on these quantities, the latter can be obtained by dynamic NMR spectroscopy. Using formal kinetics the inverse lifetimes can be linked to the rate constants, as will be demonstrated in Sect. 5.

The introduction of group exchange rate constants has not only the advantage that the group exchange pathway can be traced, but also, that possible additional symmetric exchange reactions of the type

$$A^rB^r + A^{x*}C^x \xrightarrow{k_{+2}} A^{r*}B^r + A^xC^x, \tag{4.5}$$

$$A^rB^r + B^{y*}D^y \xrightarrow{k_{+3}} A^rB^{r*} + B^yD^y, \tag{4.6}$$

$$A^rB^r + A^{r*}B^{r*} \xrightarrow{k_{+4}} A^{r*}B^r + A^rB^{r*} \tag{4.7}$$

etc. are "built" into the formalism. Consider, for example, Eq. (4.5) which can be decomposed into the group exchange reaction set

$$A^r \xrightarrow{\tau_{rx}^{-1}(A)} A^x, \quad B^r \xrightarrow{\tau_{rr}^{-1}(B)} B^r, \quad A^{r*} \xrightarrow{\tau_{rx}^{-1}(A)} A^{x*}, \quad C^x \xrightarrow{\tau_{xx}^{-1}(C)} C^x. \tag{4.8}$$

By comparison with Eq. (4.4) it follows that the quantity $\tau_{rx}^{-1}(A)$ does not depend only on the rate constant k_{+1} in Eq. (4.3) but also on the constant k_{+2} defined in Eq. (4.5). As further seen by inspection of Eq. (4.8) one needs only to include in the set of group exchange rate constants $\tau_{rx}^{-1}(A)$ the possibility that $r = x$, in order to take self exchange reactions of the type described in Eqs. (4.4) to (4.8) into account.

A further advantage of the concept of group exchange rate constants is that one can drop the above restriction to two groups in each molecule. Thus, the presence of additional dissociation and transfer reactions of the type

$$A^rB^r + C^sD^s \xrightarrow{k_{+5}} A^t + B^uC^uD^u, \tag{4.9}$$

$$A^t + C^sD^s \xrightarrow{k_{+6}} A^xC^x + D^v \tag{4.10}$$

etc. is taken into account by considering the additional group exchange reactions

$$A^r \xrightarrow{\tau_{rt}^{-1}(A)} A^t;\quad B^r \xrightarrow{\tau_{ru}^{-1}(B)} B^u;\quad C^s \xrightarrow{\tau_{su}^{-1}(C)} C^u;\quad D^s \xrightarrow{\tau_{su}^{-1}(D)} D^u;$$

$$A^t \xrightarrow{\tau_{tx}^{-1}(A)} A^x;\quad C^s \xrightarrow{\tau_{sx}^{-1}(C)} C^x;\quad S^s \xrightarrow{\tau_{sv}^{-1}(D)} D^v. \tag{4.11}$$

Thus, it is easy to set up the NMR lineshape equations for exchanging systems without assuming a particular reaction network when using pseudo first order group exchange rate constants.

When setting up lineshape equations it is useful to neglect spin–spin interactions between the different groups in a first stage. Then, the whole master equation splits up into separate sets of equations, one set for each reacting group. The matrix $\mathscr{M}_A$ of group A has the structure:

$$\mathscr{M}_A = \begin{array}{|c|c|c} \hline i\mathscr{L}_{A^r}^{CS} + \Xi_{A^r} + \mathscr{R}_{A^r} & \Xi_{A^sA^r} & \cdots \\ \hline \Xi_{A^rA^s} & i\mathscr{L}_{A^s}^{CS} + \Xi_{A^s} + \mathscr{R}_{A^s} & \cdots \\ \hline \vdots & \vdots & \ddots \end{array} \tag{4.12}$$

The elements denoted by $\Xi_{A^rA^s}$ and $\Xi_{A^sA^r}$ depend only on single group exchange rate constant $\tau_{rs}^{-1}(A)$, those denoted by Ξ_{A^r} on the sum of all group exchange rate constants. The actual elements can be calculated from Eq. (2.12) and are given by

$$\Xi^{(A)}_{\mu_r\nu_r;\kappa_s\lambda_s} = \tau_{rs}^{-1}(A)\,\delta_{\mu_r\kappa_s}\delta_{\nu_r\lambda_s} - \delta_{\mu_r\kappa_s}\delta_{\nu_r\lambda_s}\sum_t \tau_{rt}^{-1}(A), \quad t = r, s \tag{4.13}$$

The elements of $\mathscr{L}_{A^r}$ are calculated according to Eqs. (2.3), (3.2) and (3.3). In the case where A^r contains more than one interacting spin either a product base of the single spins is used or the eigenbase of the Hamiltonian $\mathscr{H}_{A^r}$. As stated above, $\mathscr{L}_{A^r}$ is diagonal in the latter case and the diagonal elements correspond to the transition frequencies.

In a second stage scalar coupling between different groups A^r and B^r of a molecule r is introduced. Note that because of the group exchange processes present one cannot use an Eigenbase of the total spin Hamiltonian $\mathscr{H}_r$ of the molecule r, but has to use the product base A^rB^r. Thus, the spin functions are written in the form $\mu_{A^r}u_{B^r} \equiv \mu_r u_r$, where a greek letter has been used for the spin functions of group A^r and a latin letter for the spin functions of B^r. Then, the Liouville operator of the molecule $r = A^rB^r$

$$\mathscr{L}_r = \mathscr{L}_{A^r} + \mathscr{L}_{B^r} + \mathscr{L}^S_{A^rB^r}, \quad \mathscr{L}_{A^r} = \mathscr{L}^{CS}_{A^r} + \mathscr{L}^S_{A^r} \tag{4.14}$$

contains non-diagonal elements connecting the two blocks $\mathscr{L}_{A^r}$ and $\mathscr{L}_{B^r}$ because of the interaction term $\mathscr{L}^S_{A^rB^r}$. In constructing the exchange operator one has to take into account that the number of spin states in the spin systems in the different groups are different. The following general exchange operator has been derived to deal with this problem [94]:

$$\begin{aligned}
&\Xi_{\mu_r u_r v_r v_r;\kappa_s x_s \lambda_s y_s} \\
&\quad = \tau_{rs}^{-1}(A)\, n_{B^r}^{-1}\, \delta_{u_r v_r}\delta_{x_s y_s}\delta_{\mu_r \kappa_s}\delta_{v_r \lambda_s} g_{x_s} - \delta_{u_r v_r}\delta_{\mu_e \kappa_s}\delta_{v_r y_s}\delta_{v_r \lambda_s}\sum_t \tau_{rt}^{-1}(A) \\
&\quad\quad + \tau_{rs}^{-1}(B)\, n_{A^r}^{-1}\, \delta_{\mu_r v_r}\delta_{\kappa_s \lambda_s}\delta_{u_r x_s}\delta_{v_r y_s} g_{\kappa_s} - \delta_{\mu_r v_r}\delta_{v_r y_s}\delta_{u_r x_s}\delta_{v_r \lambda_s}\sum_t \tau_{rt}^{-1}(B), \\
&\qquad\qquad\qquad\qquad\qquad\qquad\qquad\qquad\qquad\qquad\qquad\qquad t = r, s \cdots
\end{aligned} \tag{4.15}$$

n_{A^r} and n_{B^r} represent the number of spin states in A^r and B^r. g_{k_s} and g_{x_s} represent the statistical weights of the spin function k_s and x_s. $g_{k_s}(g_{x_s})$ is different from unity when an Eigenbase of the group spin-Hamiltonian $\mathscr{H}_{A^r}(\mathscr{H}_{B^r})$ is used [94].

4.2 Dynamic NMR Spectroscopy in the Presence of Isotopic Exchange Reactions

The above theory of NMR lineshapes in the presence of group exchange can immediately be applied to the problem of superposed isotopic exchange reactions by taking into account that the different environments r, s,... in the previous section may correspond not only to different chemical species such as AH and XH, but also to isotopically different molecules of the same species, e.g. AH and AD. Since isotopic reactions introduce, however, a higher degree of complexity it is preferable to study reaction mixtures in which only one chemical reaction is present. Before starting a systematic study of kinetic isotope effects it is, therefore, always desirable to find experimental conditions where the dynamic quantities measured are determined only by a single reaction.

Before we can set up typical NMR lineshape equations in order to determine kinetic hydrogen/deuterium isotope effects of proton transfer reactions we first have to relate the inverse lifetimes accessible by dynamic NMR to conventional kinetic quantities, as shown in the following section.

5 The NMR Proton Inventory Technique

In order to obtain information on kinetic hydrogen/deuterium isotope effects by dynamic NMR spectroscopy it is necessary to measure inverse proton and deuteron life times as a function of the deuterium fraction in the mobile proton sites. The function obtained will depend on the actual exchange mechanism and will provide information about the rate constants of the isotopic reactions. Kinetic experiments of this kind are called in the literature "proton inventories" [43]. Therefore, the term "NMR proton inventory technique" is used.

Unfortunately, because of the multitude of different mechanisms and rate laws of hydrogen transfer reactions it is not possible to derive reactions between inverse lifetimes and rate constants of isotopic reactions in a general way. In this section it will be demonstrated—using specific examples of proton and deuteron transfer reactions relevant in the context of this study—how such relations may be obtained. The extension to other reactions and isotopes is then straightforward. Intra- and intermolecular single and multiple H/D transfer reactions are treated in separate sections; some of the equations given in this section were first derived in Refs. [106–108].

5.1 *Intramolecular Hydrogen Transfer Reactions*

The term "intramolecular reaction" is used here to describe reactions in molecules or molecular systems where all groups of atoms may rearrange but stay together for an infinite time as compared to the NMR-timescale. It is clear that hydrogen transfer reactions within monomeric molecules fall into this definition, but, in principle, also intermolecular hydrogen transfer reactions in molecular aggregates such as dimers, if the lifetime of the latter is large enough, as, for example, in the solid state.

In the case of intramolecular reactions the quantities entering the elements of the exchange operator Ξ are the first order rate constants themselves. In this section, therefore, one does not need to apply the concept of group exchange proposed in Sect. 4.

5.1.1 *Single Intramolecular Hydrogen Transfer Reactions*

Let us consider first an intramolecular single hydrogen transfer between heavy atoms of a molecular fragment. Denoting the latter as A before and as X after the transfer process one can write

$$\mathrm{AH} \underset{k_-^{\mathrm{H}}}{\overset{k_+^{\mathrm{H}}}{\rightleftarrows}} \mathrm{XH}, \tag{5.1}$$

$$AD \underset{k_-^D}{\overset{k_+^D}{\rightleftarrows}} XD. \tag{5.2}$$

More precisely, AL, L = H, D may correspond to A–B–L and XL to L–A–B. The quantities k_+^L and k_-^L represent the first order forward and backward rate constants. At an arbitrary deuterium fraction the H and the D transfer reactions are superposed without mutual interference. Thus, the ^1H NMR signals of the mobile L spins only contain contributions from AH and XH molecules, whereas the ^2H NMR signals exclusively arise from the AD and XD species. Changing the overal deuterium fraction D in the mobile proton sites only changes the ratio of the protonated vs. the deuterated molecules but not their contributions to the NMR line-shape. As a consequence, the ^1H NMR signals contain information about the quantity k_+^H and the ^2H NMR signals about the quantity k_+^D. By contrast, the spectra of the A and the X spins depend on both quantities. One can, therefore, choose the most convenient way to determine the isotopic reaction rate constants. Note, that it might in some cases be difficult to derive both quantities k_+^H and k_+^D from the A/X spectra of a single sample at an intermediate deuterium fraction D because isotope effects on the A/X chemical shifts will be too small to resolve the lineshape contributions of the protonated and the deuterated molecules. These quantities can, however, easily, be derived from the A/X NMR spectra of samples with deuterium fractions of D = 0 and D = 1.

5.1.2 Double Intramolecular Hydrogen Transfer Reactions

An intramolecular double hydrogen transfer system contains two mobile hydrogen atoms. The reaction, can, therefore, be written in the form

$$AHH \underset{k_-^{HH}}{\overset{k_+^{HH}}{\rightleftarrows}} XHH, \tag{5.3}$$

$$AHH \underset{k_-^{HD}}{\overset{k_+^{HD}}{\rightleftarrows}} XHD, \tag{5.4}$$

$$ADH \underset{k_-^{HD}}{\overset{k_+^{HD}}{\rightleftarrows}} XDH, \tag{5.5}$$

$$ADD \underset{k_-^{DD}}{\overset{k_+^{DD}}{\rightleftarrows}} XDD, \tag{5.6}$$

where, for example, ALL ≡ ALBLCD and XLL ≡ ABCLDL. There are four reactant species giving contributions to the NMR lineshapes of the L spins. The species AHH, AHD and ADH contribute to the H signals which depend, therefore, on the first order rate constants k_+^{HH}, k_+^{HD} and k_+^{DH} as well as on the corresponding equilibrium constants. k_+^{HH} is most easily determined in a separate ^1H NMR experiment at a deuterium fraction of D = 0, whereas k_+^{HD} and k_+^{DH} have to be determined at higher D values. Similar arguments hold for the ^2H

NMR spectra. Thus, the whole set of kinetic isotope effects can be determined. Since, as mentioned in Sect. 3.1.1.2, liquid state ^{2}H NMR spectra are sometimes of low kinetic information content, additional experiments on the A/X spins are of help. The latter depend on all quantities k_+^{HH}, k_+^{HD}, k_+^{DH} and k_+^{DD}. In order to check the k_+^{HH} values obtained from the ^{1}H spectra, and in order to obtain the only lacking quantity k_+^{DD}, A spin NMR experiments are preferentially carried out at $D=0$ and 1. The determination of k_+^{HD} and of k_+^{DH} at intermediate deuterium fractions may represent problems because of the difficulty to distinguish the different species in the A/X spectra.

5.1.3 Multiple Intramolecular Hydrogen Transfer Reactions

The extension of the arguments given in the two previous sections to cases where more than two protons are transferred along intramolecular pathways is straightforward and will, therefore, not be considered further. In the spectral analysis one only needs to keep in mind that the lineshapes are the sum of contributions from the different isotopic species, which do not interconvert between each other. The larger the number of isotopic rate constants characterizing a given reaction system the more experiments have to be carried out using a certain combination of different multinuclear NMR experiments and deuterium fractions in the mobile proton sites.

5.2 Intermolecular Hydrogen Transfer Reactions

In this section we consider intermolecular multiple hydrogen transfer reactions of proton donors AL containing one labile hydrogen isotope L. By contrast to intramolecular reactions, the kinetic variables which determine the NMR

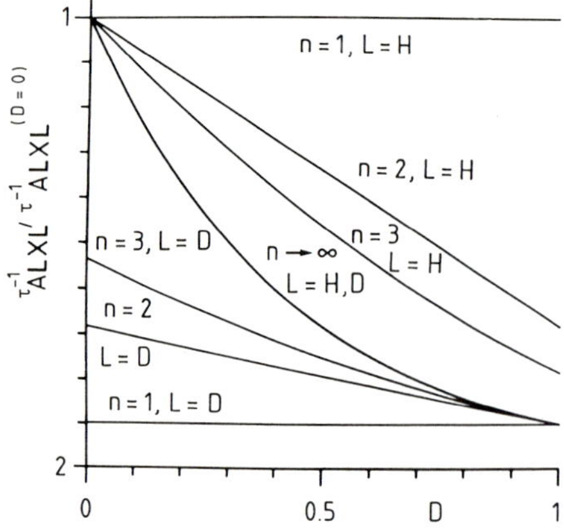

Fig. 5.1. NMR proton inventory plots as a function of the deuterium fraction D for different numbers n of protons transported in the rate limiting step for $C_B=$ constant. L is the nucleus whose NMR spectra are observed. The overall kinetic isotope effect was set arbitrarily to a value of 10. Adapted with permission from Ref. [106]

spectra will depend themselves on the deuterium fractions in the labile proton sites. In the following we will derive appropriate equations for different kinds of hydrogen transfer reactions. Some of these equations are illustrated graphically in Fig. 5.1.

5.2.1 Single Intermolecular Hydrogen Transfer Reactions

Let us first consider the basic single hydrogen transfer reactions

$$AH + X \underset{k_-^H}{\overset{k_+^H}{\rightleftarrows}} A + XH, \tag{5.7}$$

$$AD + X \underset{k_-^D}{\overset{k_+^D}{\rightleftarrows}} A + XD, \tag{5.8}$$

where k_+^L is the forward second order rate constant. Equations (5.7) and (5.8) can be decomposed into the following group exchange reaction set

$$A\mathbf{H} \xrightarrow{\tau^{-1}_{A\mathbf{H}X\mathbf{H}}} X\mathbf{H}, \quad A\mathbf{D} \xrightarrow{\tau^{-1}_{A\mathbf{D}X\mathbf{D}}} X\mathbf{D},$$

$$\mathbf{A}H \underset{}{\overset{\tau^{-1}_{\mathbf{A}H\mathbf{A}}}{\rightleftarrows}} \mathbf{A}, \quad \mathbf{A}D \underset{}{\overset{\tau^{-1}_{\mathbf{A}D\mathbf{A}}}{\rightleftarrows}} \mathbf{X}, \quad \mathbf{X} \underset{}{\overset{\tau^{-1}_{\mathbf{X}X\mathbf{H}}}{\rightleftarrows}} \mathbf{X}H, \quad \mathbf{X} \underset{}{\overset{\tau^{-1}_{\mathbf{X}X\mathbf{D}}}{\rightleftarrows}} \mathbf{X}D. \tag{5.9}$$

As compared to Sect. 3, we use here a slightly different, for the case of hydrogen transfer reactions, better adapted notation. The exchanging group is marked here in boldface, and the full chemical symbols are used as subscript. Thus, $\tau^{-1}_{A\mathbf{L}X\mathbf{L}}$ represents the average inverse lifetime of L in AL before it is incorporated into XL; $\tau^{-1}_{\mathbf{A}L\mathbf{A}}$ is the lifetime of group A in AL before the change to A. Clearly, $\tau^{-1}_{A\mathbf{L}X\mathbf{L}}$ is obtained by analysis of the L signals and $\tau^{-1}_{\mathbf{A}L\mathbf{A}}$ of the A signals.

From conventional kinetics it follows that

$$\tau^{-1}_{\mathbf{A}L\mathbf{A}} = \tau^{-1}_{A\mathbf{L}X\mathbf{L}} = -(1/C_{AL})\,dC_{AL}/dt = k_+^L\,C_X, \quad L = H, D. \tag{5.10}$$

As a consequence, in a single hydrogen transfer reaction the different inverse lifetimes are independent of the deuterium fraction in the mobile proton sites. E.g. it follows that

$$\tau^{-1}_{A\mathbf{L}X\mathbf{L}}/\tau^{-1}_{A\mathbf{L}X\mathbf{L}}(D=0) = 1. \tag{5.11}$$

Using an arbitrary kinetic isotope effect of $\tau^{-1}_{A\mathbf{H}X\mathbf{H}}(D=0)/\tau^{-1}_{A\mathbf{D}X\mathbf{D}}(D=1) = k_+^H/k_+^D = 10$ one obtains for the quantities $\tau^{-1}_{A\mathbf{L}X\mathbf{L}}/\tau^{-1}_{A\mathbf{L}X\mathbf{L}}(D=0)$, $L = H, D$ in the graph of Fig. 5.1 two horizontal lines (n = 1, L = H and n = D).

Note that in the case where the concentrations of A and X become very small, i.e. where only AL and XL can be observed one obtains a formal double proton transfer reaction treated in the following section.

5.2.2 *Double Intermolecular Hydrogen Transfer Reactions*

Let us consider here the basic hydrogen exchange reaction between two different labile proton sites AH and XH, with an equilibrium constant of K = 1:

$$AH + XH^* \underset{}{\overset{k_+^{HH}}{\rightleftarrows}} AH^* + XH, \tag{5.12}$$

$$AH + XD \underset{k_-^{HD}}{\overset{k_+^{HD}}{\rightleftarrows}} AD + XH, \tag{5.13}$$

$$AD + XH \underset{k_-^{DH}}{\overset{k_+^{DH}}{\rightleftarrows}} AH + XD, \tag{5.14}$$

$$AD + XD^* \underset{}{\overset{k_+^{DD}}{\rightleftarrows}} AD^* + XD. \tag{5.15}$$

Let k_+^{LL} be the pseudo second order rate constants of the different forward isotopic reactions whose rate constants are to be determined. Note that

$$k_+^{HD} = k_-^{DH} \quad \text{and} \quad k_+^{DH} = k_-^{HD}, \tag{5.16}$$

where k_-^{LL} are again backward pseudo second order rate constants. A star is introduced here to characterize chemically equivalent but physically different groups. The above reaction set can be decomposed into the following set of group exchange reactions:

$$AH \xrightarrow{\tau^{-1}_{AHAH^*}} AH^*, \quad AH \xrightarrow{\tau^{-1}_{AHAD}} AD, \quad AD \xrightarrow{\tau^{-1}_{ADAH}} AH, \quad AD \xrightarrow{\tau^{-1}_{ADAD^*}} AD^*, \tag{5.17}$$

$$XH \xrightarrow{\tau^{-1}_{XHXH^*}} XH^*, \quad XH \xrightarrow{\tau^{-1}_{XHXD}} XD, \quad XD \xrightarrow{\tau^{-1}_{XDXH}} XH, \quad XD \xrightarrow{\tau^{-1}_{XDXD^*}} XD^*, \tag{5.18}$$

$$AH \xrightarrow{\tau^{-1}_{AHXH}} XH, \tag{5.19}$$

$$AD \xrightarrow{\tau^{-1}_{ADXD}} XD \text{ etc.} \tag{5.20}$$

Here $\tau^{-1}_{ALAL^*}$ with LL* = HH*, HD, DH, DD* represents the observable pseudo first order exchange rate constant of group A (marked in boldface) between the environments AL and AL*, or in other words, the inverse correlation time of group A with spin L before the latter is replaced by spin L*. It is evident that $\tau^{-1}_{ALAL^*}$ can be determined only by lineshape analysis of the A-spin spectra when the Hamiltonian of A is modulated by the exchange. By contrast, τ^{-1}_{ALXL}, the inverse lifetime of spin L in the environment AL before the reaction to XL occurs, has to be determined from the NMR spectra of the L spins.

In contrast to single proton transfer reactions, the above inverse group lifetimes depend on the individual deuterium fractions D_A and D_X in the two mobile proton sites AL and XL. By measuring these lifetimes as a function of the overall deuterium fraction D, the four independent isotopic rate constants k_{+}^{HH}, k_{+}^{HD}, k_{+}^{DH}, and k_{+}^{DD} can be determined. Relations between these rate constants and the measured inverse lifetimes will be derived in the following.

We first define the inverse equilibrium isotope effect or fractionation factor

$$\Phi^{-1} = k_{+}^{HD}/k_{-}^{HD} = k_{+}^{HD}/k_{+}^{DH} = C_{AD}C_{XH}/C_{AH}C_{XD}, \tag{5.21}$$

the total concentration,

$$C_A = C_{AH} + C_{AD}, \tag{5.22}$$

the individual deuterium fraction

$$D_A = C_{AD}/C_A, \tag{5.23}$$

and the overall deuterium fraction

$$D = (C_{AD} + C_{XD})/(C_A + C_X). \tag{5.24}$$

By combination of these equations functions of the type $D_A = f(D, \Phi^{-1}, C_A/C_X)$ can be obtained. Φ, i.e. D_A and D_X, can be obtained by lineshape analysis in the slow exchange region. In the absence of equilibrium isotope effects the following relations hold:

$$\Phi = 1, \quad \text{i.e.} \quad k_{+}^{HD} = k_{+}^{DH} = k_{+}^{HD} \quad \text{or} \quad D_A = D_X = D. \tag{5.25}$$

For the observable inverse lifetimes the following relations are easily derived:

$$\tau_{AHAH^*}^{-1} = -(1/C_{AH})\,dC_{AH\to AH^*}/dt = k_{+}^{HH}C_{XH} = (1 - D_X)\,k_{+}^{HH}C_X. \tag{5.26}$$

$$\tau_{AHAD}^{-1} = -(1/C_{AH})\,dC_{AH\to AD}/dt = D_X k_{+}^{HD}C_X. \tag{5.27}$$

$$\tau_{ADAH}^{-1} = -(1/C_{AD})\,dC_{AD\to AH}/dt = (1 - D_X)\,k_{+}^{DH}C_X. \tag{5.28}$$

$$\tau_{ADAD^*}^{-1} = -(1/C_{AD})\,dC_{AD\to AD^*}/dt = D_X k_{+}^{DD}C_X \tag{5.29}$$

$$\begin{aligned}\tau_{AHXH}^{-1} &= -(1/C_{AH})\,dC_{AH\to XH}/dt \\ &= k_{+}^{HH}C_{XH} + k_{+}^{HD}C_{XD} = ((1 - D_X)k_{+}^{HH} + D_X k_{+}^{HD})C_X.\end{aligned} \tag{5.30}$$

$$\begin{aligned}\tau_{ADXD}^{-1} &= -(1/C_{AD})\,dC_{AD\to XD}/dt \\ &= k_{+}^{DH}C_{XH} + k_{+}^{DD}C_{XD} = ((1 - D_X)k_{+}^{DH} + D_X k_{+}^{DD})C_X.\end{aligned} \tag{5.31}$$

The remaining quantities are obtained by substitution of X by A and of A by X. Equations (5.26) to (5.31) indicate that the different inverse lifetimes depend linearly on the deuterium fraction D_X of the reaction partner XL. When Eq. (5.25) is valid it follows that the inverse lifetimes τ_{AHXH}^{-1} and τ_{ADXD}^{-1} depend linearly on the overall deuterium fraction D. One then obtains from Eqs. (5.26) to (5.31)

$$\tau_{AHXH}^{-1}(D)/\tau_{AHXH}^{-1}(D = 0) = 1 - [1 - k^{HD}/k^{HH}]\cdot D, \tag{5.32}$$

$$\tau_{ADXD}^{-1}(D)/\tau_{ADXD}^{-1}(D = 1) = k^{DH}/k^{DD} - [k^{DH}/k^{DD} - 1]\cdot D \tag{5.33}$$

Note that Eqs. (5.26) to (5.33) also hold for the case of self exchange between

AH molecules, i.e. where X ≡ A. In this case one can abbreviate

$$\tau^{-1}_{AHA^{*}H} \equiv \tau^{-1}_{AH} \tag{5.34}$$

and

$$\tau^{-1}_{ADA^{*}D} \equiv \tau^{-1}_{AD}. \tag{5.35}$$

Using again an arbitrary kinetic isotope effect of $k^{HH}/k^{DD} = 10$ and the geometric mean values $k^{HH}/k^{HD} = k^{DH}/k^{DD} = 10^{1/2}$ the linear curves n = 2, L = H and n = 2, L = D shown in Fig. 5.1 are obtained. Since equilibrium isotope effects were neglected in Fig. 5.1 it follows from Eqs. (5.30) and (5.31) that $\tau^{-1}_{ADXD}(D = 0) = \tau^{-1}_{AHXH}(D = 1) = k^{HD} = k^{DH}$, which—as an experimental observation—would be a proof for a double proton transfer reaction.

5.2.3 *Triple Intermolecular Hydrogen Transfer Reactions*

Let us consider in this section a triple proton transfer reaction of the kind

$$AH + BH^{*} + XH' \underset{}{\overset{k_{+}^{HHH}}{\rightleftarrows}} AH^{*} + BH' + XH, \tag{5.36}$$

$$AH + BH^{*} + XD \underset{}{\overset{k_{+}^{HHD}}{\rightleftarrows}} AH^{*} + BD + XH, \tag{5.37}$$

$$AH + BD + XH' \underset{}{\overset{k_{+}^{HDH}}{\rightleftarrows}} AD + BH' + XH, \tag{5.38}$$

$$AD + BH^{*} + XH \underset{}{\overset{k_{+}^{DHH}}{\rightleftarrows}} AH^{*} + BH + XD, \tag{5.39}$$

$$AD + BD^{*} + XH \underset{}{\overset{k_{+}^{DDH}}{\rightleftarrows}} AD^{*} + BH + XD, \tag{5.40}$$

$$AD + BH^{*} + XD' \underset{}{\overset{k_{+}^{DHD}}{\rightleftarrows}} AH^{*} + BD' + XD, \tag{5.41}$$

$$AH + BD' + XD \underset{}{\overset{k_{+}^{HDD}}{\rightleftarrows}} AD' + BD + XH, \tag{5.42}$$

$$AD + BD' + XD^{*} \underset{}{\overset{k_{+}^{DDD}}{\rightleftarrows}} AD' + BD^{*} + XD. \tag{5.43}$$

The backward reactions are characterized by the symbols k_{-}^{LLL}. As in the previous section, let us only consider the inverse average lifetimes of one of the reactants, i.e. here AL. Note that one obtains for AL exactly the same group exchange reaction set as in the case of the double proton transfer reactions in the previous section:

$$AH \xrightarrow{\tau^{-1}_{AHAH^{*}}} AH^{*},\quad AH \xrightarrow{\tau^{-1}_{AHAD}} AD,\quad AD \xrightarrow{\tau^{-1}_{ADAH}} AH,$$

$$AD \xrightarrow{\tau^{-1}_{ADAD}} AD^{*}, \tag{5.44}$$

$$AH \xrightarrow{\tau_{AHXH}^{-1}} XH, \quad AD \xrightarrow{\tau_{ADXD}^{-1}} XD \text{ etc.} \tag{5.45}$$

In a similar way as in the previous section we then obtain

$$\begin{aligned}\tau_{AHAH^*}^{-1} &= (k_+^{HHH} + k_-^{HHH})C_{BH}C_{XH} + (k_+^{HHD} + k_-^{DHH})C_{BH}C_{XD} \\ &= [(1 - D_B)(1 - D_X)(k_+^{HHH} + k_-^{HHH}) \\ &\quad + (1 - D_B)D_X(k_+^{HHD} + k_-^{DHH})]C_B C_X.\end{aligned} \tag{5.46}$$

$$\begin{aligned}\tau_{AHAD}^{-1} &= (k_+^{HDH} + k_-^{HHD})C_{BD}C_{XH} + (k_+^{HDD} + k_-^{DHD})C_{BD}C_{XD} \\ &= [D_B(1 - D_X)(k_+^{HDH} + k_-^{HHD}) + D_B D_X(k_+^{HDD} + k_-^{DHD})]C_B C_X.\end{aligned} \tag{5.47}$$

$$\begin{aligned}\tau_{ADAH}^{-1} &= (k_+^{DHH} + k_-^{HDH})C_{BH}C_{XH} + (k_+^{DHD} + k_-^{DHD})C_{BH}C_{XD} \\ &= [(1 - D_B)(1 - D_X)(k_+^{DHH} + k_-^{HDH}) \\ &\quad + (1 - D_B)D_X(k_+^{DHD} + k_-^{DDH})]C_B C_X.\end{aligned} \tag{5.48}$$

$$\begin{aligned}\tau_{ADAD^*}^{-1} &= (k_+^{DDH} + k_-^{HDD})C_{BD}C_{XH} + (k_+^{DDD} + k_-^{DDD})C_{BD}C_{XD} \\ &= [D_B(1 - D_X)(k_+^{DDH} + k_-^{HDD}) + D_B D_X(k_+^{DDD} + k_-^{DDD})]C_B C_X.\end{aligned} \tag{5.49}$$

$$\begin{aligned}\tau_{AHXH}^{-1} &= (k_+^{HHH} + k_-^{HHH})C_{BH}C_{XH} + (k_+^{HHD} + k_-^{DHH})C_{BH}C_{XD} \\ &\quad + (k_+^{HDH} + k_-^{HHD})C_{BD}C_{XH} + (k_+^{HDD} + k_-^{DHD})C_{BD}C_{XD} \\ &= [(1 - D_B)(1 - D_X)(k_+^{HHH} + k_-^{HHH}) + (1 - D_B)D_X(k_+^{HHD} + k_-^{DHH}) \\ &\quad + D_B(1 - D_X)(k_+^{HDH} + k_-^{HHD}) + (k_+^{HDD} + k_-^{DHD})D_B D_X]C_B C_X.\end{aligned} \tag{5.50}$$

$$\begin{aligned}\tau_{ADXD^*}^{-1} &= (k_+^{DHH} + k_-^{HDH})C_{BH}C_{XH} + (k_+^{DHD} + k_-^{DDH})C_{BH}C_{XD} \\ &\quad + (k_+^{DDH} + k_-^{HDD})C_{BD}C_{XH} + (k_+^{DDD} + k_-^{DDD})C_{BD}C_{XD} \\ &= [(1 - D_B)(1 - D_X)(k_+^{DHH} + k_-^{HDH}) + D_B(1 - D_X)(k_+^{DHD} + k_-^{DDH}) \\ &\quad + (1 - D_B)D_X(k_+^{DDH} + k_-^{HDD}) + D_B D_X(k_+^{DDD} + k_-^{DDD})]C_B C_X.\end{aligned} \tag{5.51}$$

Note that $k_+^{HHH} = k_-^{HHH}$ and $k_+^{DDD} = k_-^{DDD}$. Now, the inverse average lifetimes depend on products of the individual deuterium fractions. For the case where equilibrium isotope effects are absent, i.e. where

$$k_+^{HHD} = k_+^{HDH} = k_+^{DHH} = k_-^{HHD} = k_-^{HHD} = k_-^{HHD} \text{ etc.,} \tag{5.52}$$

the above equations can be simplified. Setting $k_+^{LLL} + k_-^{LLL} = k^{LLL}$ it follows that

$$\tau_{AHAH^*}^{-1} = [(1 - D)^2 k^{HHH} + (1 - D)D k^{HHD}]C_B C_X, \tag{5.53}$$

$$\tau_{\mathrm{AHAD}}^{-1} = [\mathrm{D}(1-\mathrm{D})\mathrm{k}^{\mathrm{HHD}} + \mathrm{D}^2\mathrm{k}^{\mathrm{HDD}}]\mathrm{C_B C_X}, \tag{5.54}$$

$$\tau_{\mathrm{ADAH}}^{-1} = [(1-\mathrm{D})^2\mathrm{k}^{\mathrm{HHD}} + (1-\mathrm{D})\mathrm{D}\mathrm{k}^{\mathrm{HDD}}]\mathrm{C_B C_X}, \tag{5.55}$$

$$\tau_{\mathrm{ADAD*}}^{-1} = [\mathrm{D}(1-\mathrm{D})\mathrm{k}^{\mathrm{HDD}} + \mathrm{D}^2\mathrm{k}^{\mathrm{DDD}}]\mathrm{C_B C_X}, \tag{5.56}$$

$$\tau_{\mathrm{AHXH}}^{-1} = \tau_{\mathrm{AH}}^{-1} = [(1-\mathrm{D})^2\mathrm{k}^{\mathrm{HHH}} + 2(1-\mathrm{D})\mathrm{D}\mathrm{k}^{\mathrm{HHD}} + \mathrm{D}^2\mathrm{k}^{\mathrm{HDD}}]\mathrm{C_B C_X}, \tag{5.57}$$

$$\tau_{\mathrm{ADXD}}^{-1} = \tau_{\mathrm{AD}}^{-1} = [(1-\mathrm{D})^2\mathrm{k}^{\mathrm{HHD}} + 2\mathrm{D}(1-\mathrm{D})\mathrm{k}^{\mathrm{HDD}} + \mathrm{D}^2\mathrm{k}^{\mathrm{DDD}}]\mathrm{C_B C_X}. \tag{5.58}$$

Note that this set of equations is also valid for the case where $\mathrm{A = B = C}$. Using again $\mathrm{k^{HHH}/k^{DDD}} = 10$ and $\mathrm{k^{HHD}/k^{HDD} = k^{HDD}/k^{DDD}} = 10^{1/3}$ we obtain the non-linear curves in Fig. 5.1 for the case $n = 3$, $\mathrm{L = H, D}$. Note that $\tau_{\mathrm{AHXH}}^{-1}(\mathrm{D}=1) = \mathrm{k^{HDD}} < \tau_{\mathrm{ADXD}}^{-1}(\mathrm{D}=0) = \mathrm{k^{HHD}}$.

5.2.4 Superposed Intermolecular Double and Triple Hydrogen Transfer Reactions

As one can note by comparison of Eqs. (5.26) to (5.31) with Eqs. (5.53) to (5.58), the inverse lifetimes measured are independent of the particular reaction network. Thus, a superposed double and triple proton transfer reaction can be described just by adding up the corresponding equations. For the case where

$$\mathrm{A = X}, \quad \text{i.e.} \quad \mathrm{D_A = D_X = D_B \equiv D}, \tag{5.59}$$

we obtain

$$\tau_{\mathrm{AHAH*}}^{-1} = (1-\mathrm{D})\cdot\mathrm{k}^{\mathrm{HH}}\cdot\mathrm{C_A} + [(1-\mathrm{D})^2\cdot\mathrm{k}^{\mathrm{HHH}} + \mathrm{D}\cdot(1-\mathrm{D})\cdot\mathrm{k}^{\mathrm{HHD}}]\cdot\mathrm{C_A^2}, \tag{5.60}$$

$$\tau_{\mathrm{AHAD}}^{-1} = \mathrm{D}\cdot\mathrm{k}^{\mathrm{HD}}\cdot\mathrm{C_A} + [\mathrm{D}\cdot(1-\mathrm{D})\cdot\mathrm{k}^{\mathrm{HDH}} + \mathrm{D}^2\cdot\mathrm{k}^{\mathrm{HDD}}]\cdot\mathrm{C_A^2}, \tag{5.61}$$

$$\tau_{\mathrm{ADAH}}^{-1} = (1-\mathrm{D})\cdot\mathrm{k}^{\mathrm{DH}}\cdot\mathrm{C_A} + [(1-\mathrm{D})^2\cdot\mathrm{k}^{\mathrm{DHH}} + \mathrm{D}\cdot(1-\mathrm{D})\cdot\mathrm{k}^{\mathrm{DHD}}]\mathrm{C_A^2}, \tag{5.62}$$

$$\tau_{\mathrm{ADAD*}}^{-1} = \mathrm{D}\cdot\mathrm{k}^{\mathrm{DD}}\cdot\mathrm{C_A} + [\mathrm{D}\cdot(1-\mathrm{D})\cdot\mathrm{k}^{\mathrm{DDH}} + \mathrm{D}^2\cdot\mathrm{k}^{\mathrm{DDD}}]\cdot\mathrm{C_A^2}, \tag{5.63}$$

and

$$\tau_{\mathrm{AHA*H}}^{-1} = \tau_{\mathrm{AH}}^{-1} = [(1-\mathrm{D}]\cdot\mathrm{k}^{\mathrm{HH}} + \mathrm{D}\cdot\mathrm{k}^{\mathrm{HD}}]\cdot\mathrm{C_A} + [(1-\mathrm{D})^2\cdot\mathrm{k}^{\mathrm{HHH}} + 2\cdot\mathrm{D}\cdot(1-\mathrm{D})\cdot\mathrm{k}^{\mathrm{HHD}} + \mathrm{D}^2\cdot\mathrm{k}^{\mathrm{HDD}}]\cdot\mathrm{C_A^2}, \tag{5.64}$$

$$\tau_{\mathrm{ADA*D}}^{-1} = \tau_{\mathrm{AD}}^{-1} = [(1-\mathrm{D}]\cdot\mathrm{k}^{\mathrm{DH}} + \mathrm{D}\cdot\mathrm{k}^{\mathrm{DD}}]\cdot\mathrm{C_A} + [(1-\mathrm{D})^2\cdot\mathrm{k}^{\mathrm{DHH}} + 2\cdot\mathrm{D}\cdot(1-\mathrm{D})\cdot\mathrm{k}^{\mathrm{DDH}} + \mathrm{D}^2\cdot\mathrm{k}^{\mathrm{DDD}}]\cdot\mathrm{C_A^2}. \tag{5.65}$$

The calculated values τ^{-1}_{AHA*H} and τ^{-1}_{ADA*D} will be located in Fig. 5.1 between those expected for n = 2 and 3.

5.2.5 *Multiple Hydrogen Transfer Reactions*

If more than $n = 3$ protons are transferred in the rate limiting step of the reaction the experimental data generally do not contain enough information in order to determine the different isotopic rate constants, i.e. in order to decide whether the RGM is fulfilled or not. If one assumes the validity of this rule which implies that all single kinetic isotope effects are equal i.e. given for constant total concentrations by

$$\Phi^{\ddagger} = (\tau^{-1}_{AD}(D = 1)/\tau^{-1}_{AH}(D = 0))^{1/n} \tag{5.66}$$

one obtains [106]

$$\tau^{-1}_{AH} = \tau^{-1}_{AH}(D = 0)(1 - D + D\Phi^{\ddagger})^{n-1}, \tag{5.67}$$

$$\tau^{-1}_{AD} = \tau^{-1}_{AH}(D = 0)\Phi(1 - D + D\Phi^{\ddagger})^{n-1}. \tag{5.68}$$

These equations can be easily verified for the case n = 3 by some simple transformations of Eqs. (5.67) and (5.68) using Eq. (5.66). Formally, Eqs. (5.67) and (5.68) resemble the so-called Gross–Butler equation [42–45] with the difference that Eqs. (5.67) and (5.68) have the exponent n − 1 instead of n. Therefore, Eqs. (5.67) and (5.68) contain more information than the Gross–Butler equation. The number of protons involved in the reaction can be obtained using Eqs. (5.67) and (5.68) by measuring the quantities $\tau^{-1}_{AH}(D = 0)$, $\tau^{-1}_{AD}(D = 1)$, and two values of τ^{-1}_{AH} and τ^{-1}_{AD} at the same deuterium fraction, e.g. D = 0.5. When the number of protons transferred becomes effectively infinite, as may be the case in pure protic solvents and when the overall kinetic isotope effect $(\Phi^{\ddagger})^{-n}$ is finite, $\Phi^{\ddagger}$ must be close to one and the lifetimes τ^{-1}_{AH} and τ^{-1}_{AD} measured by ^{1}H and ^{2}H NMR spectroscopy become equal, as shown in Fig. 5.1.

5.2.6 *Double Hydrogen Transfer Involving a Preequilibrium*

So far, formally simple first, second, or third order rate laws for the proton transfer processes have been treated. In practice, rate laws of chemical reactions may be more complex when a molecule A can exist in different fast exchanging environments A_i of which only A_r is reactive. The life time of A is then given by

$$\tau_A^{-1} = -\frac{1}{C_A} dc_{A_r}/dt, \quad C_A = \sum_i c_{A_i}. \tag{5.69}$$

C_A is the overall concentration of A, c_{A_i} the concentration of species A_i. In this section it is shown that the relations derived above are still valid even in this more complicated case as long as isotopic fractionation between the different environments is absent. For this purpose let us consider a double proton transfer involving a fast preequilibrium according to

$$\mathrm{AL + XL^*} \underset{\text{fast}}{\overset{K^{LL}}{\rightleftarrows}} \{\mathrm{AL, XL^*}\} \underset{\text{slow}}{\overset{k^{LL}_{rls}}{\rightleftarrows}} \{\mathrm{AL^*, XL}\} \underset{\text{fast}}{\overset{1/K^{LL}}{\rightleftarrows}} \mathrm{AL^* + XL}, \quad (5.70)$$

where

$$K^{LL} = c_{ALXL}/(c_{AL}c_{XL}), \quad LL = HH, HD, DD \quad (5.71)$$

is the preequilibrium constant of the formation of the reaction complex in which the double proton transfer takes place. k^{LL}_{rls} is the rate constant of the rate limiting step of the proton exchange in the reaction complex. With

$$c_A = c_{AH} + c_{AD}, c_{AX} = c_{AHXH} + c_{AHXD} + c_{ADXH} + c_{ADXD} \quad (5.72)$$

it follows that in the absence of isotopic fractionation between AL and ALXL

$$K^{HH} = K^{HD} = K^{DD} = K = c_{AX}/(c_A c_X). \quad (5.73)$$

Taking into account Eq. (5.69) we obtain

$$\tau^{-1}_{AHXH} = -\frac{1}{c_{AH} + c_{AHXH} + c_{AHXD}}[dc_{AHXH}/dt + dc_{AHXD}/dt]. \quad (5.74)$$

With

$$-dc_{ALXL}/dt = k^{LL}_{rls} c_{ALXL} \quad (5.75)$$

it follows that

$$\tau^{-1}_{AHXH} = \frac{1}{c_{AH} + c_{AHXH} + c_{AHXD}}[k^{HH}_{rls} c_{AHXH} + k^{HD}_{rls} c_{AHXD}]. \quad (5.76)$$

Because of Eqs. (5.72) and (5.73) it follows that

$$c_{AH} = (1-D)c_A, \quad c_{AHXH} = (1-D)^2 c_{AX}, \quad c_{AHXD} = (1-D)Dc_{AX}. \quad (5.77)$$

Combining Eq. (5.76) with (5.77) leads to

$$\tau^{-1}_{AHXH} = F[(1-D)k^{HH}_{rls} + Dk^{HD}_{rls}], \quad F = \frac{c_{AX}}{c_A + c_{AX}}. \quad (5.78)$$

In a similar way it follows that

$$\tau^{-1}_{ADXD} = F[(1-D)k^{HD}_{rls} + Dk^{DD}_{rls}], \quad F = \frac{c_{AX}}{c_A + c_{AX}}. \quad (5.79)$$

Similar equations can be obtained for τ^{-1}_{ALXL}. Note that these equations are valid not only for the case where $A \neq X$ but also for the self exchange case where $A \equiv X$. Since F is independent of the deuterium fraction D the inverse lifetimes τ^{-1}_{ALXL} depend linearly on D, as was already expressed by Eqs. (5.30) and (5.31). Recalling that the total concentrations

$$C_A = c_A + c_{AX} = c_{AH} + c_{AD} + c_{AHXH} + c_{AHXD} + c_{ADXH} + c_{ADXD} \quad (5.80)$$

$$C_X = c_X + c_{AX} = c_{XH} + c_{XD} + c_{AHXH} + c_{AHXD} + c_{ADXH} + c_{ADXD}, \tag{5.81}$$

it follows from Eqs. (5.71), (5.77) and (5.78) for the case where

$$K \text{ or } C_A \text{ small, i.e. } c_{AX} = Kc_A c_X \ll C_A$$

that

$$F = KC_X, \quad \text{i.e.} \quad c_A \cong C_A \quad \text{and} \quad c_X \cong C_X. \tag{5.82}$$

The second order rate constants defined in Eqs. (5.26) and (5.27) are then given by

$$k^{LL} = Kk^{LL}_{rls}. \tag{5.83}$$

By contrast, when

$$K \text{ or } C_{XL} \text{ large, i.e. } C_A = c_{AX}, \tag{5.84}$$

it follows that F = 1. We can then rewrite Eq. (5.78) and (5.79) in the form

$$\tau^{-1}_{AHXH}/F = \tau^{-1}_{AHXH}(C_{AL} \to \infty) = [(1-D)k^{HH}_{rls} + Dk^{HD}_{rls}] \tag{5.85}$$

and

$$\tau^{-1}_{ADXD}/F = \tau^{-1}_{ADXD}(C_{AL} \to \infty) = [(1-D)k^{HD}_{rls} + Dk^{DD}_{rls}] \tag{5.86}$$

Thus, one can obtain k^{HH}_{rls} and k^{HD}_{rls} directly from $\tau^{-1}_{AHXH}(C_A \to \infty)$ and k^{HD}_{rls} and k^{DD}_{rls} from $\tau^{-1}_{ADXD}(C_A \to \infty)$ at D = 0 and D = 1. Thus,

$$\tau^{-1}_{AHXH}(C_A \to \infty, D)/\tau^{-1}_{AHXH}(C_A \to \infty, D=0) = 1-(1-k^{HH}_{rls}/k^{HD}_{rls})D \tag{5.87}$$

and

$$\tau^{-1}_{ADXD}(C_A \to \infty, D)/\tau^{-1}_{ADXD}(C_A \to \infty, D=0) = 1-(1-k^{HD}_{rls}/k^{DD}_{rls})D. \tag{5.88}$$

At D = 0.5 the ^{1}H NMR lineshapes are determined by $\tau^{-1}_{AHXH}(C_A \to \infty) = (k^{HH}_{rls} + k^{HD}_{rls})/2$ and the ^{2}H NMR lineshapes by $\tau^{-1}_{ADXD}(C_A \to \infty) = (k^{HD}_{rls} + k^{DD}_{rls})/2$. Note, however, that this result applies only for a very rapid hydrogen bond preequilibrium as assumed in Eq. (5.70). If these processes were very slow, as for example in intramolecular double proton transfers, one would also obtain k^{HH}_{rls} and k^{DD}_{rls} at D = 0 and D = 1, but at D = 0.5 the ^{1}H NMR lineshape would have to be described by a static superposition of two independent lineshape functions as has been described in Sect. 5.1.

Expressions for F as a function of concentration can easily be derived. This will be done here for the self exchange in dimers ALAL = A_2, i.e. the case where A ≡ X. Equation (5.80) then becomes

$$C_A = c_A + 2c_{A_2}. \tag{5.89}$$

By simple arithmetics one then obtains

$$F = \frac{c_{A_2}/C_A}{1 - c_{A_2}/C_A}. \tag{5.90}$$

An expression for c_{A_2}/C_A can be obtained by combination of Eqs. (5.71) and (5.80):

$$F = \frac{4KC_A - (8 \cdot KC_A + 1)^{1/2} + 1}{4KC_A + (8 \cdot KC_A + 1)^{1/2} - 1}. \tag{5.91}$$

This equation is used together with Eqs. (5.85)–(5.86). From Eqs. (5.87) and (5.88) it follows then by setting $k_{rls} = k_{A_2}$ that

$$\tau_{AH}^{-1}(C \to \infty, D)/\tau_{AH}^{-1}(C \to \infty, D = 0) = 1 - (1 - k_{A_2}^{HD}/k_{A_2}^{HH})D. \tag{5.91}$$

6 NMR Lineshape Equations of Hydrogen Transfers in the Presence of Kinetic Isotope Effects

All elements are now in place in order to set up specific NMR line shape equations for typical cases of intra- and intermolecular proton exchange reactions in the presence of kinetic hydrogen/deuterium isotope effects. The line shape equations considerably simplify in the absence of strong scalar coupling between the remote and the labile spins. In these cases it is possible to set up two separate NMR lineshape equations for both types of spins. Only when remote and mobile spins are strongly coupled protons with similar chemical shifts it is not possible to separate the two lineshape contributions. An experimental example for this case will be given in Sect. 10.1. The purpose of this section is to demonstrate ways of determining by dynamic NMR spectroscopy the kinetic quantities defined in Sect. 5.

6.1 *Intramolecular Hydrogen Transfer Reactions*

6.1.1 *Remote Spin Studies*

Let us consider an intramolecular hydrogen transfer reaction between two states AL and XL according to Eqs. 5.1 and 5.2, characterized by the forward and backward rate constants k_+^H, k_+^D, k_-^D, k_-^D. During the exchange the chemical shift of a remote nucleus is then modulated from the value ν_{AL} to ν_{XL}. The complex matrix $\mathscr{M}$ describing the lineshape of the remote spins is then given by

$\mathscr{M}$ (remote spins) =

$-k_+^H - \pi W_0 + 2\pi i\nu_{AH}$	k_-^H		
k_+^H	$-k_-^H + \pi W_0 - 2\pi i\nu_{XH}$		
		$-k_-^D + \pi W_0 - 2\pi i\nu_{AD}$	k_-^D
		k_+^D	$-k_-^H + \pi W_0 - 2\pi i\nu_{XD}$

,

$$p = [p_{AH}, p_{XH}, p_{AD}, p_{XD}]. \tag{6.1}$$

Because of the absence of scalar coupling all non-diagonal $\mathscr{L}$ elements connecting $\mathcal{M}$(remote spins) and the counterparts $\mathcal{M}$(mobile spins) vanish. Since in Eq. (6.1) it is assumed that intermolecular exchange processes are absent, $\mathcal{M}$(remote spins) splits into two independent submatrices $\mathcal{M}_{AL}$, L = H, D given by

$$\mathcal{M}_{AL} = \begin{array}{|c|c|} \hline -k_+^L - W_0 + 2\pi i \nu_{AL} & k_-^L \\ \hline k_+^L & -k_-^L - \pi W_0 + 2\pi i \nu_{XL} \\ \hline \end{array}\,,$$

$$p = [p_{AL}, p_{XL}], \quad (6.2)$$

Measurements at deuterium fractions D = 0 and 1 will easily lead to k_+^H and k_+^D; these constants are sufficient to describe the lineshape also at other values of D. If the lineshape at a given D value cannot be simulated one has to take the possibility of intermolecular exchange processes into account, as shown in Sect. 6.3.

Let us consider now the case of a bifunctional molecule subject to an intramolecular double proton transfer between two states ALL and XLL according to Eqs. (5.3) to (5.6). The lineshape equation is obtained by replacing L = H, D in Eq. (6.2) by LL = HH, HD, DD. Thus, because of the absence of intermolecular exchange $\mathcal{M}$(remote spins) splits up into three superposed submatrices:

$$\mathcal{M}_{ALLXLL} = \begin{array}{|c|c|} \hline -k_+^{LL} - \pi W_0 + 2\pi i \nu_{ALL} & k_-^{LL} \\ \hline k_+^{LL} & -k_-^{LL} - \pi W_0 + 2\pi i \nu_{XLL} \\ \hline \end{array}\,,$$

$$LL = HH, HD, DD, \quad p = [p_{ALL}, p_{XLL}]. \quad (6.3)$$

The populations depend on the deuterium fraction D in the mobile proton sites. If the kinetic isotope effects are not very large it will be difficult to obtain k_+^{HH}, k_+^{HD}, and k_+^{DD} by analyzing spectra at intermediate deuterium fractions D. Therefore, one proceeds as follows. From experiments at D = 0 and D = 1, k_+^{HH} and k_+^{DD} can easily be obtained. Using this information one can try to obtain k_+^{DD} at D = 0.5. A better solution is, however, to obtain k_+^{HH} and k_+^{HD} by ^{1}H NMR spectroscopy of the exchanging proton signals as shown in the following.

6.1.2 *Mobile Hydrogen Isotope Studies*

Let us consider now the lineshape equations of the mobile hydrogen isotopes for the cases presented in the previous section. Again, for simplicity the presence of scalar coupling between the hydrogen isotopes and the remote spins is neglected. For the single proton transfer according to Eqs. (5.1) and (5.2) one obtains then

$$\mathscr{M}_{\mathrm{ALXL}} = \begin{bmatrix} -k_+^L - \pi W_0 + 2\pi i \nu_{AL} & k_-^L \\ k_+^L & -k_-^L - \pi W_0 + 2\pi i \nu_{XL} \end{bmatrix},$$

$$L = H, D, \quad p = [p_{AH}, p_{XH}], \qquad (6.4)$$

for the NMR lineshape of the L spins. ν_{AL} is the chemical shift of L in AL.

In the case of intramolecular double proton transfer between ALL and XLL according to Eqs. (5.3) to (5.6) the ^{1}H NMR lineshape is given by

$$\mathscr{M}_{\mathrm{AHHXHH}} = \begin{bmatrix} -k_+^{HH} - \pi W_0 + 2\pi i \nu_{AHH} & k_-^{HH} \\ k_+^{HH} & -k_-^{HH} - \pi W_0 + 2\pi i \nu_{XHH} \end{bmatrix},$$

$$p = [p_{AHH}, p_{XHH}], \qquad (6.5)$$

$$\mathscr{M}_{\mathrm{AHDXHD}} = \begin{bmatrix} -k_+^{HD} - \pi W_0 + 2\pi i \nu_{AHD} & k_-^{HD} \\ k_+^{HD} & -k_-^{HD} - \pi W_0 + 2\pi i \nu_{XHD} \end{bmatrix},$$

$$p = [p_{AHD}, p_{XHD}], \qquad (6.6)$$

Similar equations are obtained for the ^{2}H NMR lineshape. Thus, in the case of an intramolecular double proton transfer the ^{1}H NMR lineshape depends on k_+^{HH} and k_-^{HD}, whereas the ^{2}H NMR lineshape depends on k_+^{HD} and k_-^{DD}. k_+^{HH} is best obtained at a deuterium fraction of $D = 0$. Then Eq. (6.6) can be neglected. k_+^{HD} is to be determined at a value of D close to 1 by ^{1}H NMR in order to minimize the contribution of Eq. (6.5). Because of sensitivity problems one typically chooses D between 0.9 and 0.95. However, at these D values Eq. (6.5) still contributes in a non-negligible way to the lineshapes and has to be taken into account. Similar arguments hold for the determination of k_+^{HD} by ^{2}H NMR at low D-values.

Let us consider now a case of an intramolecular single proton transfer reaction between two states $r = 1, 2$ in the presence of scalar coupling between the mobile proton and remote spins according to the modified Eqs. (5.1) and (5.2):

$$1 \equiv AL \cdots B \underset{k_-^L}{\overset{k_+^L}{\rightleftarrows}} 2 \equiv A \cdots LB. \qquad (6.7)$$

Let A and B have a spin 1/2 as is the case for proton transfer in ^{15}NH$\cdots^{15}$N hydrogen bonds. Because of the intramolecular character of the process the following independent reactions take place:

$$A(\alpha) - L \cdots B(\alpha) \rightleftharpoons A(\alpha) \cdots L - B(\alpha), \qquad (6.8)$$

$$A(\alpha) - L \cdots B(\beta) \rightleftharpoons A(\alpha) \cdots L - B(\beta), \tag{6.9}$$

$$A(\beta) - L \cdots B(\alpha) \rightleftharpoons A(\beta) \cdots L - B(\alpha), \tag{6.10}$$

$$A(\beta) - L \cdots B(\beta) \rightleftharpoons A(\beta) \cdots L - B(\beta). \tag{6.11}$$

The fact that the reaction is intramolecular leads to a splitting of $\mathscr{M}$(mobile spins) into four independent superposed submatrices, one for each reaction listed in Eqs. (6.8) to (6.11). Let ν_r be the chemical shift of the mobile hydrogen isotope in r and $J_{X^rL^r}$, $X = A, B$, $r = 1, 2$ the corresponding coupling constants. Let further be $J_{A^1L^1} = J_{AL} \neq 0$, $J_{B^1L^1} = 0$, $J_{A^2L^2} = 0$, $J_{B^2L^2} = J_{BL} \neq 0$. One then obtains

$$\mathscr{M}^{\alpha\alpha}_{ALBL} = \begin{array}{|c|c|} \hline -k^L_+ - \pi W_0 + 2\pi i(\nu_1 + J_{AL}/2) & k^L_- \\ \hline k^L_+ & -k^L_- - \pi W_0 + 2\pi i(\nu_2 + J_{BL}/2) \\ \hline \end{array},$$

$$p = 0.25[p_1, p_2], \tag{6.12}$$

$$\mathscr{M}^{\alpha\beta}_{ALBL} = \begin{array}{|c|c|} \hline -k^L_+ - \pi W_0 + 2\pi i(\nu_2 + J_{AL}/2) & k^L_- \\ \hline k^L_+ & -k_- - \pi W_0 + 2\pi i(\nu_2 - J_{BL}/2) \\ \hline \end{array},$$

$$p = 0.5[p_1, p_2], \tag{6.13}$$

$$\mathscr{M}^{\beta\beta}_{ALBL} = \begin{array}{|c|c|} \hline -k^L_+ - \pi W_0 + 2\pi i(\nu_1 + J_{AL}/2) & k^L_- \\ \hline k^L_+ & -k^L_- - \pi W_0 + 2\pi i(\nu_2 - J_{BL}/2) \\ \hline \end{array}.$$

$$p = 0.25[p_1, p_2]. \tag{6.14}$$

In the slow exchange region the L signal consists of two doublets at the frequencies $\nu_1 \pm J_{AL}/2$ and $\nu_2 \pm J_{BL}/2$. In the fast exchange range the signal consists of a multiplet at the frequency $p_1\nu_1 + p_2\nu_2$ characterized by the effective coupling constants $p_1 J_{AL}$ and $p_2 J_{BL}$.

A special situation arises in a symmetric exchange case where $A = B$ in Eq. (6.7), $p_1 = p_2 = 0.5$, $k^L_+ = k^L_- = k^L$, $\nu_1 = \nu_2 = \nu$. Then $J_{AL} = J_{BL}$ and Eqs. (6.12) to (6.14) reduce to $\mathscr{M}^{\alpha\alpha}_{AL} = -\pi W_0 + 2\pi i(\nu + J_{AL}/2)$, $\mathscr{M}^{\beta\beta}_{AL} = -\pi W_0 + 2\pi i(\nu - J_{AL}/2)$,

$$\mathscr{M}^{\alpha\beta}_{AL} = \begin{array}{|c|c|} \hline -k^L - \pi W_0 + 2\pi i(\nu + J_{AL}/2) & k^L \\ \hline k^L & -k^L - \pi W_0 + 2\pi i(\nu - J_{AL}/2) \\ \hline \end{array} \tag{6.15}$$

with $p_{\alpha\alpha} = p_{\beta\beta} = 0.25$, $p_{\alpha\beta} = 0.5$. Then, in the slow exchange region there is only one doublet at $\nu \pm J_{AL}/2$. Hydrogen isotopes reacting according to Eqs. (6.8) and (6.11) do not change their transition frequencies and, therefore, contribute sharp line components at $\nu \pm J_{AL}/2$. By contrast, the lineshape contributions of protons reacting according to Eqs. (6.9) and (6.10) correspond only in the slow exchange region to a doublet at $\nu \pm J_{AL}/2$; when the exchange becomes faster these line components broaden, coalesce, and eventually give rise to one sharp line in the center of the resonance, i.e. a triplet appears. The ratio of the integrated intensities of the inner exchange broadened line component to each of the outer lines is always 2:1; however, the peak heights behave in this way only at very high temperatures, where the inner line no longer shows line broadening; i.e. peak height ratio of less then 2:1 is, therefore, indicative that the extremely fast exchange has not yet been reached. Note that for the study of kinetic isotope effects one has to perform ^{1}H and ^{2}H NMR experiments.

In the case of a double proton transfer according to

$$(AL \cdots B)_2 \underset{k_{-1}^{LL}}{\overset{k_1^{LL}}{\rightleftarrows}} (AL \cdots B)_2 \tag{6.16}$$

the same lineshape theory applies. One only replaces L in Eqs. (6.12) to (6.15) by LL = HH, HD, DD and takes into account that the ^{1}H NMR lineshape depends on the sum

$$\mathcal{M}(\text{mobile spins}) = \mathcal{M}_{LL}^{\alpha\alpha} + \mathcal{M}_{LL}^{\alpha\beta} + \mathcal{M}_{LL}^{\beta\beta}, \quad LL = HH, HD, \tag{6.17}$$

i.e. on both k_+^{HH} and k_+^{HD} as well as on the corresponding backward rate constants.

Finally, let us consider a hydrogen isotope jumping between four heavy atoms A to D with spin 1/2, i.e. exchanging between the states r = 1 to 4 as shown in Eq. (6.18):

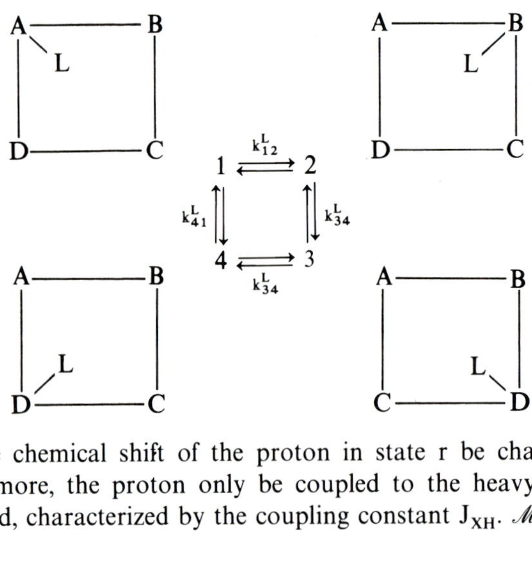

(6.18)

Let the chemical shift of the proton in state r be characterized by ν_r. Let, furthermore, the proton only be coupled to the heavy atom to which it is attached, characterized by the coupling constant J_{XH}. $\mathcal{M}$(mobile spins) is then the sum of submatrices, each given by [113, 115],

$$\mathscr{M}_L(e_1, e_2, e_3, e_4) =$$

$$\left[\begin{array}{c|c|c|c}
\begin{array}{c}-k^L_{12} - k^L_{14} - \pi W_0 \\ +2\pi i(\nu_1 + e_1 J_{AL}/2)\end{array} & k^L_{21} & & k^L_{41} \\ \hline
k^L_{12} & \begin{array}{c}-k^L_{21} - k^L_{23} - \pi W_0 \\ +2\pi i(\nu_2 + e_2 J_{BL}/2)\end{array} & k^L_{32} & \\ \hline
 & k^L_{23} & \begin{array}{c}-k^L_{32} - k^L_{34} - \pi W_0 \\ +2\pi i(\nu_3 + e_3 J_{CL}/2)\end{array} & k^L_{43} \\ \hline
k^L_{14} & & k^L_{34} & \begin{array}{c}-k^L_{41} - k^L_{43} - \pi W_0 \\ +2\pi i(\nu_4 + e_4 J_{DL}/2)\end{array}
\end{array}\right],$$

$$e_r = \pm 1, \quad p = (p_1 = p_{AL}, p_2 = p_{BL}, p_3 = p_{CL}, p_4 = p_{DL}). \qquad (6.19)$$

In the slow exchange region the signal consists of four doublets with the chemical shifts ν_r, $r = 1, 4$. In the fast exchange region Eq. (6.19) predicts a coalesced line split by coupling to all spins A to D, with the effective coupling constants given by

$$J^{eff}_{XL} = p_X J_{XL}. \qquad (6.20)$$

In the simplest case where all chemical shifts, coupling constants, and populations are equal Eq. (6.19) predicts in the slow exchange region a splitting of the L signal into a doublet by coupling with the spin to which L is bound; in the fast exchange region a pentet splitting is predicted.

If there are two protons jumping between the four atoms A to D in Eq. (6.18) the superscript L in Eqs. (6.18) and (6.19) has to be replaced by LL = HH, HD, DD. Furthermore, if ^1H measurements are performed $\mathscr{M}$(mobile spins) is the sum of the two contributions

$$\mathscr{M}(\text{mobile spins}) = \mathscr{M}^{HH} + \mathscr{M}^{HD}. \qquad (6.21)$$

6.2 *Intermolecular Hydrogen Exchange Reactions*

In the case of proton exchange according to Eqs. (5.7) and (5.8) the chemical shifts of both the remote as well as of the mobile spins are modulated. Therefore, both remote as well as mobile spin studies can be performed even in the absence of scalar spin–spin coupling between both types of spins. A scalar interaction is, however, required in the case of Eqs. (5.12) to (5.15) and (5.36) to (5.43) when performing remote spin studies. Therefore, mobile spin studies are indicated in this case because the chemical shifts of the latter are modulated by the exchange. Thus, in the case of intermolecular proton exchange between different species one does not need the scalar spin–spin interaction, and it is straightforward to set up the NMR line shape equations. A special case arises, however, when

A = X in Eqs. (5.12) to (5.15) and A = B = X in Eqs. (5.36) to (5.43). Then, the hydrogen exchange processes do not lead to a modulation of chemical shifts of either remote or of mobile spins. There are, however, often cases where the proton transfer is accompanied by a rearrangement of the group A to A′ in such a way that a remote spin in A exhibits a different chemical shift ν_{AL} before and $\nu_{A,L}$ after the exchange. Then, self exchange rates can be studied by remote spin studies. If this modulation does not occur one needs a strong scalar spin–spin coupling between remote and mobile spins. We will treat both cases in the remainder of this section.

6.2.1 *Remote Spin Studies in the Presence of Hydrogen Self Exchange*

Let us first consider the NMR line shape problem of remote spins A of monofunctional molecules AL, L = H, D as a function of the deuterium fraction D subject to intermolecular proton self exchange. Here, both species AH and AD contribute to the line shape of A. Naturally, the NMR line shape of A is only sensitive to the proton exchange when the chemical shift of A is modulated by the exchange. Let ν_{AL} be the chemical shift of A before and $\nu_{A'L}$ the chemical shift after the exchange. Using the dynamic quantities defined in Sect. 5 it follows that [108]

$$\mathscr{M}(\text{remote spins}) = \begin{bmatrix} -\tau^{-1}_{AHAH^*} - \tau^{-1}_{AHAD} - \pi W_0 + 2\pi i\nu_{AH} & \tau^{-1}_{AHAH^*} & & \tau^{-1}_{ADAH} \\ \tau^{-1}_{AHAH^*} & -\tau^{-1}_{AHAH^*} - \tau^{-1}_{AHAD} - \pi W_0 + 2\pi i\nu_{A'H} & \tau^{-1}_{ADAH} & \\ & \tau^{-1}_{AHAD} & -\tau^{-1}_{ADAH} - \tau^{-1}_{ADAD^*} - \pi W_0 + 2\pi i\nu_{AD} & \tau^{-1}_{ADAD^*} \\ \tau^{-1}_{AHAD} & & \tau^{-1}_{ADAD^*} & \tau^{-1}_{ADAH} - \tau^{-1}_{ADAD^*} - \pi W_0 + 2\pi i\nu_{A'D} \end{bmatrix},$$

$$p = [p_{AH}, p_{AH}, p_{AD}, p_{AD}], \quad \tau^{-1}_{AHAD}/\tau^{-1}_{ADAH} = p_{AD}/p_{AH} \tag{6.22}$$

and

$$p_{AH} = (1 - D) \quad \text{and} \quad p_{AD} = D. \tag{6.23}$$

Note that as compared to the intramolecular case treated in Eq. (6.1) $\mathscr{M}$ does not split any more into two submatrices because of exchange between AH and AD, besides the trivial cases at D = 0 and 1. Thus, at D = 0.5 intra- and intermolecular exchange processes can easily be distinguished. Some spectral features which follow from Eq. (6.22) will be discussed in Sect. 10.2.3

6.2.2 *Mobile Hydrogen Isotope Studies in the Presence of Hydrogen Self Exchange*

As an example relevant in the context of this study let us consider a spin system AL where A represents a spin 1/2 nucleus in the presence of intermolecular proton self exchange. Let the Larmor frequency ν_{AL} of A be so different from the Larmor frequency $\nu_{AL} = \nu$ of the exchanging L isotope that AL constitutes a first order spin system. Let us further analyze only the lineshape of the exchanging L isotope which is affected according to the previous section by the inverse life time $\tau_{ALA*L}^{-1} \equiv \tau_{AL}^{-1}$. $\mathcal{M}$(mobile spins) and the population vector are then given by

$$\mathcal{M}(\text{mobile spins}) = \begin{bmatrix} -\tau_{AL}^{-1}/2 - \pi W_0 + 2\pi i(\nu + J_{AL}/2) & \tau_{AL}^{-1}/2 \\ \tau_{AL}^{-1}/2 & -\tau_{AL}^{-1}/2 - \pi W_0 + 2\pi i(\nu - J_{AL}/2) \end{bmatrix}, \quad p = \begin{bmatrix} 1/2 \\ 1/2 \end{bmatrix}. \tag{6.24}$$

In the slow exchange range the signal of the exchanging proton consists of a doublet at the frequency ν with a splitting corresponding to the coupling constant J_{AH}. The two line components refer to the protons attached to nuclei A in the spin states α(A) and β(A). When the proton exchanges, its Larmor frequency is modulated from $\nu + J_{AH}$ to $\nu - J_{AH}$. This switch occurs, however, only every other exchange process. Thus, the inverse life times are divided by 2 in Eq. (6.24). As the proton transfer becomes faster the two lines broaden and coalesce into one single line at ν. Note that in Eq. (6.24) no assumptions concerning the proton exchange mechanism have been made. Information on the latter can be obtained by performing experiments at different deuterium fractions D as described in Sect. 5.

6.3 *Superposed Intra- and Intermolecular Hydrogen Exchange Reactions*

There may be cases where both intra- and intermolecular hydrogen exchange reactions are present. In this case one has to add up the corresponding lineshape equations. As example let us consider the degenerate superposed intramolecular and intermolecular exchange reactions

$$AH \cdots A^* \underset{}{\overset{k}{\rightleftarrows}} A \cdots HA^*, \tag{6.25}$$

$$AH \cdots A + A^*H^* \cdots A^* \overset{\tau^{-1}}{\rightleftarrows} AH^* \cdots A + A^*H \cdots A^*, \tag{6.26}$$

where A has a spin 1/2. k is the intramolecular first order rate constant, τ^{-1} the inverse lifetime characterising the intermolecular proton exchange. In both processes the Larmor frequency of the mobile proton is not changed. In the process described by Eq. (6.26) the exchanging proton jumps with the probability 0.25 to another molecule in the spin states $\alpha\alpha$ or $\alpha\beta$ or $\beta\alpha$ or $\beta\beta$. In order to describe the NMR lineshape one has to combine Eqs. (6.15) and (6.24) in the following way:

$\mathscr{M}$(mobile spins) =

$\alpha\alpha$	$\alpha\beta$	$\beta\alpha$	$\beta\beta$
$-0.75\tau^{-1} - \pi W_0$ $2\pi i(\nu + J/2)$	$0.25\tau^{-1}$	$0.25\tau^{-1}$	$0.25\tau^{-1}$
$0.25\tau^{-1}$	$-k - 0.75\tau^{-1} - \pi W_0$ $+ 2\pi i(\nu + J/2)$	$k + 0.25\tau^{-1}$	$0.25\tau^{-1}$
$0.25\tau^{-1}$	$k + 0.25\tau^{-1}$	$-k - 0.75\tau^{-1} - \pi W_0$ $+ 2\pi i(\nu + J/2)$	$0.25\tau^{-1}$
$0.25\tau^{-1}$	$0.25\tau^{-1}$	$0.25\tau^{-1}$	$-0.75\tau^{-1} - \pi W_0$ $+ 2\pi i(\nu + J/2)2$

(6.27)

In the case where $k \gg \tau^{-1}$ Eq. (6.27) predicts a doublet–triplet transition when k is increased. When increasing τ^{-1} Eq. (6.27) predicts a doublet–singlet or a triplet–singlet transition, depending on the magnitude of k.

7 Computer Program for the Calculation of NMR Lineshapes in the Presence of Kinetic Isotope Effects

In order to perform NMR lineshape calculations a computer program is necessary. Among NMR-spectroscopists the DNMR programs of Binsch have become popular [198]. They are based on the NMR lineshape theory described in Sect. 2.1. These programs are written for high-order spin systems subject to intramolecular exchange. In this case general equations for the matrix $\mathscr{M}$ can be derived and programmed. Furthermore, intramolecular exchange leads to a factorization of $\mathscr{M}$ into submatrices for each set of transitions, which greatly simplifies the diagonalization procedure indicated in Eq. (2.13). This simplification is, however, not valid in the case of intermolecular exchange. Furthermore, different exchanging species and different exchange reactions have to be considered in the case of dynamic NMR spectroscopy in the presence of kinetic isotope effects, setting which necessitates up a different matrix $\mathscr{M}$ for each problem studied. Therefore, we have written a computer program in Fortran V which uses several subroutines of the programs of Binsch, e.g. the matrix diagonalization and inversion procedures and procedures to calculate spectra from the eigenvalues and eigenvectors of $\mathscr{M}$. For each lineshape problem a separate subroutine has to be written in which $\mathcal{M}$ is calculated. The program runs on the main computer of the University of Freiburg. In a smaller version, it can also be run on a PC.

Generally, the experimental spectra are transferred from the NMR-minicomputer to a PC linked to the main computer.

8 The Preparation of NMR Samples for the Study of Kinetic Hydrogen/Deuterium Isotope Effects

When studying the kinetics of proton and deuteron transfer reactions it is of prime importance to prepare well defined NMR samples which are free from impurities which catalyze or inhibit the exchange reactions to be studied. Solute concentrations must be adjustable. The deuterium fraction in the mobile proton sites must be controlled. The latter requirement is quite easily achieved when protic solvents are employed. In this case it is only necessary to mix the non-deuterated and the deuterated liquids in the desired amounts. However, it is much more difficult to control the deuterium fraction in the mobile proton sites when aprotic media are used as solvents because of slow proton exchange with OH groups of the glass walls of the NMR tubes. Furthermore, the NMR glass tubes have to be sealed in order to prevent contamination by the environment. Sample sealing is especially important because there will always be a substantial time interval between the sample preparation and the NMR measurements. In this section, therefore, some useful sample preparation techniques which have been developed over the past years in our laboratory are described.

8.1 General Techniques for the Preparation of Sealed NMR Samples

A glass apparatus for the preparation of sealed NMR samples used in our laboratory is shown in Fig. 8.1. The apparatus is attached to a vacuum line (1); a turbomolecular pump is used because of its ease of operation and of its capability to quickly provide a vacuum of the order of 10^{-5} to 10^{-6} bar. Greaseless Teflon needle valves (2) are employed to separate the different devices from the vacuum line. Solvents are stored in solvent containers (3) over an appropriate drying agent introduced via a tube (4) sealed before condensing the solvent into (3) from another flask. The apparatus contains a system of high precision glass tubes (5) of different diameters, e.g. 0.2 mm and 2 mm, for volume measurements of liquids. The glass tubes are calibrated by filling with mercury and weighing prior to their attachment to the vacuum line. All liquid components of the NMR sample are condensed into the NMR tube (6) after volume

spectra of pure CD_2Cl_2 solutions have been measured at temperatures up to 110 °C, the boiling point of this solvent being 40 °C [122, 126].

9 Kinetic Hydrogen/Deuterium Isotope Effects of Intramolecular Double Hydrogen Transfer in the Liquid State

In this section we will give some examples of the determination of kinetic hydrogen/deuterium isotope effects of intramolecular double proton transfer reactions by dynamic liquid state NMR spectroscopy. Note that the capacity of molecules to bind a proton at different acceptor atoms has often been called proton tautomerism or just tautomerism, a term which will be used throughout this paper.

9.1 The Azophenine Tautomerism

The degenerate tautomerism of azophenine (AP, Fig. 9.1), whose systematic name is *N*, *N'*-diphenyl-3,6-bis-(phenylimino)-1,4-cyclohexadiene-1,4-diamine, was discovered some years ago [105]. The reaction has also been studied theoretically [131, 132]. A recent study [127] sought to determine the rate constants k^{HH}, k^{HD}, and k^{DD} of this process in order to obtain information on the reaction mechanism. In this section it is shown how this task was accomplished by performing various NMR experiments.

9.1.1 ^1H and ^{13}C NMR Lineshape Analysis of ^{15}N and ^{13}C Labeled Azophenine in $C_2D_2Cl_4$

Let us first discuss ^1H NMR experiments performed on the mobile protons in AP. Whereas the chemical shift of the latter is not modified when the jump process occurs, the scalar spin–spin interaction with the nitrogen atoms is

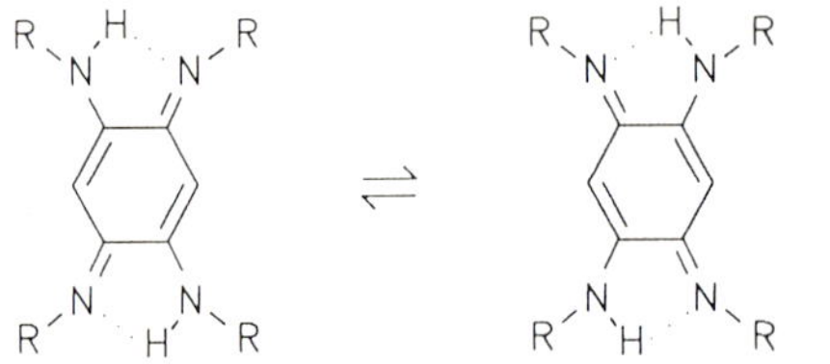

Fig. 9.1. The tautomerism of azophenine (AP)

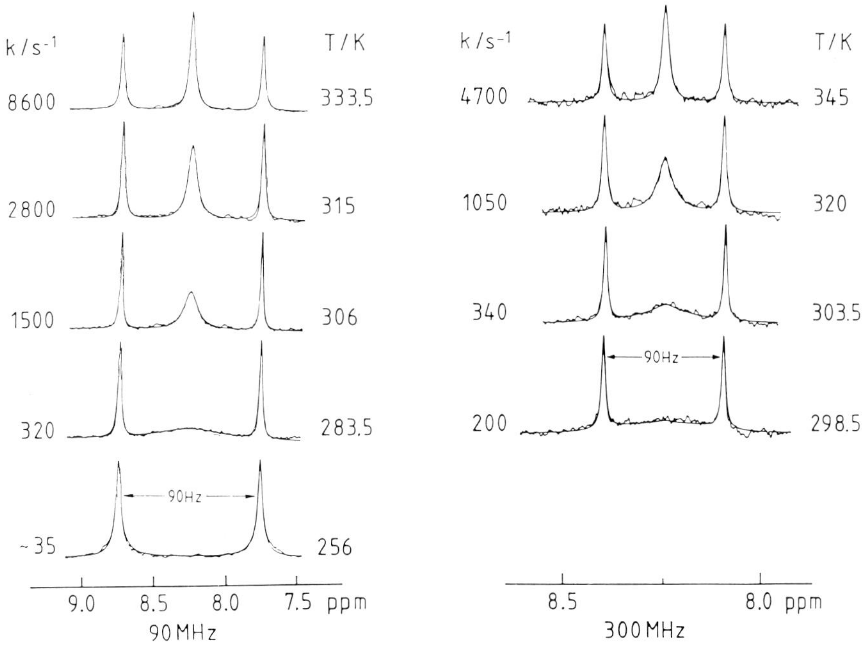

Fig. 9.2. Left: Superposed experimental and calculated 90.02 MHz ^1H NMR signals of the ^1H–^{15}N units of azophenine dissolved in $C_2D_2Cl_4$ at a deuterium fraction $D = 0$ in the NH sites as a function of temperature. The calculations were done using Eqs. (6.15) and (6.17). 200–500 scans, 60° pulses, 3 s repetition time. $k \equiv k^{HH}$ is the rate constant of the HH transfer. **Right**: Superposed experimental and calculated 300.13 MHz ^1H NMR signals of the ^1H–^{15}N units of azophenine dissolved in $C_2D_2Cl_4$. Deuterium fraction in the NH sites: $D = 0.92$. 1000–5000 scans, 90° pulses, 4 s repetition time. Reproduced with permission from Ref. [127]. $k \equiv k^{HD}$ is the rate constant of the HD transfer

modulated by the exchange. Unfortunately, the abundant ^{14}N spin has an electric quadrupole moment which leads to short ^{14}N longitudinal relaxation times, and as a consequence to a breakdown of the scalar ^1H–^{14}N coupling. Therefore, AP was labeled with the ^{15}N nucleus which has a spin 1/2.

Figure 9.2 shows the superposed experimental and calculated ^1H NMR signals of the ^1H–^{15}N units of AP dissolved in $C_2D_2Cl_4$ as a function of temperature. At low temperatures the proton signal consists of a doublet with the frequencies $\nu + J_{^1H^{15}N}/2$ and $\nu - J_{^1H^{15}N}/2$, because of coupling of the hydrogen bonded proton to one ^{15}N atom. At high temperatures the signal has a triplet structure. This observation means that the NH proton is now equally coupled to two ^{15}N atoms within the NMR time scale. The doublet–triplet transition can be understood in terms of Eq. (6.15). At low temperature the Larmor frequency ν of a mobile proton depends only on whether the ^{15}N atom to which it is bound is in the spin state α or β. As the proton transfer takes place the Larmor frequency of protons exchanging between two ^{15}N atoms in the spin states $\alpha\alpha$ or $\beta\beta$ is not altered. These protons, therefore, always contribute sharp lines to the signal at $\nu \pm J_{^1H^{15}N}/2$. However, the Larmor frequency of protons exchanging between two ^{15}N atoms in the spin states $\alpha\beta$ and $\beta\alpha$ is modulated between $\nu + J_{^1H^{15}N}/2$ and $\nu - J_{^1H^{15}N}/2$ which leads to a coalescence of the corresponding line contributions. The coalesced line then constitutes the central line of the triplet in the fast exchange regime. Note that in this regime the lineshapes can be simulated without assumptions. The line width in the absence of exchange and $J_{^1H^{15}N}$ are obtained by simulation of the outer line components, the rate constants by simulation of the inner lines. Note that within the margin of error there are no lineshape contributions due to intermolecular proton exchange as predicted by Eq. (6.26); such contributions would lead to a breakdown of the triplet into a singlet. Thus, it can be concluded that the rate constants obtained by the lineshape analysis in Fig. 9.2 do not contain terms due to intermolecular proton exchange but do arise exclusively from an intramolecular double proton transfer according to Fig. 9.1.

At a deuterium fraction of $D = 0$ in the mobile proton sites the rate constant obtained at 90 MHz by simulation correspond to k^{HH}. However, when AP is deuterated the residual ^1H–^{15}N signal dominantly stems from AP–HD and only to a minor extent from AP–HH. Thus, the rate constants obtained in this case by ^1H NMR lineshape correspond to k^{HD}. On the right side of Fig. 9.2 the superposed experimental and calculated spectra of AP in $C_2D_2Cl_4$ at a deuterium fraction of $D = 0.92$ in the NH sites are shown. Naturally, it was much more difficult to obtain these spectra because of the low concentration of the residual ^1H–^{15}N units, and the experiments had, therefore, to be carried out at 300.13 MHz. Note that this does not change the signal lineshape because the scalar ^1H–^{15}N interaction is independent of the strength of the magnetic field. A comparison of both sets of spectra in Fig. 9.2 shows that the exchange in AP–HD is slower than in AP–H_2. For example, the temperature where the peak height of the inner line component is approximately equal to the peak height of the outer line components is approximately 330 K on the left but 340 K on the right side of Fig. 9.2. At 298 K a kinetic HH/HD isotope effect of 4 was found. Note that in the spectra at D 0.92 the presence of a small amount of AP–H_2 had to be considered in the calculations. Because of the slow intermolecular proton exchange the known contributions of this species were simply added to the lineshape equation according to Eq. (6.17).

Since it was not possible to determine k^{DD} by ^2H NMR spectroscopy because of the smallness of the coupling constant $J_{^2H^{15}N}$ and because of fast deuteron quadrupole relaxation, k^{DD} had to be determined by NMR spectroscopy of a remote spin. Since the chemical shifts of the phenyl protons were not altered enough by the proton transfer, and since the longitudinal relaxation times of the ^{15}N atoms of AP–DD were too long, k^{DD} was determined by dynamic ^{13}C NMR spectroscopy. For this purpose AP was labeled with ^{13}C in the p-position of the phenyl groups [127]. Figure 9.3 shows the superposed experimental and calculated NMR signals of the p-^{13}C atoms of AP dissolved in $C_2D_2Cl_4$ as a function of temperature at a deuterium fraction of $D = 0$ and $D = 1$ in the NH sites. At low temperatures the slow exchange regime is reached, with two sharp

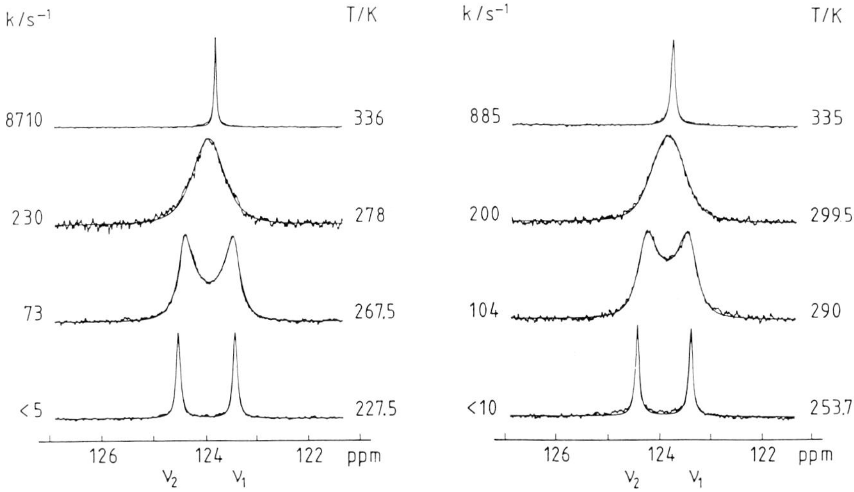

Fig. 9.3. Superposed experimental and calculated 75.47 MHz ^{13}C NMR signals of the isotopically enriched (90%) p-^{13}C atoms of AP dissolved in $C_2D_2Cl_4$ as a function of temperature. **Left**: deuterium fraction in the NH sites D = 0, $k \equiv k^{HH}$; **right**: D = 0.99, $k \equiv k^{DD}$. The calculations were done using Eq. (2.18). Reproduced with permission from Ref. [127]

signals of equal intensities. As the temperature is raised the lines broaden and finally coalesce into one sharp line, indicating a fast proton exchange. This spectral pattern indicates a symmetric exchange process with an equilibrium constant of $K_{12} = 1$. Figure 9.3 also shows that the exchange becomes much slower as the molecule is deuterated in the NH sites; the coalescence point is raised by about 20 K. The lineshape calculations were performed using Eq. (2.18). The values of $\Delta\nu = \nu_1 - \nu_2$, where ν_1 and ν_2 are the chemical shifts of the exchanging spins, could be obtained by lineshape analysis in the slow exchange regime for both AP–H_2 and AP–D_2 as a function of temperature. Note that for lineshape analyses small differences of the chemical shifts of the two isotopic species can be neglected within the margin of error. Neglecting possible kinetic $^{12}C/^{13}C$ and $^{14}N/^{15}N$ isotope effects, the ^{13}C NMR spectra at D = 0 were calculated using the rate constants k^{HH} of AP–H_2 obtained by ^{1}H NMR spectroscopy. Therefore the only adjustable parameter for the calculation of the ^{13}C NMR spectra of AP–H_2 was W_0. It was assumed that W_0 is the same for AP–HH and AP–DD. Thus, the rate constants k^{DD} could be determined without major assumptions for AP–D_2 at D = 0.99.

9.1.2 *Polarization Transfer Experiments in the Rotating Frame*

As discussed in Sect. 2.3 in the slow exchange regime lineshape analysis is not accurate. In order to overcome this difficulty, we performed low temperature

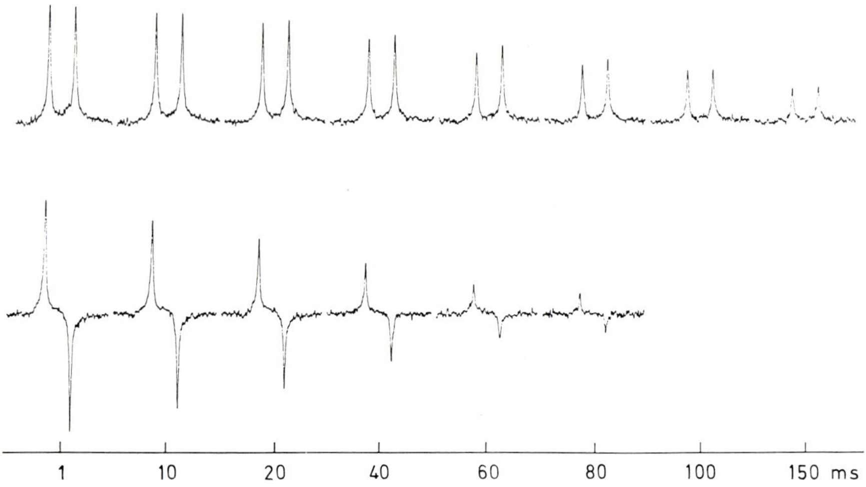

Fig. 9.4. 75.47 MHz ^{13}C NMR magnetization transfer experiment according to Fig. 2.4 in the rotating frame on AP-D_2 in $C_2D_2Cl_4$ at 265 K and a deuterium fraction in the NH sites of D = 0.99. Upper spectral set: experiment I; lower spectral set: experiment II. Reproduced with permission from Ref. [127]

^{13}C NMR polarization transfer experiments on AP–H_2 and AP–D_2 in the rotating frame. A typical experiment is shown in Fig. 9.4. From such measurements accurate kinetic HH/DD isotope effects could be obtained in the slow exchange regime. Similar experiments were performed on the ^{1}H–^{15}N signal of AP–H_2 [127]. However, both ^{1}H–^{15}N cross relaxation as well as the circumstance that only half of the exchanging doublet components exchange via proton transfer had to be taken into account.

9.1.3 Kinetic HH/HD/DD Isotope Effects of the Azophenine Tautomerism

The result of the measurements can be expressed in form of the following Arrhenius equations:

$$k^{HH} = 10^{11.6 \pm 0.14} \exp(-49.4 \pm 0.8\,\mathrm{kJ\,mol^{-1}}/RT), \quad 249 \leqslant T \leqslant 343\mathrm{K}, \quad (9.1)$$

$$k^{HD} = 10^{11.9 \pm 0.74} \exp(-54.7 \pm 0.9\,\mathrm{kJ\,mol^{-1}}/RT), \quad 299 \leqslant T \ll 352\mathrm{K}, \quad (9.2)$$

$$k^{DD} = 10^{12.1 \pm 0.25} \exp(-56.2 \pm 1.4\,\mathrm{kJ\,mol^{-1}}/RT), \quad 263 \leqslant T \leqslant 334\mathrm{K}, \quad (9.3)$$

with the kinetic isotope effects

$$k^{HH}/k^{HD} = 4.1, \quad k^{HD}/k^{DD} = 1.4, \quad \text{i.e.} \quad k^{HH}/k^{DD} = 5.6 \quad (9.4)$$

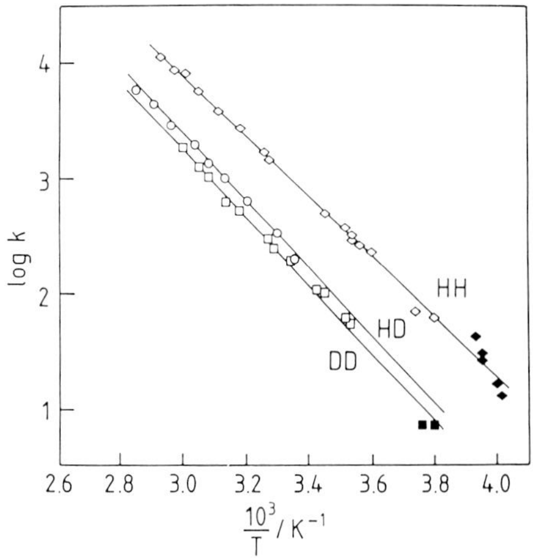

Fig. 9.5. Arrhenius curves of the azophenine tautomerism. *Open symbols*: ^{1}H and ^{13}C NMR line shape analysis; *filled symbols*: magnetization transfer in the rotating frame. Reproduced with permission from Ref. [127]

at 298 K. The corresponding Arrhenius curves are shown in Fig. 9.5. Whereas the substitution of the first mobile hydrogen by deuterium leads to a substantial primary KIE, replacement of the second H by D has only a small effect. These results constitute a considerable deviation from the rule of the geometric mean (RGM). They were taken as evidence for a stepwise proton transfer [127] between forms A and D via the intermediates B, C and the transition states J, G and K, H as shown in Fig. 9.6. This mechanism had been postulated theoretically [131, 132]. For a concerted double proton transfer via transition state E or for a pathway where B and C are transition states the RGM should have been valid in good approximation [127]. A problem with this interpretation remained in the sense that for the formation of a zwitterionic intermediate B, C = ZI in Fig. 9.6 kinetic solvent effects on the rates of proton transfer in AP should have been expected, by contrast to the experimental findings [127]. Therefore, either an unpolar singlet biradical (BR) must be formed as intermediate of the reaction or the phenyl groups in AP effectively minimize interactions between ZI and the solvent.

9.2 *The Oxalamidine Tautomerism*

The oxalamidine tautomerism (Fig. 9.7) was discovered some time ago by dynamic ^{1}H NMR spectroscopy of tetraphenyloxalamidine-$^{15}N_4$ (TPOA) in CD_2Cl_2 [120]. As shown in Fig. 9.8, the ^{1}H–^{15}N signal of TPOA consists of a doublet at low and of a triplet at room temperature which proves an intramolecular double proton transfer in this molecule, as in the case of azophenine. Figure 9.8 also indicates lineshape changes of the ^{1}H–^{15}N signal arising from a kinetic HH/HD isotope effect on the tautomerism. At higher temperatures the triplet is affected by an extra line broadening due to interconversion with

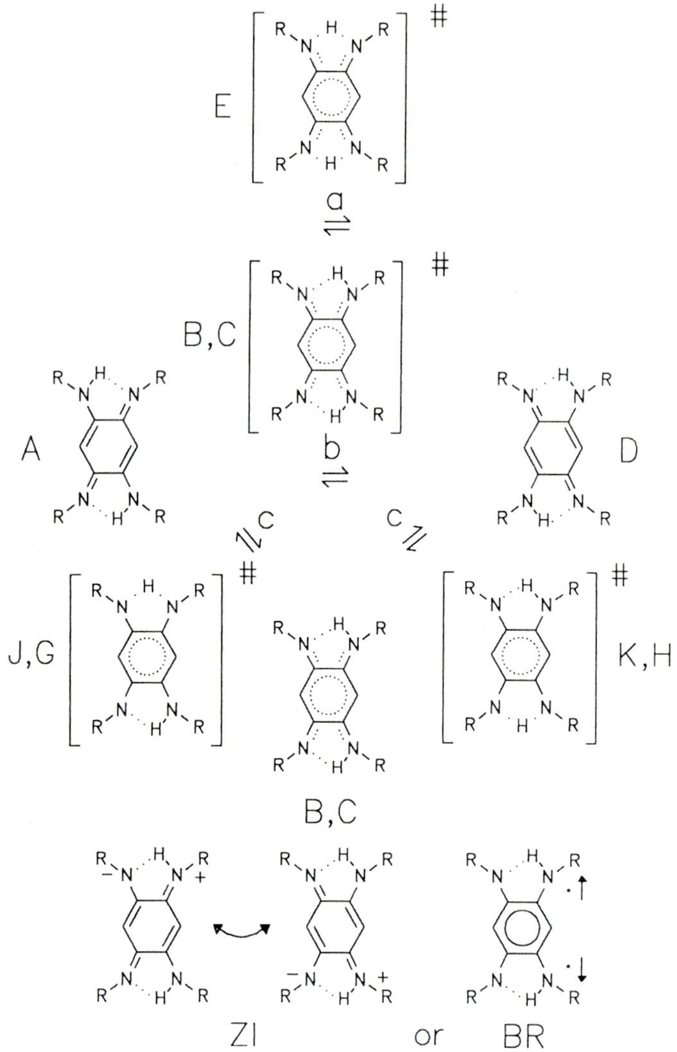

Fig. 9.6. Pathways of the azophenine (AP) tautomerism. ZI ≡ zwitterion, BR ≡ singlet biradical. Reproduced with permission from Ref. [127]

other conformers which are subject to rapid intermolecular proton exchange [120]. These spectra can be described in terms of Eq. (6.27). Unfortunately, the intermolecular exchange terms prevent the determination of the intramolecular exchange rates, especially of the k^{HD} at higher temperatures.

In order to obtain the kinetic isotope effects of the oxalamidine tautomerism ^1H NMR experiments were, therefore, performed on the bicyclic compound B7OA (Fig. 9.7), where the molecular structure prevents conformational

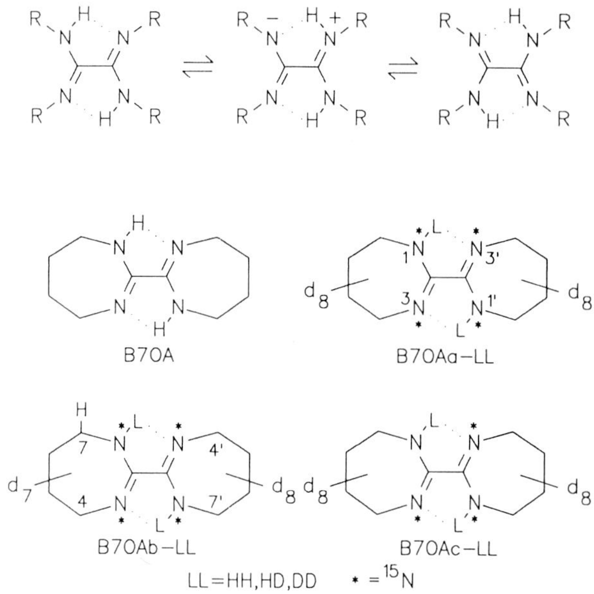

Fig. 9.7. The tautomerism of oxalamidine (OA, R = H), tetraphenyloxalamidine (TPOA, R = Ph) and B7OA, R, R = —$(CH_2)_4$—. Reproduced with permission from Ref. [129]

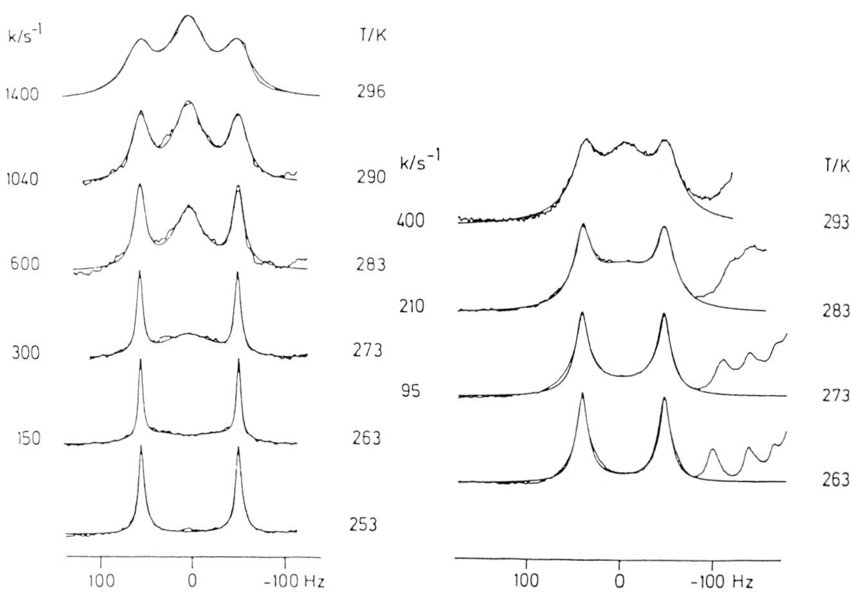

Fig. 9.8. Left: Superposed experimental and calculated 90:02 MHz ^{1}H NMR signals of the ^{1}H–^{15}N units of TPOA dissolved in CD_2Cl_2 at a deuterium fraction D = 0 in the NH sites as a function of temperature. k is the rate constant of the intramolecular HH transfer according to Fig. 9.7. The spectra were calculated using Eq. (6.27). Due to intermolecular proton exchange the average inverse lifetimes τ^{-1} are non-zero, although ≪ k. The spectrum at 296 K was recorded at 250 MHz. **Right**: As left, but D ≃ 0.9. k corresponds to the rate constants of the intramolecular HD-transfer. Reproduced with permission from Ref. [120]

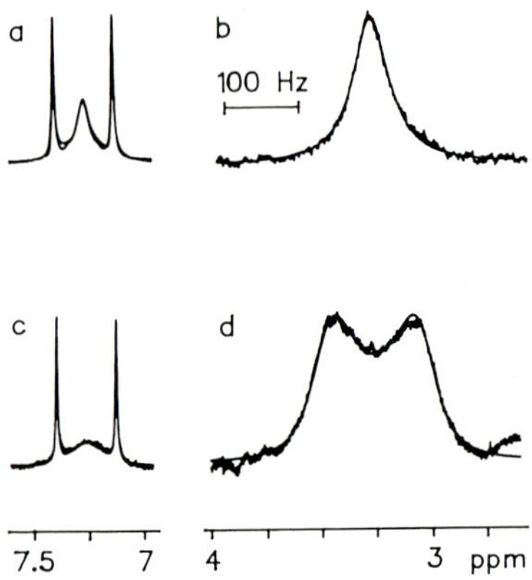

Fig. 9.9 a-d. Superposed experimental and calculated 300.13 MHz ^{1}H NMR spectra of a 0.2 molar solution of B7OA in methylcyclohexane-d$_{14}$ (MCY) at 362 K. (**a**) ^{15}N–^{1}H$\cdots$^{15}N signal of B7OAa–HH and (**b**) 4,4′ and 7,7′ proton signal of B7OAb–HH at a deuterium fraction D = 0 in the mobile proton sites; (**c**) ^{15}N–^{1}H$\cdots$^{15}N signal of B7OHa–HD and (**d**) 4,4′ and 7,7′ proton signal of B7OAb–DD at a deuterium fraction D = 0.94 in the mobile proton sites. The presence of a small amount of BCOAc was taken into account in the simulations; in addition, at D = 0.94 also the presence of B7OAa–HH, B7OAa–HH, and B7OAa– HD was taken into account. Further simulation parameters: $\nu_{^1H-^{15}N} = 7.26$ ppm, $^1J_{^1H-^{15}N} = 80.3$ Hz, $W_0 = 3.7$ Hz, and chemical shifts $\nu_4 = \nu_{7'}$ = 3.505 ppm and $\nu_7 = \nu_{4'}$ = 3.03 ppm. Reproduced with permission from Ref. [129]

isomerism [129]. The molecule was enriched with ^{15}N and deuterated on the carbon atoms. A set of experiments was performed at 362 K as shown in Figs. 9.9 and 9.10, using methylcyclohexane-d$_{14}$ (MCY) and acetonitrile-d$_3$ (AN) as solvents.

The kinetic HH/HD isotope effects could be determined by ^{1}H NMR line shape analysis of the ^{15}N–^{1}H$\cdots$^{15}N signals of B7OAa–HH and B7OAa–HD as shown in Figs. 9.9a, 9.9c, 9.10a and 9.10c. Although the temperature is elevated, no contributions from intermolecular proton exchange and conformational isomerism could be detected. The signals constitute triplets with two sharp outer line components as expected for a moderately fast degenerate intramolecular proton transfer between two ^{15}N atoms. By line shape analysis of the ^{15}N–^{1}H$\cdots$^{15}N signals using Eq. (6.15) the rate constants k^{HH} of the HH reaction in B7OA could be obtained for both solvents at D = 0 (Figs. 9.9a and 9.10a). At deuterium fractions close to unity (Fig. 9.9c and 9.10c) the residual ^{15}H–^{1}H $\cdots$ ^{15}N proton signal stems from the species B7OAa–HD, i.e. its lineshape depends on the rate constant k^{HD} of the HD process. The

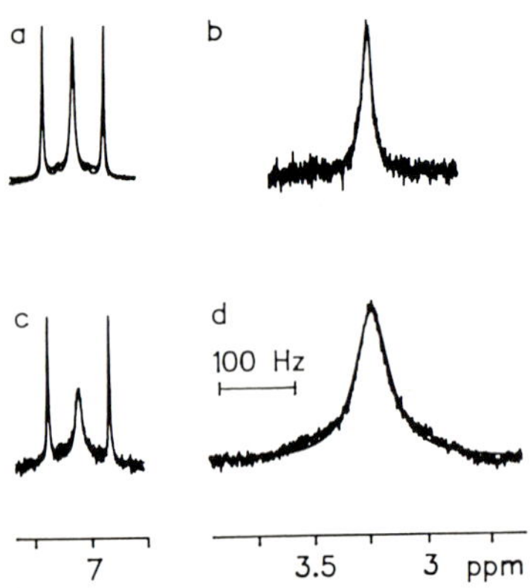

Fig. 9.10 a-d. Superposed experimental and calculated 300.13 MHz ^{1}H NMR spectra of a 0.2 molar solution of B7OA in acetonitrile-d$_3$ (AN) at 362 K. (**a**) ^{15}N–^{1}H$\cdots$^{15}N signal of B7OAa–HH and (**b**) 4,4′ and 7.7′ proton signal of B7OAb–HH at a deuterium fraction D = 0 in the mobile proton sites; (**c**) ^{15}N–^{1}H$\cdots$^{15}N signal of B7OAa–HD and (**d**) 4,4′ and 7,7′ proton signal of B7OAb–DD at a deuterium fraction D = 0.94 in the mobile proton sites. Further simulation parameters: $\nu_{^1H-^{15}N} = 7.05$ ppm, $^1J_{^1H-^{15}N} =$ 80.3 Hz, $W_0 = 3$ Hz, and chemical shifts $\nu_4 = \nu_{4'} =$ 3.492 ppm and $\nu_7 = \nu_{7'}$ = 2.965 ppm. For further description see text. Reproduced with permission from Ref. [129]

finding that the center lines of the ^{15}N–$^{1}H \cdots ^{15}N$ signals are larger for MCY as solvent than for AN and larger at $D = 0.94$ than at $D = 0$ directly indicates substantial kinetic solvent and HH/HD isotope effects on the reaction rates. Actually, it was found that $k^{HH}(AN)/k^{HH}(MCY) = 4.5$, $k^{HH}(MCY)/k^{HD}(MCY) = 2.44$, and $k^{HH}(AN)/k^{HD}(AN) = 2.63$ at 362 K.

The kinetic HH/DD isotope effects were determined by analysis of the signals of the remote 4, 4′ and 7, 7′ protons in B7OAb which interconvert by the double proton transfer. Because of the absence of homonuclear couplings the lineshape of these signals can be described in terms of Eq. (2.18). At $D = 0$ the signals constitute exchange broadened singlets (Figs. 9.9b and 9.10b); since they stem from the species B7OAb–H_2 their lineshape depends again on k^{HH}; however, this quantity as well as W_0 is already known from the simulations in Figs. 9.9a and 9.10a. Therefore, we obtain by lineshape analysis the unknown chemical shifts of the 4, 4′ and of the 7, 7′ protons. The result agrees very well with the values measured in the slow exchange region at low temperatures. By contrast, at $D = 0.94$ the signal dominantly stems from the species B7OAb–D_2 and contains information about k^{DD}. A comparison of Fig. 9.9b with 9.9d and of Fig. 9.10b with 9.10d shows then substantial kinetic HH/DD isotope effects. In the case of Fig. 9.9d even the case of slow exchange was reached where the chemical shifts could directly be obtained by lineshape analysis. The following kinetic isotope effects $k^{HD}(MCY)/k^{DD}(MCY) = 1.2$, $k^{HH}(MCY)/k^{DD}(MCY) = 3$, $k^{HD}(AN)/k^{DD}(AN) = 1.3$ and $k^{HH}(AN)/k^{DD}(AN) = 3.5$ were obtained. It was further established that the results found for both solvents were independent of the concentration of B7OA.

The observed kinetic HH/HD/DD isotope effects are similar, as in the case of AP, and can be explained with a stepwise proton transfer mechanism according to Fig. 9.7. By contrast to AP, the observed increase of the rate constants in B7OA with the increase of the dielectric constant—2 for MCY and 37.5 for AN at 298 K—establishes, however, a polar transition state of the reaction as expected for the stepwise double proton transfer via the zwitterion intermediate as shown in Fig. 9.7.

Note that in the experiments of Figs. 9.9 and 9.10 no assumptions concerning chemical shift differences, coupling constants, or line widths in the absence of exchange were necessary. Moreover, the isotopic reaction rate constants of the HD and the DD reactions could be determined by analysis of different parts of the same spectrum. Therefore, there are no temperature errors on the kinetic HD/DD isotope effects which could be introduced when the latter are derived from measurements of different samples.

9.3 The Porphyrin Tautomerism

The tautomerism of porphyrines (Fig. 2.2) was discovered in 1972 by Storm and Teklu via ^{1}H NMR spectroscopy [109]. These authors showed in a qualitative way the existence of kinetic HH/DD isotope effects on this reaction by observing different ^{1}H NMR lineshapes of the β-pyrrole protons (see Fig. 2.2) in the presence and absence of deuteriomethanol in the NMR samples. Rate constants [80, 105, 110–116, 119, 124] and kinetic HH/DD isotope effects [111, 115] have since been studied quantitatively by dynamic NMR spectroscopy. The possibility of obtaining rate constants of the tautomerism by analysis of the ^{1}H NMR lineshape of the inner proton signal of ^{15}N labeled porphyrines [112–115] also opened the possibility of studying the kinetic HH/HD isotope of the reaction. This was accomplished in 1982 [105], when complete kinetic HH/HD/DD isotope effects were reported for the first time. Since then these effects have been studied in more detail [124–126]. Recently, it has been shown by NMR that the porphyrin tautomerism can also take place in the solid state [167–177]. Note that kinetic HH/DD isotope effects were also obtained by optical methods on porphyrin embedded in solid hexane [141]. The mechanism of the porphyrin tautomerism has been subject to intense theoretical studies [133–142]. The phenomenon of solid state tautomerism of porphyrines will be dealt with in Section 11. Here, we will discuss liquid state NMR experiments devoted to the determination of the kinetic HH/HD/DD isotope effects of the porphyrin tautomerism.

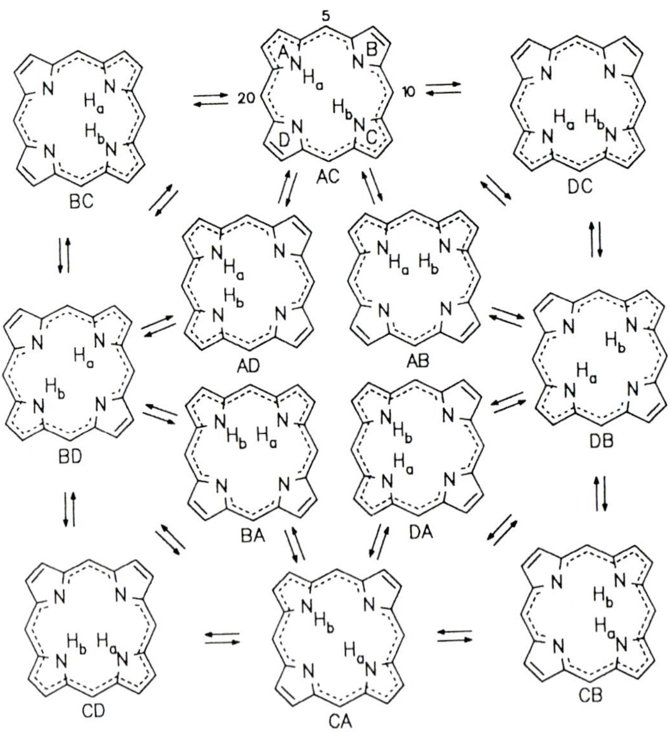

Fig. 9.11. Proton transfer pathways in porphyrins according to Ref. [123]

9.3.1 Symmetrically Substituted Porphyrines

The complete reaction pathways of symmetrically substituted porphyrines are shown in Fig. 9.11 [123]. The different tautomers are labeled according to the pyrrole rings to which the mobile protons are attached. The rate constant of the reaction of tautomer MX to NY is then written as $k_{MX \rightarrow NY}$. Observable species are the degenerate *trans*-tautomers AC, BD, DB, and CA. The remaining *cis*-tautomers are postulated intermediates of a stepwise reaction pathway. Note that in the case of porphyrins the NMR lineshapes of the mobile protons bound to ^{15}N spins and of the remote spins depend on slightly different quantities. Thus, all published rate constants k, including those given in Sects. 2.2 and 2.3, correspond to the values $2k_{AC \rightarrow DB}$. By contrast, the constants $k_{AC \rightarrow DB}$ are

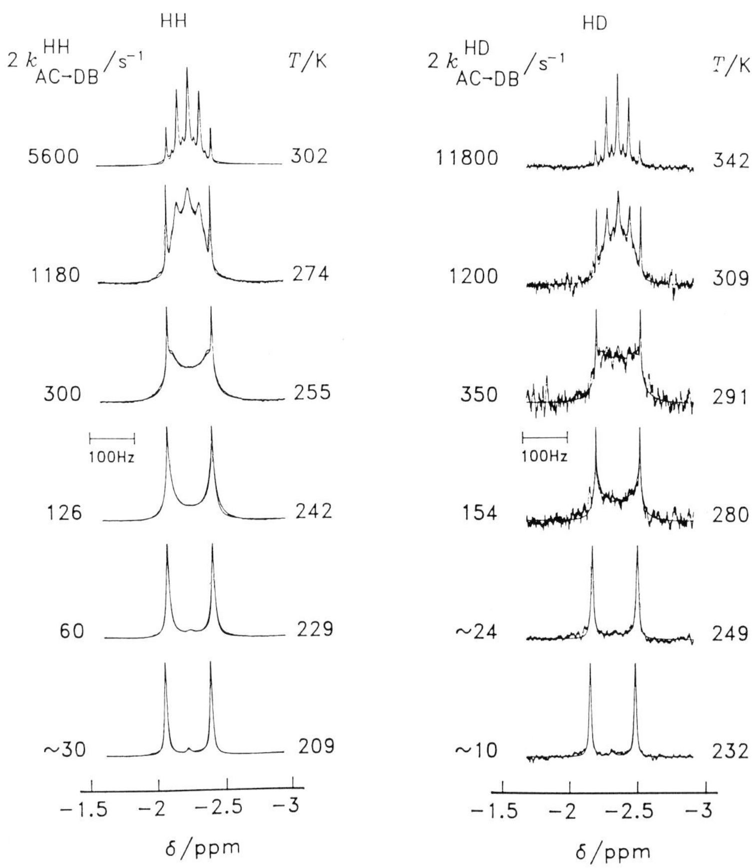

Fig. 9.12. Superposed experimental and calculated 300.13 MHz ^1H NMR spectra of the inner hydrogen atoms of *meso*-tetraphenylporphyrin-$^{15}N_4$ dissolved in toluene-d_8. **Left**: D = 0, **right**, D = 0.85. The spectra were calculated using Eqs. (6.19) and (6.20). Reproduced from Ref. [124]

obtained from the lineshape analysis of the inner proton signals. The lineshape equations of the mobile proton signals migrating between four ^{15}N atoms are set up as indicated in Sect. 6.1.2.

Since the error limits of the early $2k^{HD}_{AC\rightarrow DB}$ determination [105] were large because of experimental limitations in the signal to noise ratio, these measurements have been repeated recently [124]. The results of the 300 MHz ^{1}H NMR experiments on the inner proton signals of TTP at deuterium fractions of $D = 0$ and 0.9 in the mobile proton sites are shown in Fig. 9.12. The spectra were calculated using Eq. (6.19). As observed previously [112, 115], the intramolecular proton transfer gives rise to a doublet–pentet transition when going from the slow to the fast exchange regime. As in the case of azophenine, the coupling constant $J_{^1H-^{15}H}$ and W_0 could be obtained by simulation of the outer line components. At $D = 0$ the lineshapes depend on $2k^{HH}_{AC\rightarrow DB}$, at $D = 0.85$ mostly on $2k^{HD}_{AC\rightarrow DB}$. Naturally, the signal to noise ratio is not as good at $D = 0.85$ at to $D = 0$. In calculating the spectra the 15% of TPP–HH was taken into account.

Figure 9.13 contains the results of ^{13}C NMR experiments performed on TPP with a natural abundant ^{13}C content, dissolved in tetrahydrofuran. The o-, m-, and p-phenyl carbon signals are not affected by the exchange. At

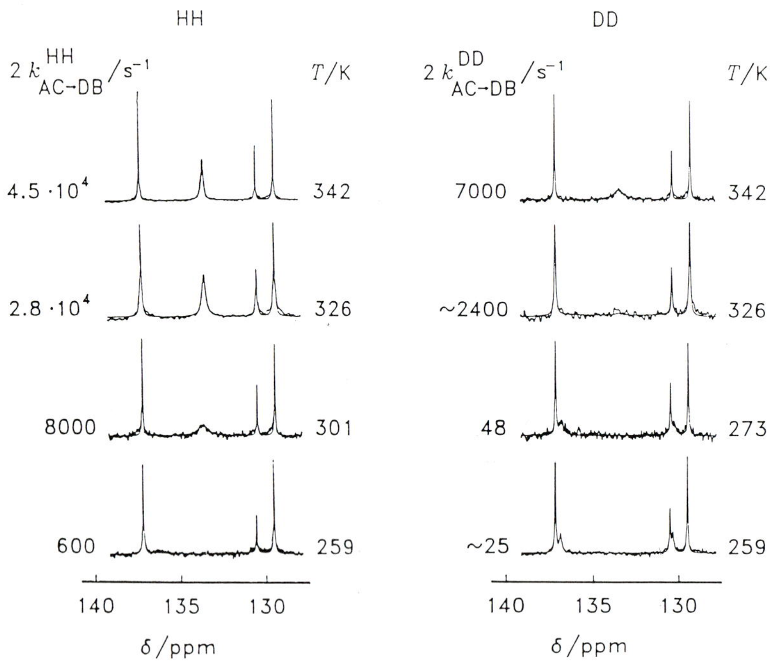

Fig. 9.13. Superposed experimental and calculated 75 MHz ^{13}C NMR spectra of TPP dissolved in tetrahydrofuran-d_8. **Left**: deuterium fraction in the mobile proton sites $D = 0$, **right**: $D = 0.85$. Reproduced from Ref. [124]

D = 0.85 and 259 K the deuteron transfer is so slow that the splitting of the two inequivalent β-pyrrole carbons 2, 3 and 7, 8 can easily be observed. In the intermediate temperature range, i.e. the coalescence region it is very difficult to detect the very broad signal hidden in the noise. The kinetic HH/DD isotope effect is clearly demonstrated in the two sets of spectra, which were calculated using Eq. (2.18).

The Arrhenius diagram of the TPP migration shows similar features to the diagram of azophenine (Fig. 9.5): a large kinetic HH/HD isotope effect is observed and only a very small kinetic HH/HD isotope effect. These results can again most easily be interpreted in terms of a stepwise proton motion via *cis*-intermediates, as predicted theoretically [133, 134, 140–142]. At low temperatures the reaction proceeds by tunneling, which has definitively been established recently using a laser temperature jump method [125, 141]. This idea was proposed a long time ago [105, 115].

9.3.2 *Chemically Perturbed Porphyrines*

So far, we have focused in this review on symmetric exchange reactions with equilibrium constants of unity. However, dynamic NMR spectroscopy can also follow asymmetric exchange reactions with equilibrium constants $\neq 1$ within certain limits. E.g., NMR measurements on asymmetric porphyrines and chlorins have been performed [121–123, 126] which open the possibility of studying kinetic hydrogen/deuterium isotope effects of asymmetric exchange reactions in this class of compounds. As an example, let us consider the tautomerism of a mono-substituted acetylporphyrin (ACP). For this purpose ACP was once triply labeled with ^{15}N (ACP–$^{15}N_3$) and once mono labeled with ^{13}C (ACP–^{13}C) as shown in Fig. 9.14. The full proton transfer pathways

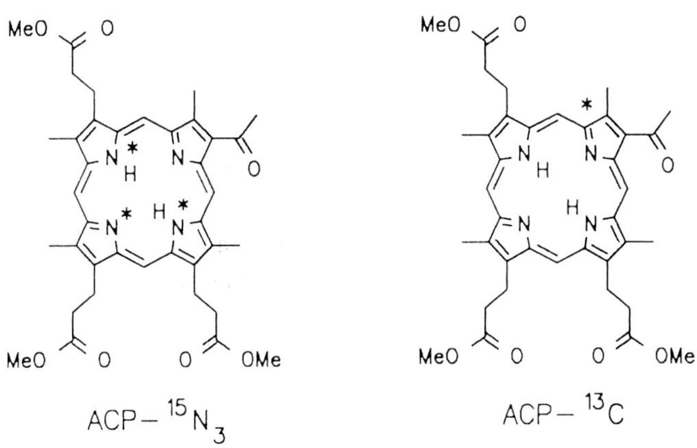

Fig. 9.14. The structure of the triply labeled ACP-$^{15}N_3$ and the ^{13}C labeled ACP-^{13}C

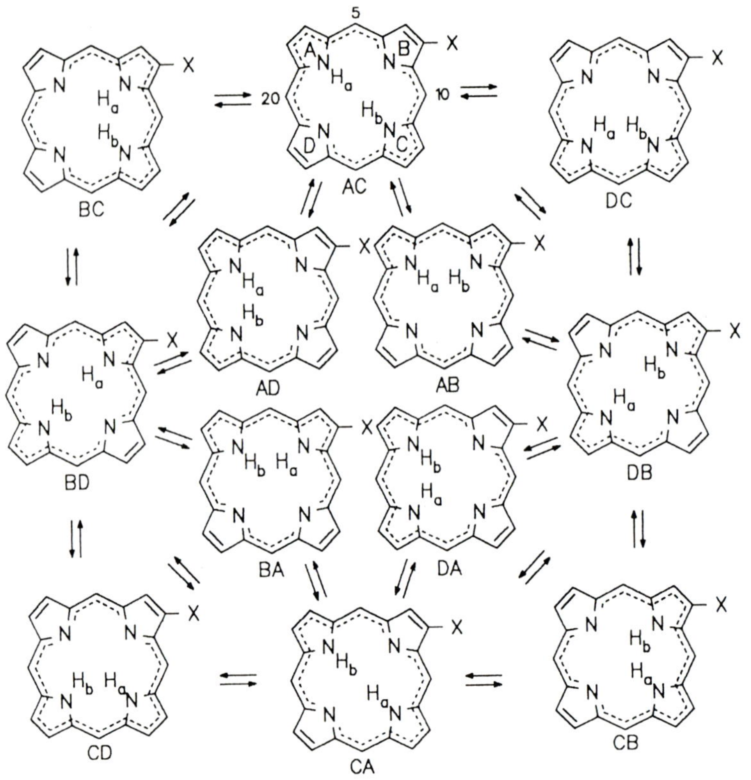

Fig. 9.15. Proton transfer pathways in ACP. Reproduced from Ref. [126]

of ACP are shown in Fig. 9.15. Preliminary HH/HD isotope effects for this compound were reported in Ref. [122]. The full isotope effects could be obtained only recently [126].

The results of the ^{13}C NMR experiments [126] on ACP–^{13}C–HH and ACP–^{13}C–DD are shown in Fig. 9.16. The spectra were calculated using Eq. (2.18). At low temperatures the labeled ^{13}C atom exhibits a strong line at 148 ppm. This line was assigned to the tautomers AC and CA. A second very weak line at ~ 132 ppm was attributed to the tautomers BD and DB (arrows in Fig. 9.16). The weak line is sharp only in the case of ACP–^{13}C–DD, but exchange broadened in the case of ACP–^{13}C–HH. As temperature is raised both lines broaden and coalesce into one sharp line. The coalescence temperature is higher for the deuterated compound as compared to the protonated compound due to a kinetic HH/DD isotope effect. The analysis of the spectra in Fig. 9.16 is complicated by a temperature dependence of the ^{13}C chemical shifts due to aggregation of ACP. However, it was found by analysis of the spectra of ACP–^{13}C–DD that the difference $\Delta\nu$ of the chemical shifts of the exchanging carbon atoms is almost independent of temperature. Thus, the chemical shifts

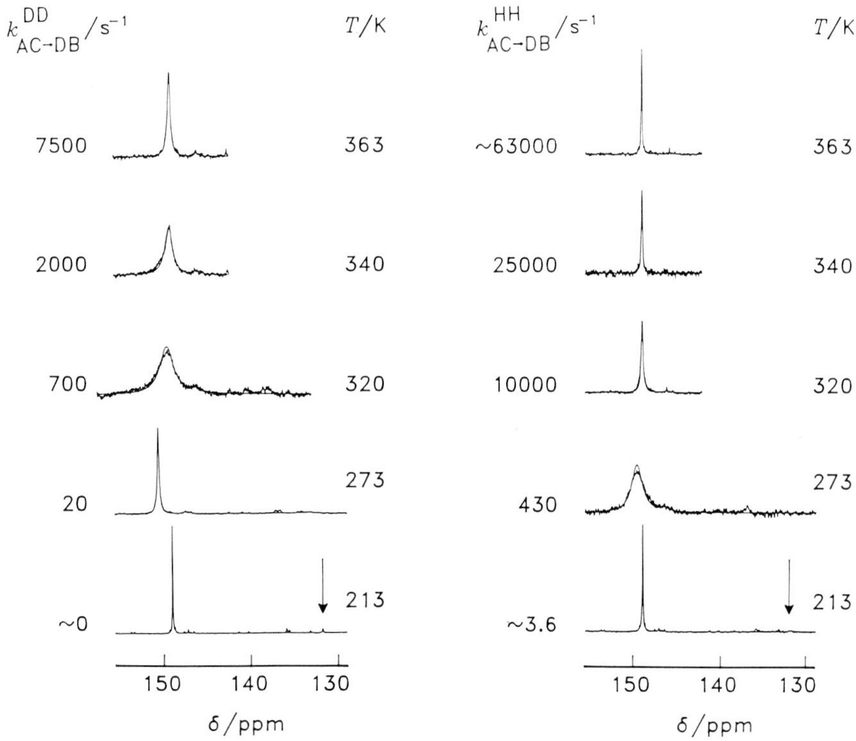

Fig. 9.16. Superposed experimental and calculated 75 MHz ^{13}C NMR signals of the C-6 carbon atom of the monoacetylporphyrin derivative ACP dissolved in highly purified CD_2Cl_2 as a function of temperature. **Left**: D = 0, **right**: D = 0.95. The *arrows* indicate the signals of tautomers DB and BD in the slow exchange limit. The spectra were calculated using Eq. (2.18). Reproduced from Ref. [126]

needed to simulate the exchange broadened spectra of ACP–^{13}C–HH could be obtained from the spectra of ACP–^{13}C–DD at a given temperature. Note that the line shapes depend here on the rate constants $2k^{LL}_{AC\to DB}$, LL = HH, DD. For comparison with the rates obtained by ^{1}H NMR, however, only the constants $k^{LL}_{AC\to DB}$ are listed in Fig. 9.16. The ratio of the two tautomers, also needed to calculate the spectra, could be obtained at low temperature directly by simulation. At higher temperatures this ratio was obtained from ^{1}H NMR experiments as shown in the following.

The latter were performed on the inner proton signals of ACP–$^{15}N_3$ as shown in Fig. 9.17 at D = 0 and D = 0.95 [126]. The spectra were simulated using Eqs. (6.19) and (6.21). The two inner protons have different isotropic chemical shifts giving rise to two inequivalent ^{1}H–^{15}N doublets with equal splittings $J_{^1H-^{15}N}$ in the slow exchange range. Unfortunately, the signals of the second tautomer could not be identified. Since ring B was not labeled with ^{15}N the observation of two ^{1}H–^{15}N doublets indicates that the protons are located

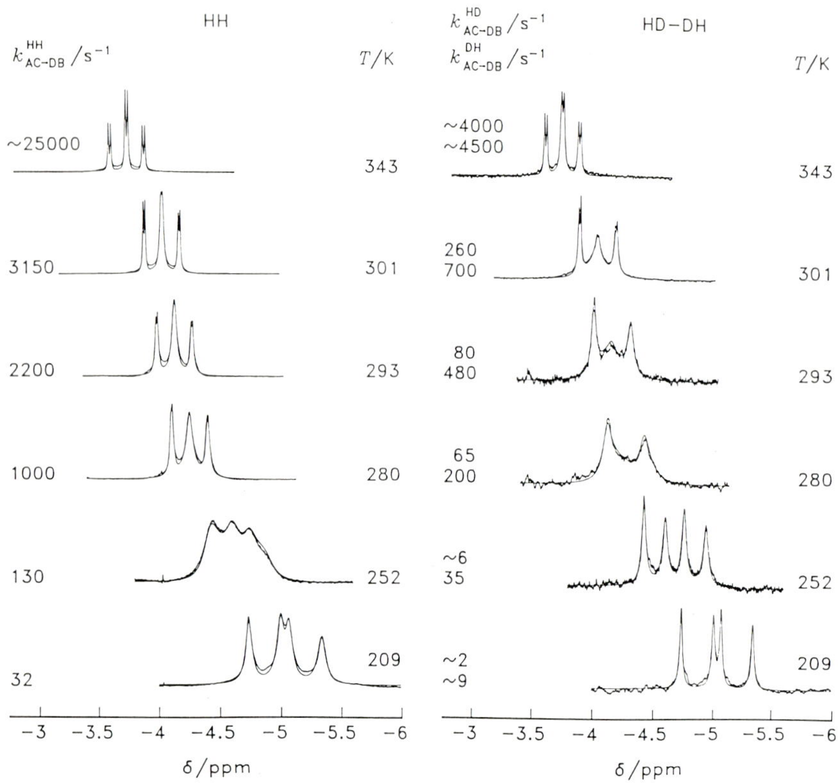

Fig. 9.17. Superposed experimental and calculated 300.13 MHz ^{1}H NMR signals of the inner hydrogen atoms of ACP dissolved at a concentration of 20 mmol l^{-1} in highly purified CD_2Cl_2 as a function of temperature. The spectra were calculated using Eqs. (6.19) and (6.21). According to Eq. (6.20), the temperature dependent ratio of the effective triplet and doublet splittings in the fast exchange region is equal to the tautomer ratio AC vs. BD. $k_{r \to s}$ is the rate constant of the interconversion r → s. **Left**: D = 0, **right**: D = 0.95. Reproduced from Ref. [126]

on the rings A and C and not on B and D in the dominant tautomers, i.e. AC and CA. At high temperatures a coalesced ^{1}H–^{15}N triplet is observed where each line component is further split into a doublet. This result indicates that AC and CA exchange rapidly via the tautomers DB and BD. The doublet splitting arises by coupling of the inner protons with the ^{15}N atom on ring D in the minor tautomers BD and DB. According to Eq. (6.20), the population of all tautomers can be obtained exactly from the effective splittings given by $J_X^{eff} = p_X J_{^1H-^{15}N}$, where p_X is the population of each of the inner protons on ring X [122]. Extrapolation of the data to low temperature showed a good agreement with the value obtained by ^{13}C NMR. A comparison of both sets of spectra in Fig. 9.17 shows significant kinetic isotope effects. Note that in the case of ACP–$^{15}N_3$–HH, i.e. at D = 0, AC and CA are degenerate. In principle, AC and

CA are not degenerate in the case of ACP–$^{15}N_3$–HD; however, they are quasi degenerate when equilibrium isotope effects can be neglected. This is indeed the case: at $D = 0.95$ where the second set of experiments was performed the lineshape almost exclusively stems from the species ACP–$^{15}N_3$–HD, i.e. of species AC and CA. In the presence of an equilibrium isotope effect the two doublets should be of unequal intensity, which is not the case. Of special importance is that the spectra of ACP–$^{15}N_3$–HD are characterized by two different rate constants, $k^{HD}_{AC \to DB} \neq k^{DH}_{AC \to DB}$. Thus, the full kinetic HH/HD/DH/DD isotope effects could be obtained for ACP, further evidence for a stepwise proton transfer pathway in this molecule [126].

Note also that recently the tautomerism of hydroporphyrins has been studied by mobile and remote spin studies [123]. In this class of compounds it was possible to significantly perturb the reaction barrier by specific hydrogenation of the pyrrole rings, which leads to the biologically important hydroporphyrins chlorin, bacteriochlorin and isobacteriochlorin. The increase or decrease of the reaction energy barrier could be explained with an increased or decreased energy of the intermediates along the reaction pathways. In future studies it will be very interesting to know how the kinetic isotope effects vary when the barrier height is altered.

10 Kinetic Hydrogen/Deuterium Isotope Effects of Intermolecular Proton Exchange Reactions in Liquids

Compared to intramolecular rearrangements, the study of intermolecular proton exchange reactions is more tedious. As in ordinary kinetics, one has to perform kinetic runs as a function of the concentration of the reactants in order to establish the rate law of the exchange. Such rate laws can be complicated in the presence of a complex reaction network. Often, it will not even be possible to obtain rate laws when the concentrations of the reacting species cannot be varied as desired because of solubility problems, or because the solvent participates in the exchange either as inhibitor or as catalyst. This is often the case in biochemical reactions. In this situation one can obtain information on the number m of protons transferred in the rate limiting step as well as on kinetic isotope effects by performing a so-called "proton inventory" [41–48]. In a proton inventory kinetic measurements are performed as a function of the deuterium fraction D in the mobile proton sites. This technique has been adapted to dynamic NMR spectroscopy [106–108] and is treated theoretically in Sect. 5 of this review. Here, we will give examples for the application of this technique both in the cases of proton self exchange as well of proton exchange between different molecules.

10.1 *Intermolecular Self Exchange: Pure CH_3OH and CH_3OH/CH_3OD*

As a first example, let us discuss the problem of proton and deuteron exchange in pure methanol. This problem has been studied in Ref. [89]. Fig. 10.1 shows the superposed experimental and calculated ^{1}H NMR spectra of pure CH_3OH (left) and of 1% CH_3OH dissolved in CH_3OD (right) as a function of temperature. CH_3OH leads to an AB_3 spin system with a doublet splitting of the CH_3 signal and a quartet splitting of the OH signal at low temperatures. Because of hydrogen bond association the OH signal shifts to low field when the temperature is lowered, a phenomenon which has been used for temperature calibration in ^{1}H NMR spectroscopy [199]. As the temperature is raised the spin–spin splitting pattern collapses into a singlet for the CH_3 and the OH signal, respectively indicating that even in pure CH_3OH a proton exchange takes place which can formally be described by

$$AL + A^*L^* \rightleftharpoons AL^* + A^*L, \tag{10.1}$$

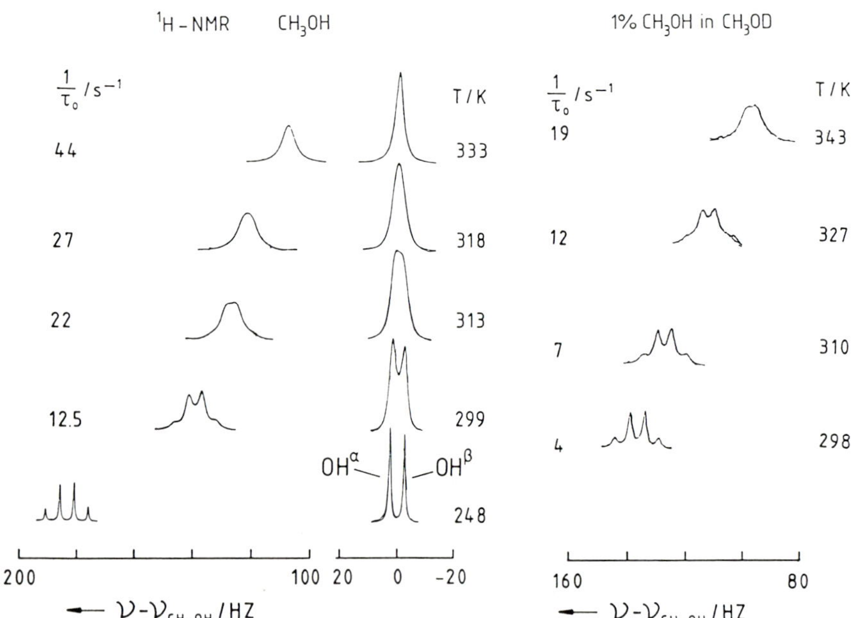

Fig. 10.1. Left: Superposed experimental and calculated ^{1}H 90.02 MHz NMR signals of pure methanol as a function of temperature. The CH_3OH signal appears at low, the CH_3OH signal at high field. The intensities of the two signals are not normalized to the corresponding number of protons (100 scans, 90° pulses, repetition time 10 s). Reproduced with permission from Ref. [89]. **Right:** Superposed experimental and calculated ^{1}H 90.02 MHz NMR signals of the residual CH_3OH protons (1 vol-%) in CH_3OD as a function of temperature (1000 scans, 90° pulses, repetition time 10 s). Reproduced with permission from Ref. [89]

with $A = CH_3O$ and $L = H$. This phenomenon has also been known for a long time and has been used to study proton exchange rates in solutions of acids in methanol [85]. The spectra in Fig. 10.1 were calculated in terms of the exchange operator given in Eq. (3.15), taking into account the magnetic equivalence of the CH_3 spins. Thus, the dimension of $\mathcal{M}$ could be reduced to a 16×16 matrix [94]. From the line shape analysis the proton lifetimes $\tau_{AHA^*H}^{-1} \equiv \tau_{AH}^{-1}(\tau_0^{-1}$ in Fig. 10.1) could be obtained as a function of temperature. Note that the coupling constant J_{CH_3OH} is only 5 Hz, which strongly limits the dynamic range in which inverse life times can be obtained. Furthermore, it was not possible to measure the deuteron exchange rates τ_{AD}^{-1} by ^{2}H NMR spectroscopy of CD_3OD, because the coupling constant $J_{CD_3OD} = 0.9$ Hz is too small to resolve a spin–spin splitting pattern, giving rise to dynamic line shape effects. The only way to obtain information on possible kinetic isotope effects was, therefore, to perform ^{1}H NMR measurements on mixtures of CH_3OH/CH_3OD, looking selectively at the OH proton signals. The results are shown for a deuterium fraction of $D = 0.99$ in Fig. 10.1 on the right side. From a comparison of both sets of spectra it immediately becomes evident that the OH-proton lifetime of an individual CH_3OH molecule is longer if it is surrounded by CH_3OD molecules. It follows from the experiments in Fig. 10.1 that $\tau_{AH}^{-1}(D = 0)/\tau_{AH}^{-1}(D = 0.99) \cong 3$ at 298 K. Arguments could be given [89] in favor of an ionic exchange mechanism, in which the proton exchange is catalyzed by the ions AL_2^+ and A^-, $L = H, D$:

$$A^*L + A^- \rightleftharpoons A^{*-} + AL, \quad A^*L + AL_2^+ \rightleftharpoons A^*L_2^+ + AL \tag{10.4}$$

generated via autoprotolysis

$$AL + AL \rightleftharpoons AL_2^+ + A^-. \tag{10.5}$$

The observed kinetic isotope effect was then attributed to a large extent to an equilibrium isotope effect of the autoprotolysis leading to a smaller concentration of ions At $D = 0.99$ as compared to $D = 0$. Unfortunately, the small dynamic range of this system did not allow to obtain more detailed information on the kinetic isotope effects of the different isotopic exchange reactions taking place in this reaction system.

Note that the presence of equilibrium isotope effects between a protic solvent like methanol and reacting acids and bases leads to complicated kinetic expressions [90].

10.2 Intermolecular Self Exchange: The System Diarylformamidine-$^{15}N_2$/Tetrahydrofuran

Much simpler self exchange reactions were found in the systems diphenylformamidine-$^{15}N_2$ (DPFA) [107] and di-*p*-flurophenylformamidine-$^{15}N_2$ (DFFA) [108] dissolved in tetrahydrofuran (THF). The structure of these compound, which can exist in an *s*-trans form A (or AL, $L = H, D$) and an *s-cis* form B (or BL), is shown in Fig. 10.2. Both conformers represent four-spin-1/2-

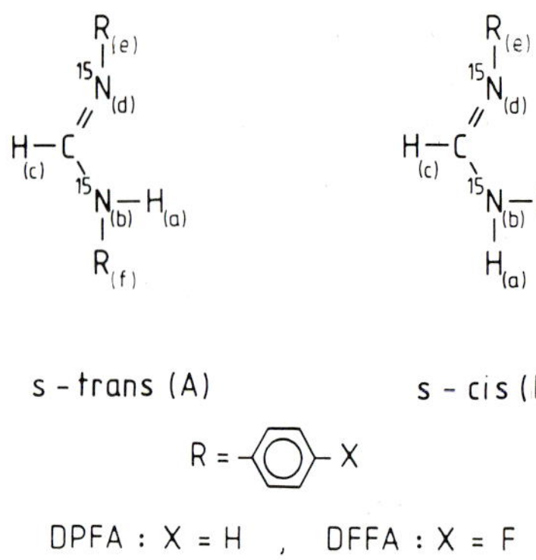

Fig. 10.2. Atom numbering of ^{15}N labeled diarylformamidines in the *s-cis* and the *s-trans* form. Reproduced with permission from Ref. [108]

systems because of the ^{15}N labeling in the nitrogen atom sites. The tautomerism of these compounds was first described by Borisov et al. [96]. In recent kinetic and thermodynamic NMR studies the complex reaction network shown in Fig. 10.3 has been elucidated [107, 108]. Both conformers form at low concentrations solvated monomers A_{1s} and B_{1s} which interconvert slowly on the NMR

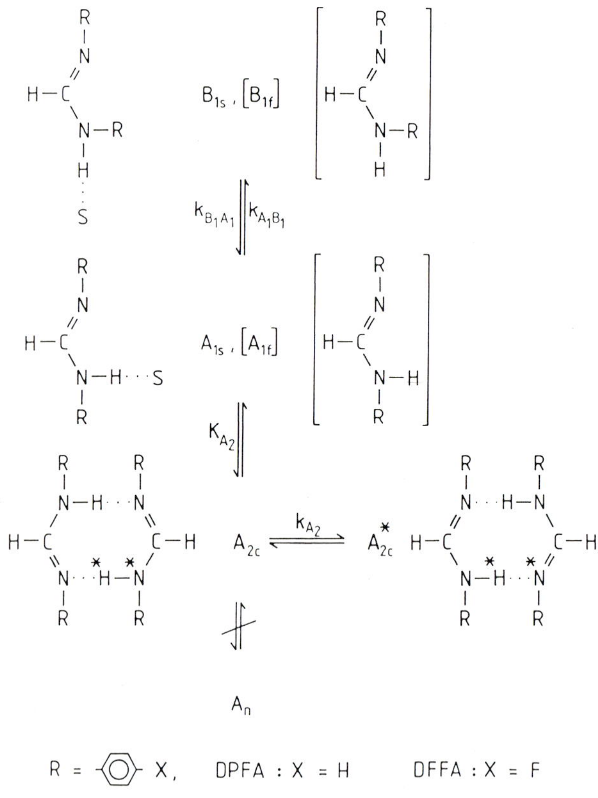

Fig. 10.3. The formamidine tautomerism. Exchange processes of DPFA and of DFFA dissolved in THF as identified by NMR. K_i and k_i are equilibrium and rate constant obtained by lineshape analysis. Brackets represent possible intermediates present in minor concentrations. S ≡ solvent. Reproduced with permission from Ref. [107]

timescale. The free monomers A_{1f} and B_{1f} which are believed to be intermediates of the conformational exchange are present only in very low concentrations. At high concentrations only the *s-trans* conformer AL is able to form cyclic dimers A_{2c} in which a double proton transfer takes place. Higher associates are formed by neither A nor B in detectable concentrations. The following will describe how this information as well as the kinetic HH/HD/DD isotope effects of the double proton transfer were obtained.

10.2.1 Dynamic ^{1}H NMR Spectroscopy of DPFA in Tetrahydrofuran

Figure 10.4 shows a superposition of experimental and calculated ^{1}H NMR spectra of DPFA in THF at a very low concentration. At room temperature only one $^{1}H-^{15}N$ singlet A_a, B_a and one CH-proton signal A_c, B_c are found. By contrast, at low temperatures two CH multiples A_c and B_c as well as two $^{1}H-^{15}N$ signals A_a and B_a are observed, indicating the presence of two conformers A and B which interconvert at high temperature. Both $^{1}H-^{15}N$ signals are split into doublets by scalar coupling with the ^{15}N spin b; each component is further split by coupling with c. For B_a an additional long range coupling to d is resolved, leading to an apparent subtriplet structure. As temperature is raised first signal A_a is broadened; at higher temperatures all signals are affected by the rotational isomerism.

The concentration dependent superposed experimental and calculated ^{1}H NMR spectra of DPFA in THF at 179 K (Fig. 10.5) give further insights

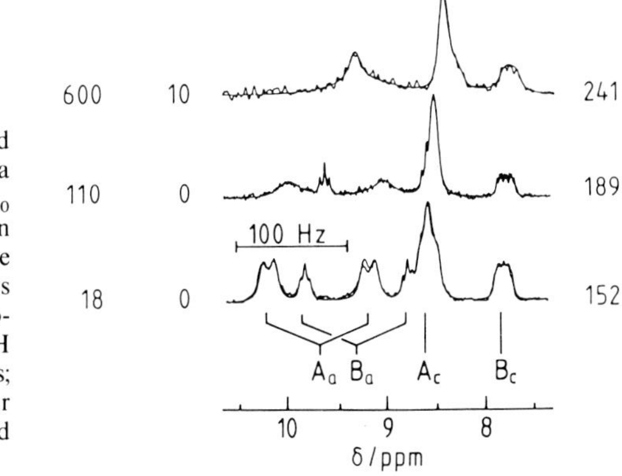

Fig. 10.4. Superposed experimental and calculated ^{1}H NMR 90.02 MHz spectra of ^{15}N, $^{15}N'$-diphenylformamidine-d_{10} (DPFA) in THF-d_8 for a concentration of 0.004 M as a function of the temperature. k_{AB}: forward rate constants of the hindered rotation between monomeric conformers A and B; A_a, B_a: ^{1}H NMR signals of the mobile $^{1}H-^{15}N$ sites; A_c, B_c: ^{1}H signals of the CH sites. For further description see text. Reproduced with permission from Ref. [107]

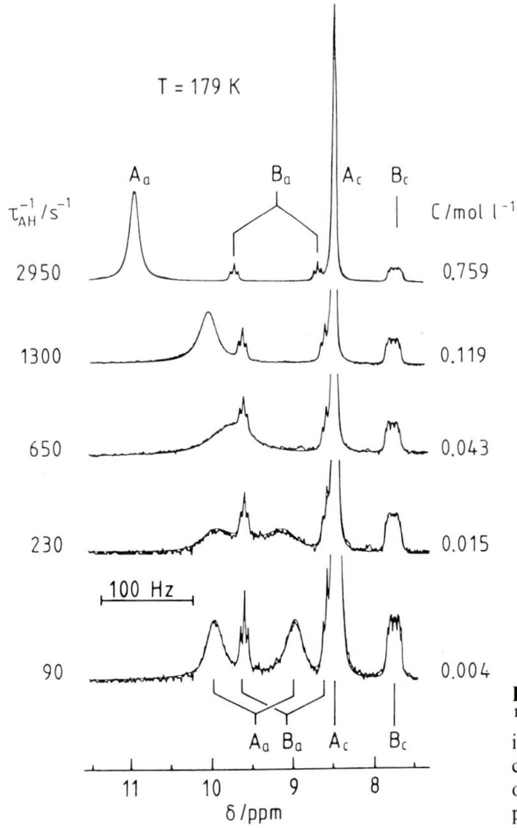

Fig. 10.5. Experimental and simulated ^{1}H NMR 90.02 MHz spectra of DPFA in THF-d_8 at 179 K as a function of the concentration. τ_{AH}^{-1} is the inverse lifetime of the mobile proton in AH. Reproduced with permission from Ref. [107]

into the reaction network of this molecule. The spectra were calculated using a combination of Eqs. (2.12) and (4.15) according to Ref. [107]. Neither the position nor the shape of the B_a signal changes with concentration, indicating absence of self association and proton exchange of B. By contrast, the $^1H-^{15}N$ proton signal A_a of the *s-trans* from A (i) strongly shifts to lower field and (ii) collapses from a doublet into a singlet as concentration is raised because of (i) self association and (ii) proton self exchange. This is because fast intermolecular proton exchange modulates the Larmor frequency of the mobile proton every other exchange process (see Eq. (6.24)). Because of the self association the AH/BH ratio increases with concentration. When the chemical shifts of A_a are plotted as a function of concentration a non-linear behavior is observed (Fig. 10.6, top); the dependence could be fitted assuming only the formation of dimers, with the monomer and the dimer chemical shifts and the equilibrium constant K_{A_2} (Fig. 9.3) of the dimerization as a parameter. Thus, at low concentration the solvated monomer dominates, at higher concentration the dimer. Note that the dynamic range in which proton exchange rates can be measured is much greater than

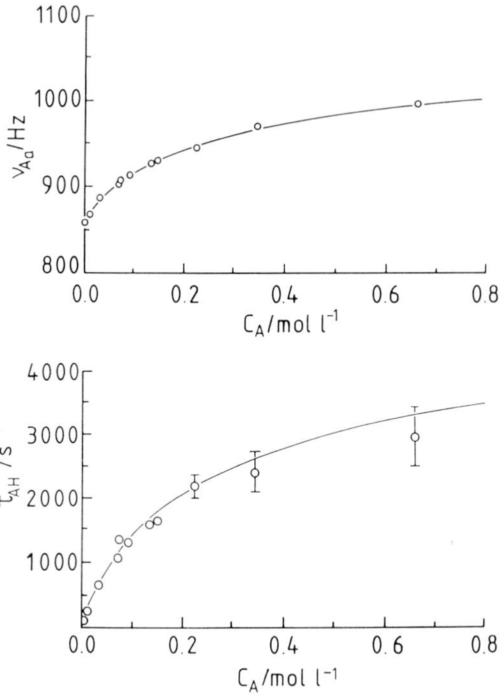

Fig. 10.6. Top: Experimental and calculated chemical shifts ν_{A_a} of the $^1H-^{15}N$ signal A_a of DPFA in THF as a function of the concentration C_A of the conformer A at 179 K. **Bottom**: Experimental and calculated inverse $^1H-^{15}N$ proton lifetimes τ_{AH}^{-1} in conformer A as a function of the concentration C_A of conformer A at 179 K. The *solid line* was calculated using Eqs. (5.85) and (5.91) from the known values of the constants $K \equiv K_{A_2}$ and $k_{A_2} = k_{rls}$. Reproduced with permission from Ref. [107]

in to the case of methanol because the coupling constant $J_{^1H-^{15}N} \approx 92$ Hz. From the calculations the average inverse life times $\tau_{AHAH}^{-1} \equiv \tau_{AH}^{-1}$ are obtained. If the τ_{AH}^{-1} are plotted vs. the concentration C_A also a non-linear dependence is also observed (Fig. 10.6, bottom). Using the known value of K_{A_2} the data could be fitted to Eqs. (5.85) and (5.91) by adapting the exchange rate constant $\tau_{rls}^{-1} = k_{A_2}$ in the cyclic dimer and setting $D = 0$ in Eq. (5.85). At low concentrations τ_{AH}^{-1} is proportional to the concentration C_A of the *s-trans* form A as expected for a second order rate law; at higher concentrations the reaction becomes independent of C_A. This is because A then dominantly forms dimers in which the proton exchange takes place. The ^{1}H NMR spectra of DFFA in THF exhibit similar characteristics [107].

10.2.2 ^{1}H NMR Proton Inventory of the System DFFA/Tetrahydrofuran

Having established the reaction rate law the problem arose how to count the number of protons transferred and how to determine the kinetic isotope effects of the double proton transfer. For this purpose, ^{1}H NMR proton inventories of DPFA and DFFA were performed, i.e. proton exchange rates were measured as a function of the deuterium fraction D in the NH sites. An example is shown in Fig. 10.7 for DFFA/THF. Only the superposed experimental and calculated

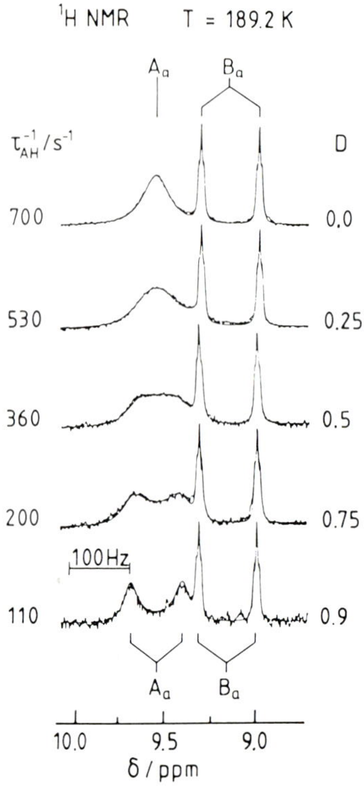

Fig. 10.7. Superposed experimental and calculated 300.13 MHz $^1H-^{15}N$ NMR signals of a 0.02 molar solution of DFFA at 189.2 K as a function of the deuterium fraction D in the $^1H-^{15}N$ position. The $^1H-^{15}N$ signals of the *s-trans* form A are labeled as A_a and of the *s-cis* form B as B_a according to Fig. 10.2. The inverse proton lifetimes in X = A, B are labeled as τ_{XH}^{-1}. τ_{BH}^{-1} was set to zero. Reproduced with permission from Ref. [108]

$^1H-^{15}N$ signals A_a and B_a of DFFA are shown as a function of the deuterium fraction D in the mobile proton sites. Whereas the lineshape of signal B_a is not affected by deuteration the exchange broadened singlet A_a splits up into an exchange broadened doublet when D is increased. Thus, the proton life times τ_{AH}^{-1} decrease with increasing D, which is proof of a multiple proton transfer process.

In Fig. 10.8 the ratio of the inverse proton lifetimes in $A \equiv AH$, $\tau_{AHA*H}^{-1}(D)/\tau_{AHA*H}^{-1}(D=0) = \tau_{AH}^{-1}(D)/\tau_{AH}^{-1}(D=0)$ is plotted as a function of D. A

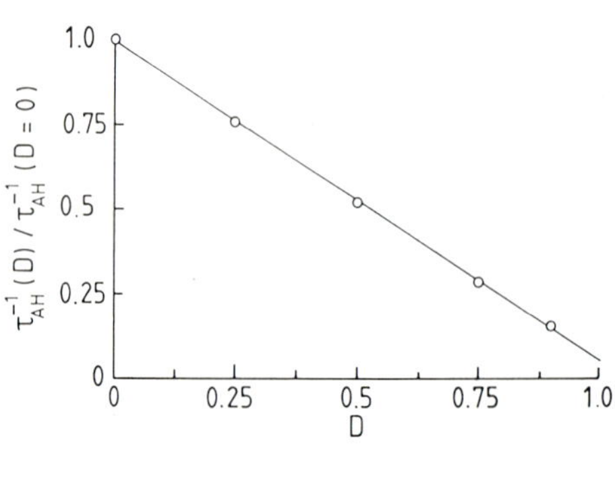

Fig. 10.8. ^{1}H NMR proton inventory plot of the data in Fig. 10.7. According to Eq. (5.32) a linear dependence $\tau_{AH}^{-1} = f(D)$ is expected for a double proton transfer process. The ratio $\tau_{AH}^{-1}(D=0)/\tau_{AH}^{-1}(D=1)$ represents the kinetic HH/HD isotope effect of the exchange. Reproduced with permission from Ref. [108]

linear decrease as predicted by Eq. (5.32) for a double proton transfer process is observed. The inverse lifetimes $\tau_{AH}^{-1}(D=0)$ and $\tau_{AH}^{-1}(D=1)$ can then be identified as the pseudo first order rate constants k^{HH} and k^{HD} of the processes

$$AH + A^*H^* \xrightarrow{k^{HH}} AH^* + A^*H,$$

$$AH + A^*D \xrightarrow{k^{HD}} AD + A^*H.$$

By linear regression analysis we obtain $k^{HH} = 700 \pm 20\,s^{-1}$ and $k^{HD} = 37 \pm 5\,s^{-1}$, i.e. a kinetic HH/HD isotope effect of $k^{HH}/k^{HD} = 19 \pm 3$. Assuming that equilibrium isotope effects on the dimerization are small this value is equal to the intrinsic isotope effect $k_{A_2}^{HH}/k_{A_2}^{HD}$ of the exchange in the cyclic dimer.

Thus, the ^1H proton inventory supports the interpretation of the rate law in terms of the reaction network of Fig. 10.3. These results show the power of the proton inventory technique for the analysis of dynamic processes in the liquid state; this technique is especially important when the concentrations of the reactants cannot be varied.

10.2.3 ^{19}F NMR Proton Inventory of the System DFFA/Tetrahydrofuran

Unfortunately, although the experiments in Fig. 10.8 revealed the kinetic HH/HD isotope effects, the kinetic HD/DD isotope effects could not be determined by ^2H NMR spectroscopy. This is because of the smallness of the scalar coupling constant $J_{^2H-^{15}N} = \gamma_{^2H}/_{^{15}H-^{15}N} \approx 14$ Hz. In view of the much larger natural ^2H line width arising from quadrupole relaxation at low temperatures this value did not provide a sufficient dynamic range. Using the information that $m = 2$ protons are transferred during the exchange the full kinetic HH/HD/DD isotope effects of the exchange could, however, be determined for the system DFFA/THF by ^{19}F NMR spectroscopy [108] as shown in the following.

Figure 10.9 shows the superposed experimental and calculated ^{19}F NMR spectra of DFFA at 189.2 K and a concentration of $C_{298K} = 0.02\,mol\,l^{-1}$ as a function of the deuterium fraction D. In fact, the spectra in Fig. 10.9 stem from the same samples whose ^1H NMR signals were shown in Fig. 10.7, with exception of the highest deuterated sample where $D = 0.99$. All samples were taken under exactly the same experimental conditions, each with a total number of 1500 scans.

One main difference between the spectra in Figs. 10.7 and 10.9 is that in Fig. 10.7 only the protonated molecules AH and BH are observed, whereas in Fig. 10.9 both $A = \{AH, AD\}$ and $B = \{BH, BD\}$ contribute to the spectra. Two singlets $B_e = \{BH_e, BD_e\}$ and $B_f = \{BH_e, BD_e\}$ (see atom numbering in Fig. 10.3) are observed for the *s-cis* form B, indicating two inequivalent fluorine atom positions in this species, which is conform with the molecular structure. Again, the signals B_e and B_f do not depend on D. By contrast, the lineshape of the

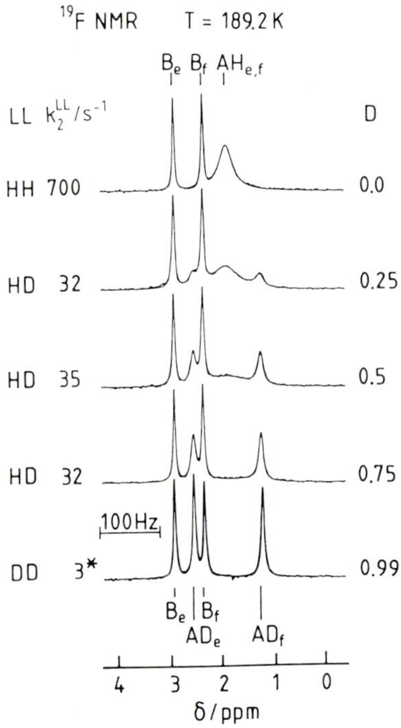

Fig. 10.9. Superposed experimental and calculated ^{19}F–$\{^1H\}$–NMR-spectra of 0.02 molar solutions of DFFA in THF as a function of the deuterium fraction D. For the explanation of the atom numbering see Fig. 10.2. The pseudo first order rate constants of the proton exchange are labeled as k_2^{LL}, LL = HH, HD, DD. In the fitting procedure k_2^{DD} was set constant to a value of $3\,s^{-1}$ (see Figs. 10.10 and 10.11); by non-linear regression analysis of all spectra the values $k_2^{HH} = 700\,s^{-1}$ and $k_2^{HD} = 33\,s^{-1}$ were obtained. The line width in absence of exchange $W_0 \approx 5$ Hz was obtained from the B_e and B_f signals. Reproduced with permission from Ref. [108]

fluorine signals of A depends strongly on the deuterium fraction D. Let us first take a look at the spectrum with D = 0.99. Two sharp signals AD_e and AD_f are observed which indicate two inequivalent fluorine positions in the AD molecule, as expected for a slow deuterium exchange. Now, at D = 0 one broadened coalesced fluorine signal AH_e, AH_f is observed, indicating that a fast process takes place between protonated AH molecules which renders the two inequivalent AH_e and AH_f fluorine atoms equivalent. In view of the molecular structure of DFFA and the dependence of this process on the deuterium fraction it must correspond to the intermolecular double proton transfer involving two AH molecules as observed in the previous section by ^1H, of the type shown in Fig. 10.3, where the proton exchange is associated to an intramolecular exchange of ^{19}F atoms.

Now, a very interesting phenomenon is observed at intermediate deuterium fractions: the fluorine atoms of the protonated AH and the deuterated AD molecules can be separated; they give rise to a **trio**: Whereas the outer two singlets of the **trio** stem from the AD_e and AD_f, the inner broad coalesced line stems from the AH_e, AH_f fluorine atoms. As D increases the latter signal becomes broader, indicating longer lifetimes of the mobile proton in AH. By contrast, the lifetime of the mobile deuteron in AD shortens as D increases, leading to a broadening of the outer two singlets. Note that the line intensities AH_e and

AH_f are determined by the concentration $C_A(1-D)/2$, whereas the line intensities AD_e and AD_f are given by $C_AD/2$.

The lineshape analysis of the spectra in Fig. 10.9 was performed using Eq. (6.22). Unfortunately, not all kinetic quantities $\tau^{-1}_{AHAH^*}(D)$, $\tau^{-1}_{AHAD}(D)$, $\tau^{-1}_{AHAD}(D)$, and $\tau^{-1}_{ADAD^*}(D)$ can be obtained at all values of D. This is clear because AD does not contribute to the line shape at $D=0$. The same is true for AH at $D=1$. Therefore, it was a great help to know from the ^1H spectra that $m=2$, i.e. that Eqs. (5.26)–(5.31) apply. In other words, the deuterium fraction dependent inverse life times in Eq. (6.22) could be replaced by the three pseudo first order rate constants k^{HH}, k^{HD}, and k^{DD} in the following way:

$$\tau^{-1}_{AHAH^*} = (1-D)\cdot k^{HH}, \quad \tau^{-1}_{AHAD} = D\cdot k^{HD}, \tag{10.2}$$

$$\tau^{-1}_{ADAH} = (1-D)\cdot k^{DH}, \quad \tau^{-1}_{ADAD^*} = D\cdot k^{DD}. \tag{10.3}$$

The value of $k^{HH}=700\,s^{-1}$ was determined from the spectrum at $D=0$; this value concides well with the value obtained by ^1H NMR line shape analysis. Unfortunately, the value of $k^{DD}=3\,s^{-1}$ was too small to induce a measurable line broadening. The latter value was, therefore, determined by the method of magnetization transfer in the rotating frame (see Sect. 2.4) between the lines AD_e and AD_f at $D=0.99$ as shown below. From the spectra in the intermediate D range we obtained then values of $k^{HD}=33\,s^{-1}$ by non-linear least squares fit of the simulated to the experimental spectra. Again, W_0 was determined by simulation of the signals B_e and B_f.

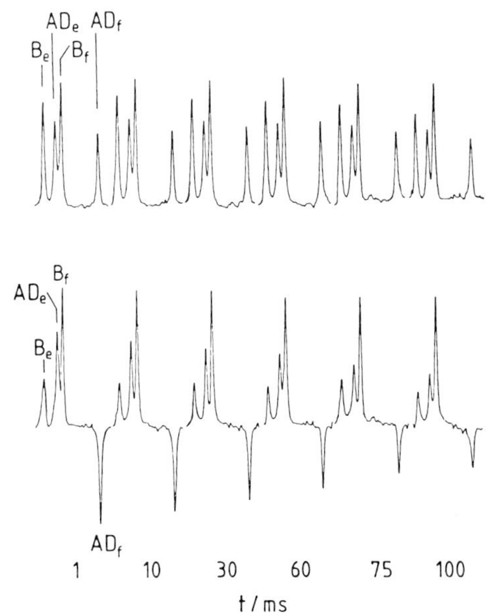

Fig. 10.10. ^{19}F-magnetization transfer experiments in the rotating frame of a 0.02 molar solution of DFFA in THF at a deuterium fraction of $D=0.99$ in the mobile proton sites. The carrier frequency was set to the position of the AD_f line. **Upper curves**: experiment (i) corresponding to a usual $T_{1\rho}$ experiment; **lower curves**: experiment (ii) with a delay $\tau = 1/(2\Delta\nu) = 4.2$ ms between the first 90° pulse and the spin locking pulse. $\Delta\nu$ is the frequency difference between the signals AD_e and AD_f. B_e, B_f: non-exchanging, AD_e, AD_f: exchanging magnetizations. Repetition time 3.8 s, 4.2 μs 90° pulses, strength of the spin locking field $t_{180^\circ} = 165$ μs. Reproduced with permission from Ref. [108]

The results of the magnetization transfer experiments are shown in Fig. 10.10. As described in Sect. 2.3 two sets of experiments were carried out. In the "parallel" experiment (i) the decay of the magnetizations AD_e and AD_f is solely governed by the longitudinal relaxation time in the rotating frame, (Eq. (2.20)):

$$M_{AD_f} = M_{AD_f}(t=0)\exp(-t/T_{1\rho}), \tag{10.4}$$

whereas in the "antiparallel" experiment (ii), (Eq. (2.21))

$$M_{AD_f} = M_{AD_f}(t=0)\exp(-t(2k_2^{DD} + T_{1\rho}^{-1})). \tag{10.5}$$

Since it was difficult to measure the line intensity of AD_e due to signal overlap, only the integrated intensities of AD_f were used to determine k_2^{DD} by non-linear least-squares fit of the data shown in Fig. 10.11.

The following full kinetic isotope effects at 189.2 K and a concentration of $0.02\,\text{mol}\,l^{-1}$ were then obtained:

$$k_2^{HH}/k_2^{dd} = 233 \pm 20,\ k_2^{HH}/k_2^{HD} = 21 \pm 3 \text{ and } k_2^{HD}/k_2^{DD} = 11 \pm 2.$$

The kinetic HH/HD-isotope effect of 21 agrees well within the margin of error with the value of 19 ± 3 determined in the ^1H experiments. Note that the kinetic HH/DD isotope effect is the largest kinetic hydrogen/deuterium isotope effect measured so far by dynamic NMR spectroscopy. Note also that the rule of the geometric mean (RGM) would predict on the basis of the known HH/DD isotope effect for the HH/HD and the HD/DD isotope effects equal values of 15.

It was further shown [200] that the kinetic isotope effects of the proton exchange did not depend in a significant way on concentration, i.e. that equilibrium isotope effects of the dimerization of DFFA are absent. Thus, after determination of the equilibrium constants of dimerization it was possible to obtain the rate constants k_{A_2} of proton exchange in the cyclic dimer as a function of temperature. The Arrhenius diagram is shown in Fig. 10.12. By contrast to the intramolecular proton transfer systems large kinetic HD/DD isotope effects are observed. This has been interpreted with a concerted proton transfer process in the cyclic formamidine dimer [108] which has also been supported theoretically [44]. The deviation from the rule of the geometric mean value was attributed

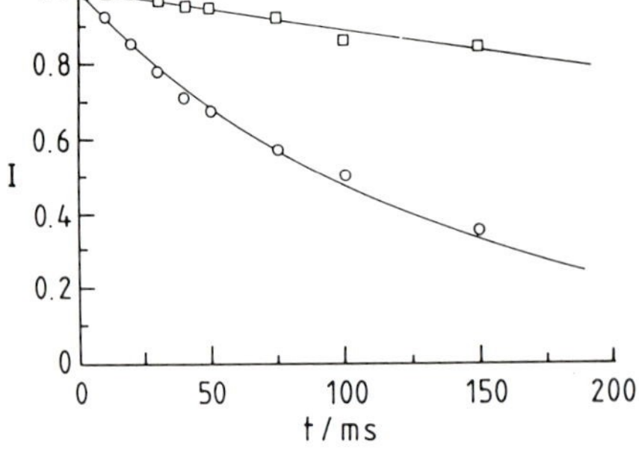

Fig. 10.11. Analysis of the data in Fig. 10.10 in terms of Eqs. (10.4) and (10.5). Parameters of the non-linear regression analysis: $T_{1\rho} = 808$ ms and $k_2^{DD} = 3.05\,s^{-1}$. Reproduced with permission from Ref. [108]

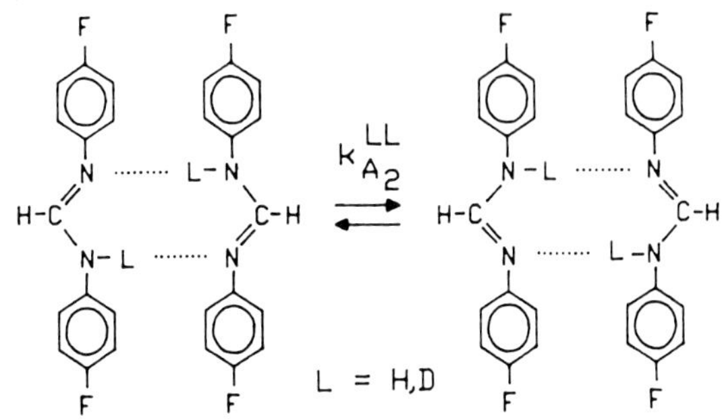

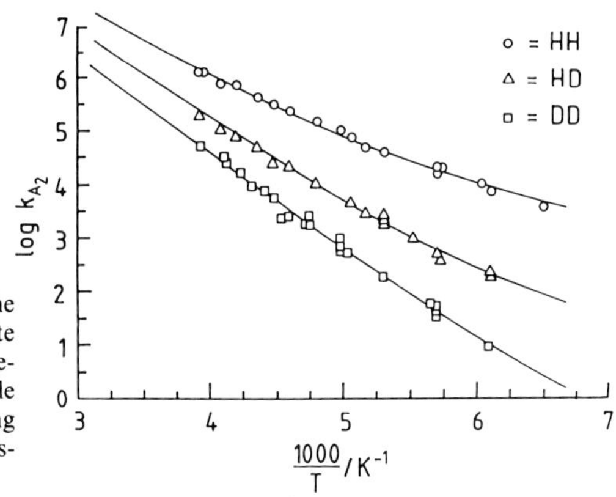

Fig. 10.12. Arrhenius diagram of the DFFA tautomerism in THF. The rate constants correspond to the unimolecular rate constants k_{A_2} of the double proton transfer in the dimer according to Fig. 10.3. Reproduced with permission from Ref. [200]

to tunneling [108]. The calculation of the Arrhenius curves in Fig. 10.12 was performed with a modified Bell model described in Refs. [106] and [200].

10.3 Exchange Between Different Sites: The System Acetic Acid/Methanol/Tetrahydrofuran

In this section an example of an intermolecular proton and deuteron exchange between two proton donors with different spin Hamiltonians is presented: the system acetic acid (AL)/methanol (BL)/tetrahydrofuran (THF). A 1:1 cyclic proton transfer between both molecules in THF shown in Fig. 10.13 was first established by ^{1}H NMR line shape analysis at a deuterium fraction of D = 0 in the mobile proton sites [94, 97]. Later, the number of protons transferred as well as the multiple kinetic hydrogen/deuterium isotope effects were studied by a combination of ^{1}H and ^{2}H NMR spectroscopy [102–106] as a function of temperature, concentration, and D; thus, an additional 2:1 proton exchange according to Fig. 10.13 was detected at higher concentrations of acetic acid.

In order to understand how the results of this study were obtained let us first have a look at the superposed experimental and calculated ^{1}H NMR spectra

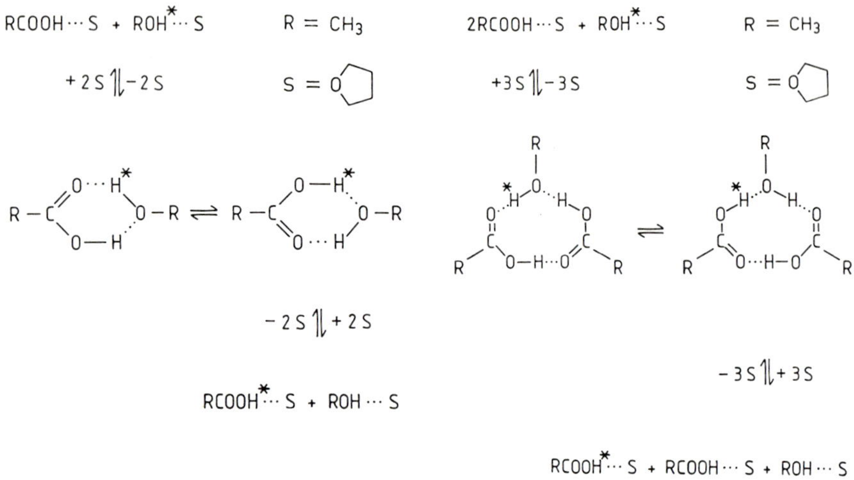

Fig. 10.13. Cyclic 1:1 (**left**) and 2:1 (**right**) proton transfer between acetic acid and methanol in THF; S ≡ THF. Reproduced with permission from Ref. [106]

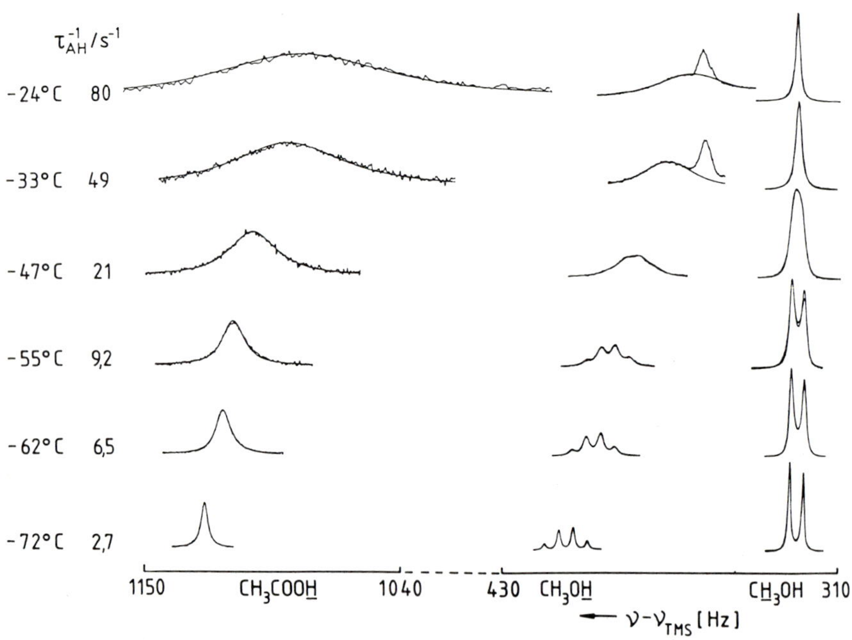

Fig. 10.14. Superposed experimental and calculated 100 MHz ^{1}H CW NMR spectra of a solution of 0.29 mol l^{-1} CH_3COOH(AH) and 0.8 mol L^{-1} CH_3OH(BH) in THF-d_8 as a function of temperature. The peak at 370 Hz arises from residual solvent protons. $\tau_{AH}^{-1} \equiv \tau_{AHBH}^{-1}$. The τ_{BHBH}^{-1} values were found to be zero. Reproduced with permission from Ref. [97]

of a solution of 0.19 mol l^{-1} CH_3COOH(AH) and of 0.8 mol l^{-1} CH_3(BH) in THF-d_8 shown in Fig. 10.14 as a function of temperature. At low temperatures the slow exchange range is reached where the COOH signal represents an exchange broadened singlet; the two signals of CH_3OH are those expected for an AB_3 spin system as found for pure methanol, i.e. a quartet for the OH and a doublet for the CH_3 group. From the simulation of these spectra using the exchange operator of Eq. (4.15) [94, 97] the inverse lifetimes $\tau_{AH}^{-1} = \tau_{AHBH}^{-1} = \tau_{BHAH}^{-1} C_B/C_A$, and τ_{BHBH}^{-1} could be obtained. C_A and C_B are the concentrations of the reactants. The values of τ_{AHAH}^{-1} characterizing the inverse lifetime of a carboxylic proton before the exchange with another carboxyl group occurs could not be obtained because this process does not modulate the spin Hamiltonian of the COOH proton, which is not coupled to a remote spin. It was found that the quantities τ_{BHBH}^{-1} were zero in pure samples [94, 97]; this means that within the errorlimits no proton jumps directly from one methanol molecule to another.

Because of the limited dynamic range of the NMR receiver all methyl groups in this system had to be deuterated in order to obtain exchange broadened NMR spectra in the region of the coalescence of the COOH and the OH signals [106]. The results are shown in Fig. 10.15 which contains on the right side the superposed experimental and calculated FT ^{1}H NMR spectra of a 0.4 molar 1:1 solution of CD_3COOH/CD_3OH/THF-d_8. The residual solvent signals were minimized by subtracting the spectra from those of the pure solvent recorded under the same conditions. Although this procedure leads to some artifacts at high field, they do not hinder the lineshape analysis. Thus, the spectra were easily calculated using Eq. (2.18).

In order to obtain the kinetic isotope effects ^{2}H NMR spectra of a mixture of 0.29 mol l^{-1} CH_3COOD and 0.19 mol l^{-1} CH_3OD in THF were recorded [106]. The superposed experimental and calculated spectra are shown in Fig. 10.15 on the left side. Since isotope effects on chemical shifts, measured in ppm, are small, $\delta_{COOH} - \delta_{OH} \cong \delta_{COOD} - \delta_{OD}$ to a good approximation. However, the frequency differences in Hz are related by $\nu_{COOH} - \nu_{OH} \cong \gamma_{^1H}/\gamma_{^2H}(\nu_{COOD} - \nu_{OD})$, with $\gamma_{^1H}\gamma_{^2H} = 6.51$. This means that the COOD and the OD signals coalesce at much smaller exchange rate constants than the COOH and the OH signals. Thus, the observation that the external aspect of both sets of spectra in Fig. 10.15 is similar signifies a large kinetic hydrogen/deuterium isotope effect.

These spectra and the kinetic isotope effects obtained so far do not yet indicate the rate law of the exchange or the number of protons transferred in the rate limiting step. Before performing experiments in a large concentration range an NMR proton inventory was performed which consisted of ^{1}H and ^{2}H NMR measurements as a function of the deuterium fraction D in the mobile proton sites at fixed concentrations and a fixed temperature of $T = 298$ K. The ^{2}H NMR results are shown in Fig. 10.16. The slow deuteron exchange range is reached at values of $D \approx 1$. As the deuterium content is lowered, the inverse deuteron life times increase. Coalescence of the signals is reached at about $D = 0.2$. Since for a single proton transfer reaction the deuterium life times

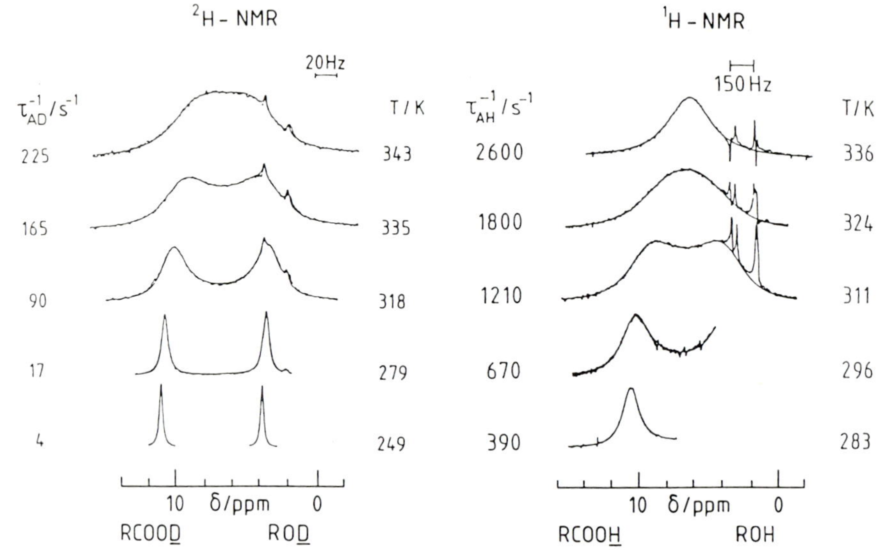

Fig. 10.15. Right: Superposed experimental and calculated ¹H 90.02-MHz FT NMR difference spectra of a solution of 0.4 mol L^{-1} CD_3COOH and 0.4 mol L^{-1} CH_3OH in THF-d_8 and of pure THF-d_8 at different temperatures using a 10-mm probe head (1000 scans, 60° pulses, 10-s repetition time, non-spinning sample). The sealed sample tube had a diameter of 8 mm, a height of 40 mm and was placed in a 10 mm NMR tube. The gas phase in the sample was about 20% of the total sample volume. The small sharp peaks arise from the residual aliphatic protons of the solvent and the solutes. The calculations were performed using Eq. (2.18). **Left**: Superposed experimental and calculated ²H 13.82-MHz FT NMR spectra of a solution of 0.29 mol L^{-1} CH_3COOD and 0.29 mol L^{-1} CH_3OD in THF at different temperatures. The deuterium fraction of the exchangeable protons was D = 0.99. The small two sharp lines arise from the natural deuterium content of THF. Reproduced with permission from Ref. [106]

should be independent of D according to Fig. 5.1, this observation is proof of a multiple proton transfer reaction. On the other hand, the proton exchange rates measured by ¹H NMR spectroscopy decrease with increasing deuterium fraction. The resulting NMR proton inventory plot is shown in Fig. 10.17. Note that it was possible to obtain both $\tau_{AH}^{-1} = \tau_{AHBH}^{-1}$ (filled circles) and $\tau_{AD}^{-1} = \tau_{ADBD}^{-1}$ (open circles) from successive ¹H and ²H NMR measurements on the same samples of CH_3COOL/CH_3OL in THF at D = 0.5 and 0.7. Thus, within a given sample the protons exchange faster than the deuterons! In first order, straight lines for τ_{AH}^{-1} and τ_{AD}^{-1} vs. D in Fig. 10.17 indicate that m = 2 according to Eqs. (5.32) and (5.33). This signifies the presence of the cyclic 1:1 process depicted in Fig. 10.13. Moreover, the intercepts $\tau_{AH}^{-1}(D=0)$ and $\tau_{AD}^{-1}(D=1)$ are approximately equal and can, therefore, be identified with the pseudo first order rate constant of the 1:1 process k^{HD} in Fig. 10.13. A closer look shows that $\tau_{AD}^{-1}(D=0) > \tau_{AH}^{-1}(D=1)$. This result can be explained with superposed double and triple proton transfers according to Fig. 10.13. For this case, the predicted dependence of the proton and deuteron exchange rates as a function of D was derived in Eqs. (5.64) and (5.65). Then, according to Eq. (5.65), $\tau_{AD}^{-1}(D=0)$ is strongly influenced by k^{HHD}, which is larger than k^{HDD}. Since $\tau_{AH}^{-1}(D=1)$ is

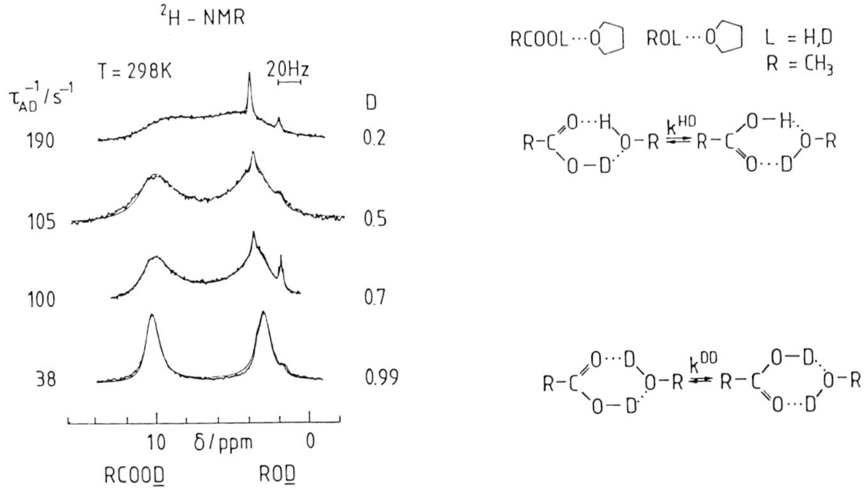

Fig. 10.16. Superposed experimental and calculated ^2H 13.82-MHz FT NMR spectra of solutions of 0.3 mol L^{-1} CH_3COOL and 0.3 mol L^{-1} CH_3OL (L = H, D) in THF at 298 K as a function of the deuterium fraction D. The calculations were performed using Eq. (2.18). Reproduced with permission from Ref. [106].

influenced by the latter the result that $\tau_{AD}^{-1}(D=0) > \tau_{AH}^{-1}(D=1)$ is understandable. As a consequence, the solid lines in Fig. 10.17 were calculated using Eqs. (5.64) and (5.65). Further concentration dependent experiments supported this interpretation. Actually, it was found that the pseudo first order rate constants increased linearly with the acetic concentration. This result was consistent with the previous finding of absence of methanol self exchange in acetic

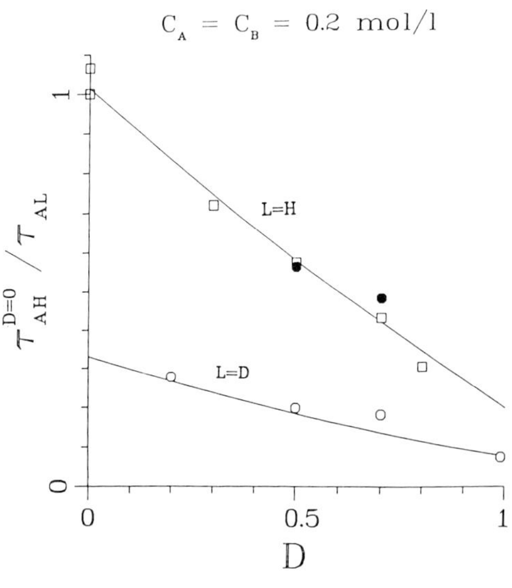

Fig. 10.17. Experimental NMR proton inventory plot at 298 K and a concentration of 0.2 mol L^{-1} for the system acetic acid/methanol/THF. (□): data obtained by ^1H NMR of the system $CD_3COOL/CD_3OL/THF\text{-}d_8$; (●): data obtained by ^1H NMR of the system $CH_3COOL/CH_3OL/THF$ (selective pulses on the COOH signal); (○): data obtained by ^2H NMR of the system $CH_3COOL/CH_3OL/THF$. The curves were calculated according to Eqs. (5.64) and (5.65) using the known rate constants and a value of $k^{HDD} = 634$ L mol^{-2} s^{-2} calculated from k^{DDD} using the RGM. Reproduced with permission from Ref. [106]

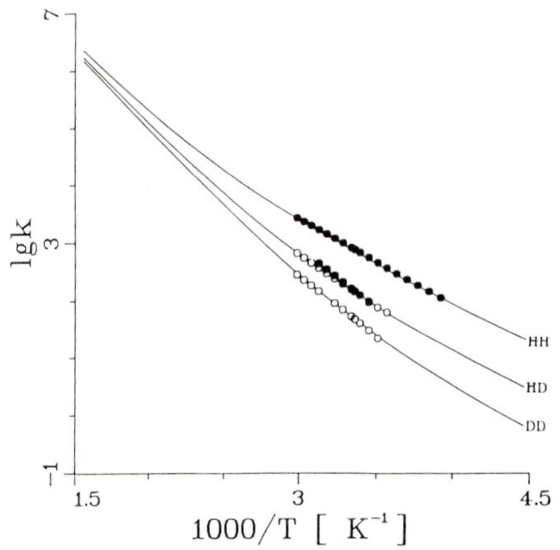

Fig. 10.18. Arrhenius curves of the intermolecular double HH, HD, and DD exchange involving one molecule each of acetic acid and methanol in tetrahydrofuran. The three curves were calculated according to a tunnel model described in Ref. [106] using only four parameters. (●): values obtained by ^{1}H NMR spectroscopy; (○): values obtained by ^{2}H NMR spectroscopy. Reproduced with permission from Ref. [106]

acid/methanol/THF mixtures, i.e. with the exchange mechanism shown in Fig. 10.13.

From the analysis of all kinetic data the following multiple kinetic isotope effects of the processes shown in Fig. 10.13 could be derived:

$$(k^{HH}/k^{DD})_{298K} = 15.5 \pm 4.4, \; (k^{HH}/k^{HD})_{298K} = 5.1 \pm 1.4,$$
$$(k^{HD}/k^{DD})_{298K} = 3.1 \pm 0.9, \quad (10.6)$$
$$(k^{HHH}/k^{DDD})_{298K} = 11.5 \pm 2.7, \; (k^{HHH}/k^{HHD})_{298K} = 2.1 \pm 0.5. \quad (10.7)$$

As in the case of the formamidines these data can not easily be explained with a stepwise exchange mechanism. The deviation from the geometric means values were interpreted in terms of a model described in Ref. [106]. With this model the Arrhenius curves of the 1:1 exchange (Fig. 10.18) and of the 2:1 exchange could well be reproduced [106].

11 NMR Studies of Kinetic Hydrogen/Deuterium Isotope Effects in the Solid State

In the previous sections we have discussed the possibilities of dynamic NMR spectroscopy to elucidate kinetic hydrogen/deuterium isotope effects of reversible chemical reactions in the liquid state. As discussed already in Sects. 1 and 3, dynamic NMR spectroscopy is also able to follow proton and deuteron transfer dynamics in the solid state. The easiest way to do so is to take advantage of a modulation of chemical shifts of remote spin 1/2 nuclei by the processes studied and to perform high resolution solid state CPMAS NMR measurements on microcrystalline powders or amorphous solids. In this method, proposed by Schaefer and Stejskal [154], all orientation dependent spin interactions are removed by ^{1}H decoupling and magic angle spinning (MAS) [155–157]. The sensitivity is increased by ^{1}H–X cross polarization (CP) [152, 153]. Thus, CPMAS NMR spectra of e.g. ^{13}C or ^{15}N spins are comparable to those obtained by liquid state NMR spectroscopy. As a consequence magnetization transfer and exchange broadening phenomena can be observed in the presence of dynamic processes which modulate the isotropic chemical shifts [162–179]. In the nomenclature defined above such experiments correspond to remote spin studies and all possibilities to determine kinetic isotope effects by looking at remote spins apply. Note that other solid state interactions may also be used [158–161, 180–189] in order to follow the dynamics of proton transfers; generally, such studies require, however, single crystals and the determination of kinetic isotope effects is not easy.

In this section we do not intend to give a comprehensive overview of dynamic CPMAS spectroscopy of solid state proton transfers. Instead, we would like to demonstrate how kinetic hydrogen/deuterium isotope effects can be measured using this method. In addition, other solid state NMR techniques will only be discussed briefly.

11.1 ^{15}N CPMAS NMR Studies of Kinetic Isotope Effects of Hydrogen Transfer Between Nitrogen Atoms

Usually, in CPMAS studies of organic compounds natural abundance ^{13}C atoms are monitored [154, 161, 162]. Thus, several solid state hydrogen transfer systems have been studied by natural abundance ^{13}C CPMAS NMR [163, 166, 175–178]. However, with the exception of hydride transfers in carbonium ions [166], carbon atoms are not directly involved in hydrogen transfers and their NMR lines may not always be sensitive to these processes. Therefore, nitrogen NMR is a more suitable method for organic dyes, where nitrogen atoms often act as proton donors or acceptors. Because of the quadrupole moment of the ^{14}N nucleus, it is, at present, necessary to enrich the molecules studied with the less abundant ^{15}N isotope. Using the ^{15}N CPMAS technique of ^{15}N enriched compounds fast proton tautomerism in a number of organic crystalline dyes was detected [164, 167–174]. In this section we show two examples of the determination of kinetic hydrogen/deuterium isotope effects using this method.

11.1.1 Intramolecular Proton and Deuteron Transfer in Solid meso-Tetratolylporphyrin

As an example for an intramolecular proton transfer reaction let us consider the tautomerism of *meso*-tetraarylporphyrins. As already mentioned in Sects 2 and 9.3, porphyrins are subject in the liquid state to a degenerate tautomerism according to Figs. 2.2 and 9.11. This tautomerism also takes place in the solid state. Generally, the two tautomers are no more degenerate by contrast to the liquid state because of a reduction of the molecular site symmetry [167]. In special cases, e.g. porphyrin [171, 176] and *meso*-tetratolylporphyrin (TTP) [167], it was found, however, that the two tautomers are quasi-degenerate. This is demonstrated in Fig. 11.1 on the left side which shows the variable temperature ^{15}N CPMAS NMR spectra of ^{15}N enriched polycrystalline TTP, at a deuterium fraction of $D = 0$ and $D \cong 1$. As expected from the liquid state studies, two singlets for the protonated and the non-protonated nitrogen atoms are obtained at low temperature which indicates that the proton migration is slow on the NMR time scale. In the fast exchange regime one coalesced line is observed

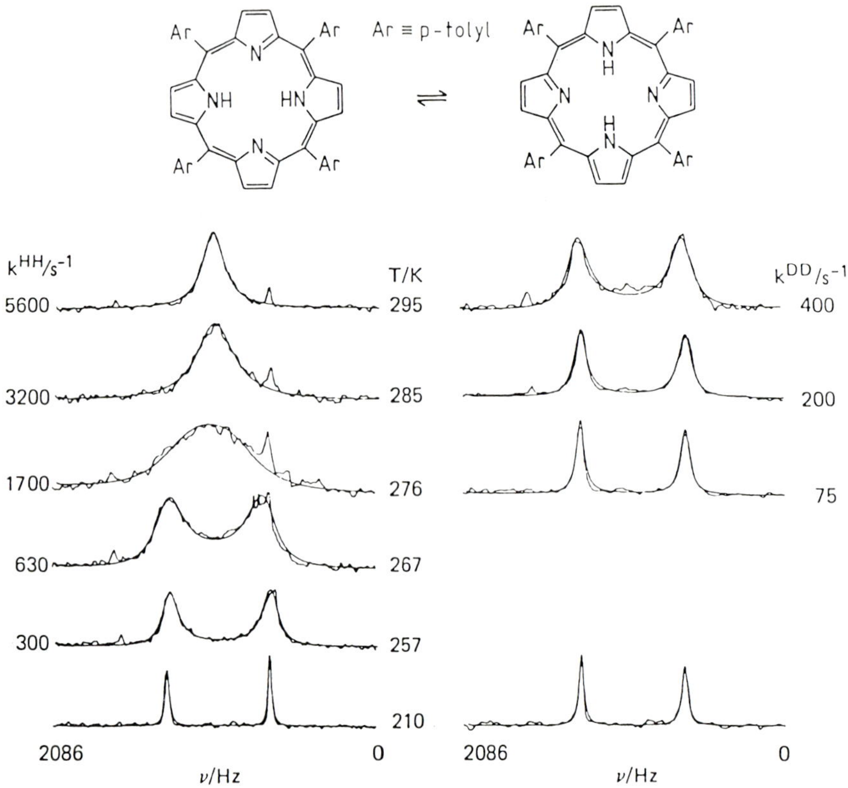

Fig. 11.1. Superposed experimental (6.082 MHz) and calculated ^{15}N CPMAS NMR spectra of 95% ^{15}N-enriched *meso*-tetratolylporphyrin (TTP) as a function of temperature. The calculations were performed using Eq. (2.18). **Left**: deuterium fraction in the mobile proton sites $D = 0$ (reproduced with permission from Ref. [167]); **right**: deuterium fraction in the mobile proton sites $D = 1$ (HH Limbach, CS Yannoni, unpublished). 10 Hz line broadening, 8000 Hz sweep width, 1.2 s repetition time, 9 μs ^{1}H-$\pi/2$ pulses, quadature detection, 25000 scans on the average; reference, external $^{15}NH_4NO_3$. The line width in the absence of exchange was taken from the spectrum at 210 K. The equilibrium constant of the tautomerism was assumed to be unity. k^{LL} is the LL = HH, DD migration rate constant. The sharp small high-field line stems from the NH signal of a low-level impurity of nonexchanging meso-tetratolylchlorin, a byproduct of the synthesis

indicating equivalent nitrogen atoms with an average proton density 1/2 within the NMR timescale. The lineshape analysis was carried out using Eq. (2.18) with an equilibrium constant of $K = k_{12}/k_{21} = 1$. The rate constants observed are very close to those obtained for the liquid solution [167].

Figure 11.1 shows on the right side what happens when the compound is deuterated in the mobile proton sites. The kinetic HH/DD isotope effect is manifest in view of the observation that at room temperature the slow exchange regime is observed. There is an additional line broadening of the high-field ^{15}N signal for the ^{15}N atom bound to deuterium. This broadening arises both from small non-averaged dipolar coupling with deuterium, which is not averaged by MAS [201–204] because of the quadrupole moment of ^{2}H, and from a nonresolved scalar ^{2}H–^{15}N coupling. These effects were taken into account by using a larger value of W_0, the line width in the absence of exchange, for the simulation of this signal. Unfortunately, the rate constants of the HD reaction in the solid state have not yet been determined via ^{15}N CPMAS NMR spectroscopy. One can expect that such measurements will be possible in the future by lowering the deuterium fractions in the mobile proton sites. In such studies one has to take into account that the ^{15}N signal intensities of protonated nitrogen are enhanced as compared to deuterated ^{15}N atoms, due to different ^{1}H–^{15}N cross polarization dynamics.

Since lineshape analysis yields unsatisfactory results in the slow exchange regime polarization transfer experiments in the rotating and the laboratory frame were performed in order to obtain kinetic isotope effects at low temperatures [164]. Here, attention had to be paid to the possibility that polarization transfer between dipolar coupled spins can also arise from "spin diffusion".

The kinetic HH/DD isotope effect found here for the solid state is similar to that observed for the tautomerism of TPP in the liquid state [124]. Thus, in the case of TTP liquid/solid state effects on the tautomerism cannot be observed. This result is very important when combining kinetic results obtained by laser methods [141] on porphyrin in solid hexane and data obtained by dynamic liquid state NMR spectroscopy.

11.1.2 Intermolecular Triple Proton and Deuteron Transfer in Solid Dimethylpyrazole

As an example of an intermolecular proton transfer we discuss in this section the ^{15}N CPMAS spectra of ^{15}N enriched solid 3,5-dimethylpyrazole (DMP) which forms in the solid state a cyclic trimer [178, 179] (Fig. 11.2). In this trimer a moderately fast proton transfer takes place. A superposition of the experimental and calculated ^{15}N CPMAS NMR spectra of DMP are shown in Fig. 11.3 for deuterium fractions $D = 0$ and 1 in the mobile proton sites. Let us first discuss the spectra at $D = 0$. At 223 K two sharp lines are observed (Fig. 11.3a) indicating the presence of two types of nitrogen atom, one

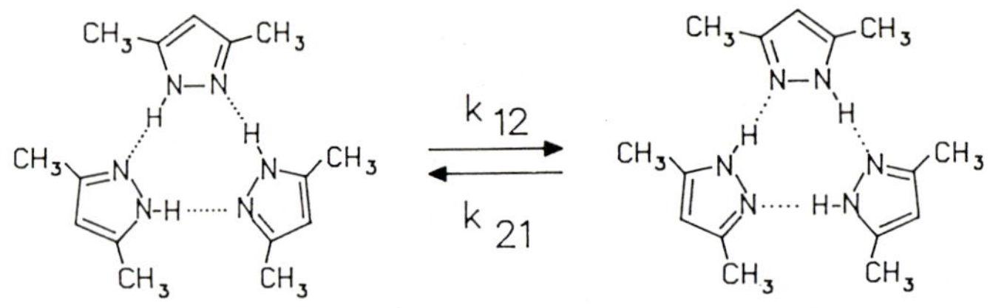

Fig. 11.2. The tautomerism of 3, 5-dimethylpyrazole (DMP) in the solid state where the molecule forms cyclic trimers. Reproduced with permission from Ref. [179]

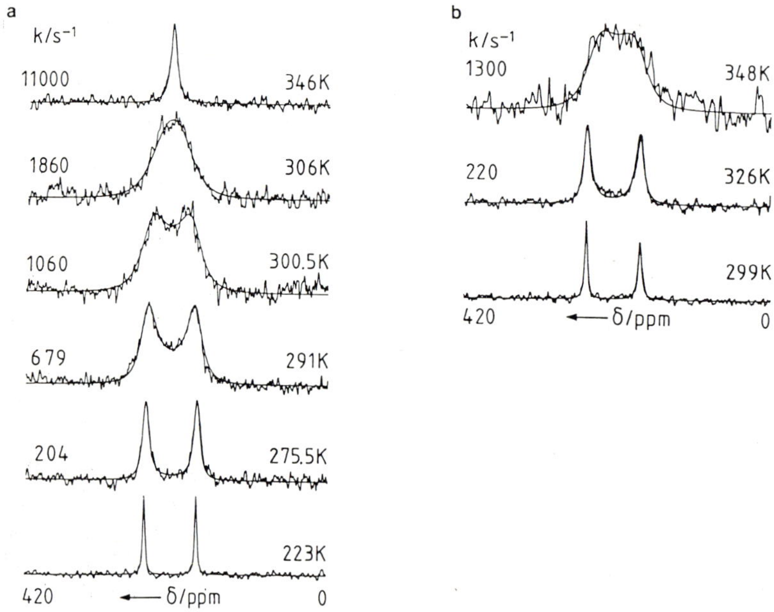

Fig. 11.3 a, b. ^{15}N CPMAS NMR spectra of 95% ^{15}N enriched 3,5-dimethylpyrazole (DMP) at 9.12 MHz as a function of temperature. (a) deuterium fraction in the mobile proton sites D = 0; (b) deuterium fraction in the mobile proton sites D = 1. Reference: external $^{15}NH_4Cl$. The spectra were calculated using Eq. (2.18). Parameters of the calculation: $\nu_{NH} = 180$ ppm; $\nu_N = 254$ ppm (at 9.12 MHz), i.e. $\Delta\nu = 544$ Hz; a: $W_0 = 35$ Hz including 15 Hz line broadening; b: $W_{0_{ND}} = 54$ Hz and $W_{0_N} = 55$ Hz. k is the rate constant. Rotation frequencies between 2.6 and 3.5 kHz. Reproduced with permission from Ref. [179]

protonated (N(1)) and one non-protonated (N(2)). As the temperature is raised, the two lines broaden and coalesce into one sharp line at 346 K. The fact that the line appears in the center of the two low temperature singlets indicates again that all nitrogen atoms have an equal proton density of 0.5, i.e. that the equilibrium constant $K_{12} \cong 1$ within the margin of error of ^{15}N CPMAS NMR spectroscopy. The rate constants obtained can be represented by the Arrhenius equation

$$k_{12}^{H} \approx A_{12} \exp(-E_{a_{12}}/RT), \quad A_{12} \approx 10^{11}, \quad E_{a_{12}} \approx 45.7\,\mathrm{kJ\,mol^{-1}}. \quad (11.1)$$

Because of the relatively small temperature range covered so far the activation parameters in Eq. (11.1) might be subject to changes when rate constants can be obtained in the future over a larger temperature range.

At $D \approx 1$ the proton transfer rates are drastically reduced as shown in Fig. 11.3b. Thus, at room temperature the slow exchange regime is realized. Whereas the width of the low field line is almost as sharp as in the 223 K spectrum of DMP at $D = 0$, the high field line is again broadened by residual scalar and dipolar coupling with deuterium. By line shape analysis a kinetic hydrogen/deuterium isotope effect of about 8 at 347 K is obtained. One can estimate that at 298 K the rate of deuteron exchange cannot exceed a value of $50\,s^{-1}$. This leads to a kinetic HHH/DDD isotope effect of $\geqslant 20$ at 298 K.

Again, it has not yet been possible to study the mixed rate constants k^{HHD} and k^{HDD} at intermediate D values. Note that hydrogen bond exchange and diffusion is extremely slow in the solid state by contrast to the liquid state. Therefore, the ^{15}N CPMAS NMR lineshape of DMP at intermediate D values is a static superposition of line shapes arising from the isotopic HHH, HHD, HDD, and DDD species. The mobile protons jump then always between the same nitrogen atoms, a situation which is equal to the one encountered in intramolecular proton transfer systems in liquid solution.

Although the value of $10^{11}\,s^{-1}$ found here for the frequency factor of this process might be subject to changes when kinetic data are obtained over a larger temperature range, one can already say that this value indicates the absence of a large negative activation entropy. This finding is expected for an intramolecular reaction. Future experiments will have to establish the origin of the large kinetic isotope effect on the solid state tautomerism of DMP.

11.2 The Use of Orientation Dependent Nuclear Spin Interactions for the Study of Kinetic Hydrogen/Deuterium Isotope Effects

There are a number of fast proton transfer processes which do not lead to a modulation of isotropic chemical shifts in the solid state. In these cases, it is not possible to obtain proton transfer rates by the CPMAS technique. As an example, consider the tautomerism of carboxylic acid dimers shown in Fig. 11.4. However, as discussed in Sect. 3, in the solid state there are orientation dependent interactions which are modulated in the presence of molecular motion. Among these are dipolar coupling, the nuclear quadrupole interaction and the chemical shift anisotropy. The first two interactions have been used in order to study the dynamics of proton and deuterium transfer in single crystals of carboxylic acids [180–189] because both the dipolar interaction between the

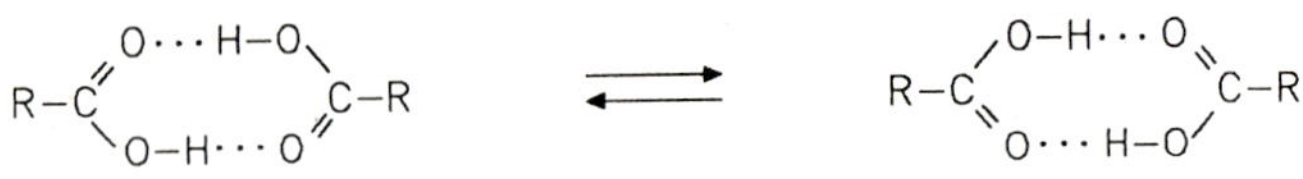

Fig. 11.4. Double proton transfer in carboxylic acid dimers

two mobile protons as well as the quadrupole interaction of the mobile deuterons are modulated when the jump processes occur.

Unfortunately, because of a reduction of the molecular site symmetry in the solid state the two tautomers in Fig. 11.4 have slightly different energies. Therefore, at low temperatures where one could hope to "freeze out" the exchange process on the NMR timescale only the energetically favored tautomer is populated and dynamic line broadening does not occur. Therefore, rate constants could not directly be obtained by lineshape analysis. However, as has been shown by Graf et al. [180], the double proton transfer provides an efficient longitudinal relaxation mechanism for the mobile protons in carboxylic acids. Thus, by measuring T_1 of the mobile protons of benzoic acid single crystals it has been possible to determine rate constants k^{HH} of the carboxylic acid tautomerism as a function of temperature. In a similar way, rate constants k^{DD} of the deuteron tautomerism of carboxylic acid dimers have been measured where advantage of the quadrupole interaction of the ^2H isotope was taken into account [180–186]. One can show that these different methods can be combined in order to obtain the full kinetic HH/HD/DD isotope effects of the carboxylic acid tautomerism [206].

12 Discussion of Kinetic Isotope Effects

We have described strategies to measure kinetic hydrogen/deuterium isotope effects of multiple intra- and intermolecular proton transfer reactions by dynamic NMR spectroscopy, including lineshape analysis and polarization transfer experiments in the liquid and the solid state. In such studies either mobile protons or deuterons or remote spins such as immobile protons or heavy atoms are monitored whose spin Hamiltonians are modulated by the exchange. In order to detect the kinetic isotope effects as well as the number of protons in flight in the rate limiting step of the processes studied dynamic NMR experiments are performed as a function of the deuterium fraction in the mobile proton sites. Such experiments are called "proton inventories". The theory of exchange broadened NMR lineshapes in connection with the proton inventory technique has been described in Sects. 2–6. Examples for this technique coming from the field of intra- and intermolecular proton transfer reactions in the liquid and the solid state were given in Sects. 9–11, after the experimental Sects. 7 and 8.

In the case of pure CH_3OH and CH_3OD evidence for an ionic proton exchange mechanism was obtained where the ionic catalysts $CH_3OH_2^+$ and CH_3O^- are created by autoprotolysis of the solvent. The apparent kinetic isotope effect is then also influenced by the equilibrium isotope effect of the autoprotolysis [89]. Such ionic proton transfer processes can be suppressed when proton donors are studied in organic solvents. Thus, for proton exchange between acetic acid and methanol in tetrahydrofuran [105, 106] the number of protons transported was measured and multiple kinetic HH/HD/DD isotope effects were obtained for these reactions. This was possible because the chemical shifts of the exchanging protons in the two environments are different. By contrast, the determination of kinetic isotope effects of proton self exchange reactions where the chemical shifts of the exchanging protons is not modulated is more difficult. Here, one has to take advantage of the modulation of the scalar spin–spin coupling of the mobile hydrogen isotopes to remote spins, e.g. scalar $^1H-^{15}N$ coupling and/or the modulation of chemical shifts of remote spins by the exchange. An example where this concept was successfully applied is the double proton transfer between formamidine molecules (Fig. 10.3) dissolved in tetrahydrofuran (Sect. 10.2 and Refs. [107, 108]). Concentration dependent studies even allowed the determination of rate constants of exchange in the cyclic dimer which are no more affected by the equilibrium constants of the preequilibrium. In addition, kinetic HH/HD/DD isotope effects were obtained for several intramolecular double proton transfer reactions in azophenine [127], oxalamidine [120, 129] and porphyrines [105, 124–126] which were labeled with ^{15}N for this purpose.

Although the number of systems where multiple kinetic isotope effects have been studied so far is not yet very large, in the following the question will be discussed what the observed kinetic isotope effects determined can tell us about the reaction mechanisms, without going into different theories of multiple kinetic isotope effects.

The kinetic HH/HD/DD isotope effects of symmetric intramolecular double proton transfer reactions determined so far heavily violate the so called "rule of the geometric mean" [43] which states that there are two primary kinetic isotope effects P_1 and P_2 of similar size:

$$P_1 = k^{HH}/k^{HD} \cong P_2 = k^{HD}/k^{DD}. \tag{12.1}$$

In fact, for the azophenine, oxalamidine, and the porphyrin tautomerism (Sect. 9) it is found that the rate constants of the different isotopic reactions are related by

$$k^{HH} \gg k^{HD} \cong k^{DD}. \tag{12.2}$$

Furthermore, in the case of the oxalamidine tautomerism a strong solvent dependence of the rate constants was observed [129] indicating a highly polar transition state, as expected for a stepwise proton transfer via a zwitterionic intermediate and transition states where one proton is bound and the other in flight as indicated Fig. 9.7. Thus, it is clear that these reactions can not be characterized by two but only by one primary kinetic isotope effect. The substitution of the bound proton by deuterium contributes then only a small secondary isotope effect S to the reaction rates. Actually, for degenerate double hydrogen transfer reactions the following equations have been derived without making any assumptions concerning an over barrier or a tunneling pathway [127]:

$$\frac{k^{HD}}{k^{DD}} = \frac{2}{S^{-1} + P^{-1}} \quad \text{and} \quad \frac{k^{HH}}{k^{DD}} = PS. \tag{12.3}$$

Since S is of the order of 1 and $P \gg 1$ Eq. (12.3) represents a very large deviation from the RGM since in this case $k^{HH}/k^{HD} \approx P$ and $k^{HD}/k^{DD} \approx 2$. The Arrhenius curves of the intramolecular HH reactions (e.g. Fig. 9.5) can easily be calculated in terms of Eq. (12.3). These findings give experimental evidence for stepwise reaction pathways in the case of the azophenine and the oxalamidine tautomerism. In the azophenine case the kinetic data were consistent with a reaction over the barrier and the kinetic isotope effects were interpreted in terms of Eq. (12.3) with values of $P = 7.2$, $S = 0.78$ at 298 K [127]. Novel independent theoretical [140–142] and experimental [123–125, 141] studies favor also a stepwise mechanism for the tautomerism of porphyrines, i.e. an interpretation of the kinetic HH/HD/DD isotope effects in terms of Eq. (12.3). Further information on the mechanism of the porphyrin tautomerism comes from a comparison of the rates of the porphyrin tautomerism in a very wide temperature range. Deviations from a non-Arrhenius behavior at low temperatures [125, 141] give then evidence for a thermally activated single proton tunneling mechanism in this temperature range.

The kinetic isotope effects of intermolecular symmetric double proton transfers differ in a significant way from the intramolecular proton transfers. In the intermolecular case the rule of the geometric mean (Eq. (12.1)) is much better fulfilled—although small deviations were still observed—indicating a more or less concerted double proton transfer mechanism. Thus, Eq. (12.3) is not able to accommodate neither the acetic acid/methanol nor the formamidine data. E.g., for the formamidines a kinetic HH/DD isotope effect of 233 is found at 189.2 K and the partial kinetic HH/HD and HD/DD isotope effects are 21 and 11. The rule of the geometric mean (RGM) would predict for both quantities equal values of 15. Note that the value of 233 is the largest kinetic hydrogen/deuterium isotope effect found so far by dynamic NMR spectroscopy, using a combination of lineshape analysis and polarization transfer methods. In the case of the double proton transfer between acetic acid and methanol it was found that $k^{HH}/k^{HD} \approx 5$ and $k^{HD}/k^{DD} \approx 3$ at 298 K [105, 106]. It is clear that these values are not consistent with a stepwise double proton transfer. This means that in the intermolecular case both protons contribute to the kinetic isotope effects. Therefore, it is understandable that the overall kinetic HH/DD isotope effects are larger in intermolecular than in intramolecular double proton transfer reactions. The largest value found at 298 K by NMR is the kinetic HHH/DDD isotope effect of the intermolecular triple proton transfer process in 3,5-dimethylpryazole (Fig. 11.2). For the formamidine reaction evidence for a concerted proton transfer [108] is supported by ab initio calculations [144]. How can one then rationalize the small deviations from the RGM in the case of intermolecular exchange reactions?

In the case of the systems formamidine/THF and acetic acid/methanol/THF these deviations could be interpreted in terms of thermally activated tunneling assuming a concerted proton motion. Such deviations from the RGM in the presence of tunneling are well understood on theoretical grounds [105, 106]. The deviations may arise because tunneling enhances the reaction rates especially of the light hydrogen isotopes. Tunneling also induces a curvature of the Arrhenius curves in a wide temperature range. This is demonstrated in the calculated Arrhenius curves of the formamidine and the acetic acid/methanol tautomerism shown in Figs. 10.12 and 10.18. These curves were calculated for a modified one-dimensional Bell-tunneling model [106]. Since the concave curvatures of the calculated Arrhenius curves is not easily detected experimentally the deviation of the HH/HD/DD isotope effects from the RGM in the case of concerted double degenerate proton transfer reactions may be taken as an additional criterion for tunneling.

Let us try now to give a qualitative explanation for the reaction pathways of intra- and intermolecular double proton transfer systems. This explanation is based on the observation that intramolecular proton transfer systems such as porphyrin and azophenine lack the usual flexibility of hydrogen bonded systems, i.e. the usual low frequency hydrogen bond stretching vibration [21] which modulates the hydrogen bond distance. Thus, the molecular frame of heavy atoms in these compounds is relatively rigid and a high energy would be required to reduce the hydrogen bond distance in such systems. This feature is expressed at the bottom of Fig. 12.1 by an outer square which schematically represents the molecular frame. It is understandable that it costs too much energy to break the bonds of both protons to their neighboring heavy atoms at the same time, and the proton transfer will be stepwise. Note that proton tunneling in this case will always require a minimum energy of activation corresponding to the energy difference between the intermediate and the initial state.

By contrast, the presence of low frequency hydrogen bond stretching vibrations in the flexible intermolecular proton transfer systems allows a comparatively easy compression of the hydrogen bond as schematically shown in the model of Fig. 12.1 top. This model has been used for the calculation of the Arrhenius curves of the proton transfer between acetic acid and methanol [105, 106]. In this case the hydrogen bond lengths are variable, i.e. the energy of activation of the proton transfer is pooled into the hydrogen bond stretching vibration which shortens the hydrogen bond length. A consequence of the shorter bond length is a smaller barrier for the proton transfer. At extreme short hydrogen bond lengths the barrier for proton transfer vanishes and, therefore, the difference between a stepwise and a concerted proton transfer mechanism. The imaginary frequency required for a transition state corresponds then to the hydrogen bond stretching rather than to the AH-stretching

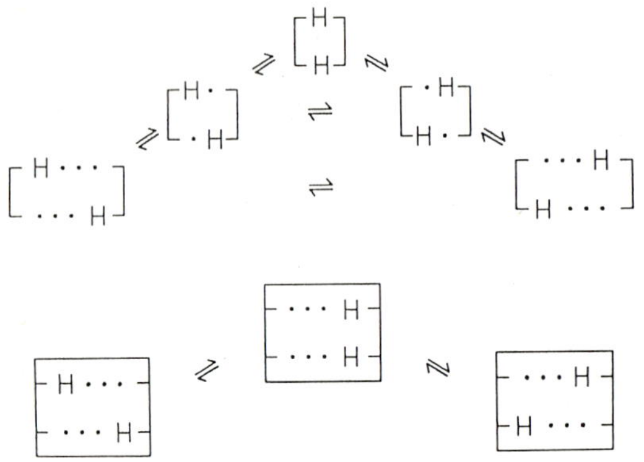

Fig. 12.1. Bottom: Stepwise double proton transfer in the case of a fixed molecular frame of heavy atoms. Top: Double proton transfer in the case of variable hydrogen bond lengths according to a model proposed in Ref. [106]. For further explanation see text. Reproduced with permission from Ref. [108]

vibrations. Now, it is well known that the latter are shifted to lower frequencies when the hydrogen bond distance is shortened [22]. Therefore, there will be a considerable loss of zero point energy of both vibrations in the highly compressed transition state and the RGM will be fulfilled at high temperatures. At lower temperatures the transfer may occur by tunneling leading to a violation of the RGM at low temperatures, as has been proposed previously [105, 106]. Thus, for the intermolecular proton transfer systems a reaction mechanism according to Fig. 12.1 top has been proposed [108]. Note that one might find intermolecular proton transfer systems with rigid hydrogen bond distances and intramolecular proton transfer systems with flexible hydrogen bonds which could lead to an inverse behavior of kinetic isotope effects.

13 Conclusions

In this review it has been shown that kinetic hydrogen/deuterium isotope effects of fast intra- and intermolecular proton transfer reactions in liquids and solids can today be determined using advanced methods of dynamic NMR spectroscopy. Thus, new insights have been obtained which could influence future experimental and theoretical development of the physics and chemistry of isotope effects. A theoretical framework has been given which could guide future research. So far, the method has mostly be applied to small organic molecules dissolved in liquid aprotic solvents where it is very difficult to obtain kinetic isotope effects with other methods.

One might ask now whether there are additional fields where dynamic NMR spectroscopy in the presence of isotope effects might be important. So far, only proton transfer reactions between nitrogen and oxygen atoms have been studied. Recent experiments show that also hydrogen transfers between carbon atoms can also be studied by NMR [207] as well as the corresponding kinetic hydrogen/deuterium isotope effects [208]. Proton, hydrogen and hydride transfer reactions also play a role in metallorganic chemistry where they are related to the problems of homogeneous catalysis. Such reactions can be followed by dynamic NMR spectroscopy; thus, NMR studies of kinetic hydrogen/deuterium isotope effects might also be useful for a better understanding of inorganic reaction mechanism in the future. NMR spectroscopy in one and two dimensions has also contributed significantly in the past to the study of biologically important molecules, such as nucleic acids, proteins, and enzymes. So far, the scope of these studies was mostly the elucidation of the structure of the biomolecules. Since interesting insights into the mechanisms of enzyme reactions have been obtained from conventional studies of kinetic hydrogen/deuterium isotope effects of enzyme reactions in H_2O/D_2O mixtures [43], one can anticipate that dynamic NMR experiments as a function of the deuterium fractions in the mobile proton sites will also contribute to an understanding of the dynamics of enzyme reactions in the future.

Thirdly, as has been shown in Sect. 11, the whole potential of NMR in the study of kinetic isotope effects of proton transfer reactions in the solid state has not yet been fully developed and future efforts in this direction can be anticipated. Such investigations are especially important because of the possibility of studying solids at low temperatures, i.e. the possibility of obtaining kinetic isotope effects in a very wide temperature range.

14 Acknowledgements

I would like to thank all my collaborators which have contributed to the work described in this review. Dr. J. Hennig was involved with the early work on porphyrines and on polarization transfer. Dr. D. Gerritzen performed the kinetic isotope effect studies on the system methanol and methanol/acetic acid/ THF. Dr. H. Rumpel studied the azophenine tautomerism, Dr. L. Meschede the formamidines, Dr. M. Schlabach the hydroporphyrins. Dr. B. Wehrle, F. Aguilar-Parilla and J. Braun were responsible for most of the solid state studies. The initial solid state studies were performed during the year 1983 in the laboratory of Dr. C. S. Yannoni, IBM, San Jose, to whom I am greatly indebted for introducing me into variable temperature high resolution solid state NMR spectroscopy. The acetylporphyrin studies were performed in collaboration with Professor C. Djerassi and his coworkers.

I also would like to thank many friends who have been involved for a long time in problems of proton transfer and isotope research. They have not only encouraged me to pursue the study of proton transfer kinetics by dynamic NMR spectroscopy but have also contributed many ideas to this work. Among these are Prof. H. W. Zimmermann, Prof. R. P. Bell, Prof. E. F. Caldin, Prof. M. M. Kreevoy and Prof. R. L. Schowen.

Finally, I have to thank the Deutsche Forschungsgemeinschaft, Bonn–Bad Godesberg for a Heisenberg grant during the years 1981–1986 as well as for continuous financial support. In addition, I would like to thank the Fonds der Chemischen Industrie, Frankfurt, for a Dozentenstipendium and other financial support.

15 References

1. Urey HC, Brickwedde FG, Murphy GM (1931) Phys Rev [2] 39: 164; (1932) Phys Rev [2] 40: 1
2. Washbourn EW, Urey HC (1932) Proc Natl Acad Sci 18: 4998
3. Melander L, Saunders WH (1980) Reaction rates of isotopic molecules, New York, Toronto: John Wiley & Sons
4. Urey HC (1947) J Chem Soc 569
5. Bigeleisen J, Goeppert Mayer M (1947) J Chem Phys 15: 261; Bigeleisen J (1955) J Chem Phys 23: 2264
6. The May issue of Z Naturfosch 44a, 1989 is devoted to the 70th birthday of Professor J Bigeleisen and contains recent developments in isotope chemistry
7. Glasstone S, Laidler KJ, Eyring H (1941) The theory of rate processes, New York, McGraw-Hill
8. Bigeleisen J (1949) J Chem Phys 17: 675
9. Bell RP (1980) The Proton in Chemistry, 2nd ed, London, Chapman and Hall
10. Bell RP (1980) The tunnel effect in chemistry, London, Chapman and Hall
11. German ED, Kusnetzov AM, Dogonatze RR (1980) J Chem Soc Faraday 2 76: 1128 and references cited herein
12. Bruniche-Olsen N, Ulstrup J (1979) J Chem Soc Faraday 1 75: 205
13. German ED, Kusnetsov AM, Dogonadze RR (1980) J Chem Soc Faraday 2 76: 128 and references cited therein
14. Kuznetsov AM, Ulstrup J (1982) J Chem Soc Far Trans 2 78: 1497
15. Siebrand W, Wildman TA, Zgierski MZ (1984) J Am Chem Soc 106: 4083
16. Siebrand W, Wildman TA, Zgierski MZ (1984) J Am Chem Soc 106: 4089
17. Brickmann J, Zimmermann H (1966) Ber Bunsenges Phys Chem 70: 157
18. Brickmann J, Zimmermann H (1966) Ber Bunsenges Phys Chem 70: 521
19. Brickmann J, Zimmermann H (1967) Ber Bunsenges Phys Chem 71: 160
20. Brickmann J, Zimmermann H (1969) J Chem Phys 50: 1608
21. Schuster P, Zundel G, Sandorfy C (eds) (1976) The hydrogen bond, vols 1–3, Amsterdam, North Holland Publ Comp
22. Novak A (1974) Struct Bond 14: 177
23. Brickmann J (1980) Ber Bunsenges Phys Chem 84: 186
24. Klöffler M, Brickmann J (1982) Ber Bunsenges Phys Chem 86: 203
25. Borgis DC, Lee S, Hynes JT (1989) Chem Phys Lett 162: 19
26. Barbara PF, Trommsdoff (eds) (1989) Chem Phys 136 p 153–360
27. Caldin E, Gold V (1975) Proton Transfer, Chapman and Hall, London
28. Swain CG, Brown JF (1952) J Am Chem Soc 74: 2534
29. Swain CG, Brown JF (1952) J Am Chem Soc 74: 2538
30. Cox MH, Jencks WP (1981) J Am Chem Soc 103: 580
31. Bell RP, Critchlow JE (1971) Proc Roy Soc A 325: 35
32. Ek M, Ahlberg P (1980) Chemica Scripta 16: 62; Ahlberg P, Janne K, Löfas S, Nettelblad F, Swahn L (1989) J Phys Org Chem 2: 429
33. Klaer AM, Nilsen H, Sorensen PE, Ulstrup J (1980) A Chem Scand A34: 281
34. Bensaude O, Chevrier M, Dubois, JE (1979) J Am Chem Soc 101: 2423
35. Bensaude O, Dreyfus H, Dodin G, Dubois JE (1977) J Am Chem Soc 99: 4438
36. Bensaude O, Chevrier M, Dubois JE (1979) J Am Chem Soc 101: 2423
37. Albery WJ (1982) J Chem Soc Far I 78: 1579

38. Albery WJ (1982) Far Disc Chem Soc 74: 245
39. Caldin EF, Mateo S (1975) J Chem Soc Faraday 2 71: 1876
40. Ek M, Ahlberg P (1980) Chemica Scripta 16: 62
41. Engdahl KA, Bivehed H, Ahlberg P, Saunders Jr WH (1982) J Chem Soc Chem Comm 423
42. Elrod JP, Gandour RD, Hogg JL, Kise M, Maggiora GM, Schowen RL, Venkatasubban KS (1975) Faraday Symp Chem Soc 10: 145
43. Gandour RD, Schowen RL (1978) Transition states of biochemical processes, New York, Plenum Press
44. Gross P, Steiner H, Krauss F (1936) Trans Faraday Soc 32: 877
45. Hornell JC, Butler JAV (1936) J Chem Soc 1361
46. Kresge AJ (1964) Pure Appl Chem 8: 243
47. Gold V (1960) Trans Far Soc 56: 255
48. Gold V (1969) Adv Phys Org Chem 7: 259
49. Albery WJ, Limbach HH (1982) J Chem Soc Faraday Disc 24: 291
50. Albery WJ (1986) J Phys Chem 90: 3773
51. Hermes JD, Cleland WW (1984) J Am Chem Soc 106: 7263
52. Hermes JD, Morrical SW, O'Leary MH, Cleland WW (1984) Biochemistry 23: 5479
53. Belasco JG, Albery WJ, Knowles JR (1986) Biochemistry 25: 2529
54. Belasco JG, Albery WJ, Knowles JR (1986) Biochemistry 25: 2552
55. Taylor CA, El Bayoumi MA, Kasha M (1969) Proc Natl Acad Sci USA 63: 253
56. Tokumura K, Watanabe A, Udagawa M, Itoh M (1987) J Am Chem Soc 109: 1346 and references cited therein
57. Moerner WE (1986) J Molecular Electronics 1: 55
58. Völker S, van der Waals JH (1976) Mol Phys 32: 1703
59. Voelker S, Macfarlane R (1979) IBM Res Develop 23: 547
60. Friedrich J, Haarer D (1984) Ang Chem 96, 96; (1984) Ang Chem Int Ed Engl 23: 113
61. Eigen M (1963) Ang Chem 75: 489; (1964) Ang Chem Int Ed Engl 3: 1
62. Eigen M, De Maeyer L (1963) In Friess SL, Lewis ES, Weissberger A (eds) Techniques of organic chemistry, vol 8, Part 2, New York, Interscience
63. Gettins WJ, Wyn-Jones E (eds) (1979) Techniques and applications of fast reactions in fast solution, D Reidel
64. Gutowsky HS, McCall DW, Slichter CP (1953) J Chem Phys 21: 279
65. Kubo R (1957) Nuovo Cimento Suppl 6: 1063
66. Sack RA (1958) Mol Phys 1: 163
67. Abragam A (1961) The principles of nuclear magnetism, Oxford: Clarendon Press
68. Alexander S (1962) J Chem Phys 37: 971
69. Kaplan JI (1958) J Chem Phys 28: 278
70. Kaplan JI, Fraenkel G (1972) J Am Chem Soc 94: 2907
71. Binsch G (1969) J Am Chem Soc 91: 1304
72. Farrar TC, Becker E (1971) Pulse and fourier transform NMR, New York, Academic Press
73. Ernst RR, Bodenhausen G, Wokaun A (1987) Principles of nuclear magnetic resonance in one and two dimensions, Oxford: Clarendon Press
74. Doddrell DM, Bendall MR, Barron PF, Pegg DT (1979) J Chem Soc Chem Comm 2, 77; Doddrell DM, Barron PF (1980) Field J Org Magn Reson 13: 119
75. Forsén S, Hofmann RA (1963) J Chem Phys 39: 2892; Forsen S, Hofmann RA (1964) J Chem Phys 40: 1189
76. Hofmann RA, Forsén S (1966) J Chem Phys 45: 2049
77. Campbell ID, Dobson CM, Ratcliffe RG, Williams RJP (1978) J Magn Reson 29: 397
78. Morris GA, Freeman R (1978) J Magn Reson 29: 433
79. Jeener J, Meier BH, Bachmann P, Ernst RR (1976) J Chem Phys 71: 4546
80. Hennig J, Limbach HH (1982) J Magn Reson 49: 322
81. Bleich H, Wilde J (1984) J Magn Reson 56: 149
82. Martin GJ, Martin ML (1981) Tetrahedron Lett 22: 3525
83. Pascal RA, Jr, Baum MW, Wagner CK, Rodgers LR (1984) J Am Chem Soc 106: 5377; Pascal RA, Jr, Baum MW, Wagner CK, Rodgers LR, Huang DS (1986) J Am Chem Soc 108 6477
84. Martin GJ, Martin ML (1990) In: NMR Basic Principles and Progress, vol 23, Springer, Berlin Heidelberg New York
85. Grunwald E, see Referece 27, p 103
86. Denisov GS, Bureiko SF, Golubev NS, Tokhadse KG (1980) In: Molecular interactions, Ratajczak H, Orville-Thomas WJ (eds) vol 2, Chichester, Wiley, Chap 2, p 107

87. Limbach HH (1983) The use of NMR spectroscopy in the study of hydrogen bonding in solution, in aggregation processes In: Gormally J, Wyn-Jones E (eds) Elsevier Amsterdam, Chap 16
88. Grunwald E, Fong DW (1972) J Am Chem Soc 94: 7371
89. Gerritzen D, Limbach HH (1981) Ber Bunsenges Phys Chem 85: 527
90. Lankhorst D, Schriever J, Leyte JC (1983) Chem Phys 77: 319
91. Limbach HH, Seiffert W (1974) Ber Bunsenges Phys Chem 78: 532
92. Limbach HH, Seiffert W (1974) Ber Bunsenges Phys Chem 78: 641
93. Halliday JD, Symons EA, Bindner PE (1978) Can J Chem 56: 1470
94. Limbach HH (1979) J Magn Reson 36: 287
95. Lunazzi L, Panciera G (1980) J Chem Soc Perkin 2: 52
96. Borisov EV, Kratsov DN, Peregudov AS, Fedin EI (1980) Izv Akad Nauk SSSR, Ser Khim 2151
97. Limbach HH, Seiffert (1980) W J Am Chem Soc 102: 538
98. Litchman WM (1979) J Am Chem Soc 101, 545
99. Chenon MT, Coupry C, Grant DM, Pugmire R (1977) J Org Chem 42: 659
100. Nesmeyanov AN, Zavelovitch EB, Babin VN, Kochetkova NS, Fedin EI (1976) Tetrahedron 31: 1461
101. Nesmeyanov AN, Babin VN, Zavelovitch EB, Kochetkova NS, Fedin EI (1976) Chem Phys Lett 37: 184
102. Bureiko SF, Denisov GS, Golubev NS, Lange IY (1979) React Kinet Cat Lett 11: 35
103. Limbach HH, Gerritzen D, Seiffert W (1980) Bull Magn Reson 2: 315
104. Gerritzen D, Limbach HH (1980) J Phys Chem 84: 799
105. Limbach HH, Hennig J, Gerritzen D, Rumpel H (1982) Far Discuss Chem Soc 74: 822
106. Gerritzen D, Limbach HH (1984) J Am Chem Soc 106: 869
107. Meschede L, Gerritzen D, Limbach HH (1988) Ber Bunsenges Phys Chem 92: 469
108. Limbach HH, Meschede L, Scherer G (1989) Z Naturforsch 44a, 459
109. Storm CB, Teklu Y (1974) J Am Chem Soc 94: 1745; Storm CB, Teklu Y (1973) Ann NY Acad Sci 206: 631
110. Abraham RJ, Hawkes GE, Smith KM (1974) Tetrahedron Lett 1483
111. Eaton SS, Eaton GR (1977) J Am Chem Soc 99: 160
112. Yeh HJC (1977) J Magn Reson 28: 365
113. Irving CS, Lapidot A (1977) J Chem Soc Chem Comm 184
114. Gust D, Roberts JD (1977) J Am Chem Soc 99: 3637
115. Hennig J, Limbach HH (1979) J Chem Soc Faraday 2 75: 752
116. Stilbs P, Moseley ME (1980) J Chem Soc Faraday 2 76: 729; Stilbs P (1984) J Magn Reson 58: 152
117. Graf F (1979) Chem Phys Lett 62: 291
118. Bren VA, Chernoivanov VA, Konstantinovskii, LE, Nivorozhkin LE, Zhdanov YA, Minkin VI (1980) Dokl Akad Nauk SSSR 251: 1129
119. Hennig J, Limbach HH (1984) J Am Chem Soc 106: 292
120. Otting G, Rumpel H, Meschede L, Scherer G, Limbach HH (1986) Ber Bunsenges Phys Chem 90: 1122
121. Crosswell MJ, Field LD, Harding MM, Sternhell S (1987) J Am Chem Soc 109: 2335
122. Schlabach M, Wehrle B, Limbach H, Bunnenberg E, Knierzinger A, Shu AYL, Tolf BR, Djerassi C (1986) J Am Chem Soc 108: 3856
123. Schlabach M, Rumpel H, Limbach HH (1989) Angew Chem 101: 84; (1989) Ang Chem Int Ed Engl 28: 76; Schlabach M, Scherer G, Limbach HH (1990) J Am Chem Soc (submitted for publication)
124. Schlabach M, Wehrle B, Braun J, Scherer G, Limbach HH (manuscript in preparation)
125. Schlabach M, Braun J, Wehrle B, Limbach HH, Köcher M, Vogel E (manuscript in preparation)
126. Schlabach M, Limbach HH, Bunnenberg E, Knierzinger A, Shu AYL, Tolf BR, Djerassi C (1990) J Am Chem Soc (submitted for publication)
127. Rumpel H, Limbach HH (1989) J Am Chem Soc 111: 5429
128. Rumpel H, Limbach HH, Zachmann G (1989) J Phys Chem 93: 1812
129. Scherer G, Limbach HH (1989) J Am Chem Soc 111: 5946
130. Ady E, Brickmann J (1971) Chem Phys Lett 11: 302
131. Dewar MJS, Merz KM (1985) J Mol Struct (Theochem) 124: 183
132. Holloway KM, Reynolds CH, Merz KM (1989) J Am Chem Soc 111: 3466
133. Sarai A (1981) Chem Phys Lett 83: 50
134. Sarai A (1982) J Chem Phys 76: 5554

135. Sarai A (1984) J Chem Phys 80: 5341
136. Bersuker GI, Polinger VZ (1984) Chem Phys 86: 57
137. Limbach HH, Hennig J (1979) J Chem Phys 71: 3120
138. Limbach HH, Hennig J, Stulz J (1983) J Chem Phys 78: 5432
139. Limbach HH (1984) J Chem Phys 80: 5343
140. Merz KM, Reynolds CH (1988) J Chem Soc Chem Comm 90
141. Butenhoff T, Moore CB (1988) J Am Chem Soc 110: 8336
142. Smedarchina Z, Siebrand W, Zerbetto F (1989) Chem Phys 136: 285
143. Yanabe T, Yamashita K, Kaminoyama M, Koizumi M, Tachibana A, Fukui K (1984) J Phys Chem 88: 1459
144. Svensson P, Bergmann NÅ, Ahlberg P (1989) Z Naturf 44a: 473
145. Prokofiev AI, Bubnov NN, Solovnikov SP, Kabachnik HI (1973) Tetrahedron Lett 2479
146. Prokofiev AI, Masalimov AS, Bubnov NN, Solodnikov SP, Kabachnik MI (1978) Izv Akad Nauk SSSR Ser Khim 310
147. Bubnov NN, Solodnikov SP, Prokofiev AI, Kabachnik MI (1978) Russian Chem Rev 47: 1048
148. Loth K, Graf F, Andris H, Gunthardt HsH (1976) Chem Phys Lett 29: 163
149. Loth K, Graf F, Gunthardt HsH (1976) Chem Phys 13: 95
150. Limbach HH, Gerritzen D (1982) J Chem Soc Far Disc 74: 279
151. Beinhölzl R, Grampp G, Jaenicke W (1989) Z Phys Chem NF 160: 169
152. Hartmann SR, Hahn EL (1952) Phys Rev 128: 2042
153. Pines A, Gibby MG, Waugh IS (1973) J Chem Phys 59: 569
154. Schaefer J, Stejskal EO (1976) J Am Chem Soc 98: 1031
155. Andrew ER, Bradbury A, Eades RG (1958) Nature (London) 182: 1659
156. Andrew ER, Bradbury A, Eades RG (1959) Nature (London) 183: 1802
157. Lowe IJ (1959) Phys Rev Lett 2: 285
158. Haeberlen U (1976) Advan Magn Reson supplement, New York, Academic Press
159. Spiess HW (1978) Rotation of molecules and nuclear spin relaxation In: NMR-Basic Principles and Progress, vol 15, p 55 Springer, Berlin Heidelberg New York
160. Mehring M (1978) High resolution NMR in solids, NMR-Basic Principles and Progress, vol 11, 2nd edn, Springer, Berlin Heidelberg New York
161. Fyfe CA (1983) Solid state NMR for Chemists, Guelph: CFC Press
162. Lyerla JR, Yannoni CS, Fyfe CA (1982) Acc Chem Res 15: 208
163. Szeverenyi NM, Sullivan MJ, Maciel GE (1984) J Magn Reson 56: 149
164. Limbach HH, Wehrle B, Schlabach M, Kendrick RD, Yannoni CS (1988) J Magn Reson 77: 84
165. Schmidt A, Vega S (1987) J Chem Phys 87: 6895
166. Myrrhe PC, Kruger JD, Hammond BL, Lok SM, Yannoni CS, Macho V, Limbach HH, Vieth HM (1984) J Am Chem Soc 106: 6079
167. Limbach HH, Hennig J, Kendrick RD, Yannoni CS (1984) J Am Chem Soc 106: 4059
168. Kendrick RD, Friedrich S, Wehrle B, Limbach HH, Yannoni CS (1985) J Magn Reson 65: 159
169. Limbach HH, Gerritzen D, Rumpel H, Wehrle B, Otting G, Zimmermann H, Kendrick RD, Yannoni CS (1985) In: Photoreaktive Festkörper, Sixl H, Friedrich J, Bräuchle C (eds) Karlsruhe, M Wahl, p 19–43
170. Limbach HH, Wehrle B, Zimmermann H, Kendrick RD, Yannoni CS (1987) J Am Chem Soc 109: 929
171. Limbach HH, Wehrle B, Zimmermann H, Kendrick RD, Yannoni CS (1987) Ang Chem 99: 241; (1987) Ang Chem Int Ed Engl 26: 247
172. Wehrle B, Limbach HH, Köcher M, Ermer O, Vogel E (1987) Ang Chem 99: 914; (1987) Ang Chem Int Ed Engl 26: 934
173. Wehrle B, Zimmermann H, Limbach HH (1987) Ber Bunsenges Phys Chem 91: 941
174. Wehrle B, Limbach HH, Zimmermann H (1988) J Am Chem Soc 110: 7014
175. Meier BH, Storm CB, Earl WL (1986) J Am Chem Soc 108: 6072
176. Frydman L, Olivieri AC, Diaz LE, Frydman B, Morin FG, Mayne CL, Grant DM, Adler AD (1988) J Am Chem Soc 110: 8336; Frydman L, Olivieri AC, Diaz LE, Frydman B, Kustanovich I, Vega S (1989) J Am Chem Soc 111: 7001
177. Frydman L, Olivieri AC, Diaz LE, Valasinas A, Frydman B (1988) J Am Chem Soc 110: 5651
178. Baldy A, Elguero J, Faure R, Pierrot M, Vincent EJ (1985) J Am Chem Soc 107: 5290
179. Smith JAS, Wehrle B, Aguilar-Parrilla, Limbach HH, Foces-Foces M, Xano FH, Elguero J, Baldy A, Pierrot M, Khurshid MMT, Larcombe-McDuall JB (1989) J Am Chem Soc 111: 7304
180. Graf F, Meyer R, Ha TK, Ernst RR (1981) J Chem Phys 75: 2914
181. Meier BJ, Graf F, Ernst RR (1982) J Chem Phys 76: 767 (1982)

182. Nagoka S, Terao T, Imashiro F, Saika A, Hirota N, Hayashi S (1981) Chem Phys Lett 80: 580
183. Nagoka S, Terao T, Imashiro F, Saika A, Hirota N, Hayashi S (1983) J Chem Phys 79: 4694
184. Schajor W, Post H, Grosescu R, Haeberlen U (1983) J Magn Reson 53: 213
185. Benz S, Haeberlen U, Tegenfeldt J (1986) J Magn Reson 66: 125
186. Scheubel W, Zimmermann H, Haeberlen U (1988) J Magn Reson 80: 401
187. Idziak S Pislewski N (1987) Chem Phys 111: 439
188. Jarvie TP, Thayer AM, Millar JM, Pines A (1987) J Phys Chem 91: 2240
189. Agaki T, Imashiro F, Terao T, Hirota H, Hayashi S (1987) Chem Phys Lett 139: 331
190. Jonsen P, Luzar M, Pines A, Mehring (1986) MJ Chem Phys 85: 4873
191. Meier P, Kothe G, Jonsen P, Trekoske M, Pines A (1987) J Chem Phys 87: 6867
192. Bothner-By AA, Stephens RL, Lee JM, Jeanloz RW (1984) J Am Chem Soc 106: 811
193. Bax A, Davis DG (1985) J Magn Reson 63: 207
194. Braunschweiler L, Ernst RR (1983) J Magn Reson 53: 521
195. Davis DG, Bax A (1985) J Am Chem Soc 107: 2820
196. Chambers VMA, Evans EA, Elvidge JA, Jones JR (1978) Tritium Nuclear Magnetic Resonance Spectroscopy, The Radiochemical Centre, Amersham, Bucks, England, Review No 19
197. Bloxsidge JP, Elvidge JA (1983) Progr NMR Spect 16: 99
198. Binsch G, Kleier A, Quantum Chem Progr Exchange, Indiana University, program DNMR2, DNMR3, and DNMR5
199. Van Geet AL (1970) Anal Chem 42: 679
200. Meschede L (1989) Dissertation Freiburg; Meschede L, Limbach HH (1990) J Phys Chem (submitted for publication)
201. Hexem JG, Frey MH, Opella SJ (1982) J Chem Phys 77: 3847
202. Olivieri AC, Frydman L, Diaz LE (1987) J Magn Reson 75: 50
203. Olivieri AC, Frydman L, Grasselli M, Diaz LE (1988) Mag Res Chem 26: 615
204. Swanson SD, Ganapathy S, Bryant RG (1987) J Magn Reson 73: 239
205. Martin GJ, Martin ML, Gouesnard JP (1981) ^{15}N-NMR spectroscopy, In: NMR Basic Principles and Progress, vol 18, Springer, Berlin Heidelberg New York
206. Meyer R, Meier BH, Ernst RR (private communication)
207. Ahlberg P, Davidsson Ö (1986) J Chem Soc Chem Comm 623
208. Ahlberg P (private communication)

The IGLO-Method: Ab-initio Calculation and Interpretation of NMR Chemical Shifts and Magnetic Susceptibilities

Werner Kutzelnigg, Ulrich Fleischer, and Michael Schindler[1]

Lehrstuhl für Theoretische Chemie, Ruhr-Universität Bochum, D-4630 Bochum, FRG

Table of Contents

[1] Present address: Bayer AG D-5090 Leverkusen, Central Research, FRG

NMR Basic Principles and Progress, Vol. 23

In the first part, the theory of molecules in a magnetic field is reviewed with special attention to chemical shifts. The central role of the current density is pointed out. The gauge origin problem and its solution by means of the IGLO method is explained. Various methods for the calculation of chemical shifts are compared and the IGLO method is presented in detail both as far as computations and interpretation of the result are concerned.

The second part is devoted to a systematic study of protons and the nuclei of the first two rows of the periodic system. Rather detailed discussions are given on the contributions of the various localized molecular orbitals (LMOs) to the chemical shifts, on their interpretation and the question of transferability. Comparison between theory and experiment are given for a wealth of molecules, with, on the whole, satisfactory agreement. In the few cases where there is not good agreement, the reasons for this are understood. It is pointed out that in case of large gas-to-liquid shifts only gas-phase data should be compared to the IGLO results. For some selected classes of molecules principal values of the full shift tensors are studied. Several examples are given where IGLO calculations have helped to elucidate molecular structures, mainly for carbocations. Many data are presented here for the first time. At the end a short review on magnetic susceptibilities is given.

Glossary of Acronyms

AO atomic orbital
CHF coupled Hartree–Fock
DIIS 'direct inversion in the iterative space', a method for convergence acceleration of SCF-like methods
DZ 'double zeta', a rather modest AO basis
GIAO gauge-invariant atomic orbital
IGLO individual gauge for localized orbitals
LMO localized molecular orbital
LORG localized orbital, localized origin
MNDO 'modified neglect of differential overlap', a semi-empirical quantum chemical method
MO molecular orbital
SCF 'self-consistent field', essentially synonymous to 'Hartree–Fock'
TMS tetramethylsilane

1 Introduction

Nuclear magnetic resonance (NMR) has turned out to be one of the most powerful methods in structural chemistry. An important part of the information contained in an NMR spectrum are the chemical shifts. Relations between chemical shifts and structure have so far largely been based on empirical rules, but the use of ab-initio calculations of chemical shifts in combination with measured NMR spectra is becoming increasingly important. Especially in the field of carbonium ions the comparison of measured chemical shifts with ab-initio results for various competitive structures has often led to unambiguous elucidations [1] which would not have been possible otherwise. In such cases various possible structures often differ so little in energy, that ab-initio calculations of the structures and their energies [2] do not allow definite conclusions as to the structure realized under the experimental conditions, in particular since these calculations cannot take environmental effects into consideration.

The just-mentioned developments were only possible after efficient methods for the ab-initio calculation of chemical shifts of molecules of chemical interest within a tolerable amount of computer time had become available. Among these methods the IGLO (*i*ndividual *g*auge for *l*ocalized *o*rbitals) approach [3, 4] plays a prominent role, and it appears in order to present this method to a wider audience.

This review consists of two main parts, one dealing with the general theory, the other with a summary of results, both published and unpublished. We have tried, in the general part, to be as elementary as possible and to leave out all complicated aspects of the theory. Readers who want to learn more details are referred to a recent review [5] in which the formal aspects of the IGLO method and other problems of the ab-initio calculation of molecular properties are discussed, and of course, to the original papers [3, 4].

Nevertheless, a minimum of formalism is unavoidable even in a review like the present one in which the emphasis is on applications. In order to appreciate the IGLO calculations one must know a little bit of the physical background, on current densities, on the meaning of the terms 'diamagnetic' and 'paramagnetic', on the vector potential of a magnetic field, and on gauge transformations. The basic concepts of perturbation theory are required as well. This review is aimed at being self-contained. For readers with too little background or for those who are interested in more details textbooks or review papers on the basic theory of molecules in a magnetic field are available [6].

It is certainly not necessary to read and understand the whole of Part 2 before one starts with Part 3. On a first reading one may e.g. skip Sect. 2.4 or Sect. 2.8.

Although this paper is mainly a review, a lot of unpublished results have been included. Among these we mention the study of the effect of hydrogen

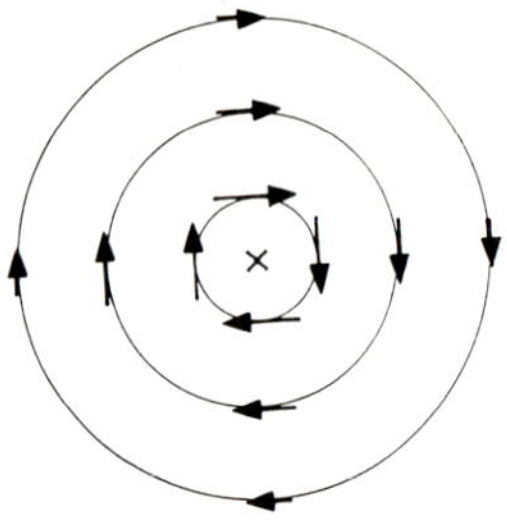

Fig. 2. The current density in a closed-shell atom in a plane that contains the nucleus, perpendicular to the direction of the field

potential of the magnetic field, to which we come back later in this section. All quantities are measured in the Gaussian system of units.

The current density satisfies a so-called continuity equation (which is an immediate consequence of the Schrödinger equation and its conjugate equation)

$$\operatorname{div} \vec{j} = \frac{\partial j_z}{\partial x} + \frac{\partial j_y}{\partial y} + \frac{\partial j_z}{\partial z} = -4\pi \frac{\partial \varrho}{\partial t} \tag{2.2.2}$$

$$\varrho = \psi^* \psi \tag{2.2.3}$$

ϱ is the electron density. For stationary states (we only consider those) ϱ is time-independent and the r.h.s. of (2.2.2) vanishes such that

$$\operatorname{div} \vec{j} = \vec{0} \tag{2.2.4}$$

The current density is hence a source-free vector field.

Current density plots are displayed in Figs. 2, 4, 5, 22, 23.

According to (2.2.1a) $\vec{j}(\vec{r})$ consists of two parts, a 'diamagnetic' contribution defined by (2.2.1b) and a 'paramagnetic' one defined by (2.2.1c). If ψ is a complex wave function it can be written as

$$\psi = \psi_R + i\psi_I \tag{2.2.5}$$

where ψ_R is the real and $i\psi_I$ the imaginary part (both ψ_R and ψ_I being real). The complex conjugate is then

$$\psi^* = \psi_R - i\psi_I \tag{2.2.6}$$

If ψ is real (i.e. $\psi_I \equiv 0$ and $\psi^* = \psi$) the paramagnetic contribution $\vec{j}_p$ (2.2.1c) obviously vanishes. The diamagnetic contribution $\vec{j}_d$ is proportional to $\vec{A}$ times the electron density (2.2.3).

For an n-electron system the expressions for $\vec{j}_d$ and $\vec{j}_p$ are expressible through the one-particle density matrix. If one describes this system by a single Slater-determinant wave function (built up from as many spin-orbitals ψ_k as there are electrons) one has simply to sum over these spin-orbitals

$$\vec{j}_d = -\frac{e^2}{mc} \vec{A} \sum_k \psi_k^* \psi_k \tag{2.2.7}$$

$$\vec{j}_p = \frac{ie\hbar}{2m} \sum_k \{\psi_k^* \operatorname{grad} \psi_k - \psi_k \operatorname{grad} \psi_k^*\} \tag{2.2.8}$$

Since the exact wave function is only approximated by a Slater determinant, the expressions (2.2.7, 8) can only be approximate. However, even the Hartree–Fock current density satisfies (2.2.4), provided that the Hartree–Fock equations are solved exactly [8], which is only possible in the limit of a complete basis.

The vector potential $\vec{A}$ is related to the magnetic field strength $\vec{B}$ as

$$\vec{B} = \mathrm{rot}\,\vec{A} = \mathrm{curl}\ \vec{A} \tag{2.2.9}$$

or written explicitly for the cartesian components

$$B_x = \frac{\partial A_z}{\partial y} - \frac{\partial A_y}{\partial z}$$

$$B_y = \frac{\partial A_x}{\partial z} - \frac{\partial A_z}{\partial x}$$

$$B_z = \frac{\partial A_y}{\partial x} - \frac{\partial A_x}{\partial y} \tag{2.2.10}$$

While $\vec{B}$ is always determined uniquely if $\vec{A}$ is given, the converse is not true. For a homogeneous magnetic field, where $\vec{B}$ is constant and independent of $\vec{r}$, a possible choice of $\vec{A}$ compatible with (2.2.9) is

$$\vec{A} = \tfrac{1}{2}\vec{B} \times \vec{r} \tag{2.2.11}$$

or explicitly

$$\begin{aligned} A_x &= \tfrac{1}{2}(B_y z - B_z y) \\ A_y &= \tfrac{1}{2}(B_z x - B_x z) \\ A_z &= \tfrac{1}{2}(B_x y - B_y x) \end{aligned} \tag{2.2.12}$$

as one can verify by inserting (2.2.11) into (2.2.9) or (2.2.12) into (2.2.10). However another possible choice is

$$\vec{A} = \tfrac{1}{2}\vec{B} \times (\vec{r} - \vec{R}) \tag{2.2.13}$$

with any constant $\vec{R}$. One may hence choose the origin $\vec{R}$ arbitrarily, with respect to which the position vector $\vec{r}$ is measured. We call $\vec{R}$ in (2.2.13) the 'gauge origin'. There is even more freedom in the choice of $\vec{A}$. One can add to $\vec{A}$ the gradient of any scalar function χ without affecting the validity of (2.2.9).

$$\vec{A} \to \vec{A}'' = \vec{A} + \mathrm{grad}\,\chi(\vec{r}) \tag{2.2.14}$$

It namely holds that rot grad $\chi = 0$ for any $\chi(\vec{r})$. The 'gauge function' χ corresponding to a change of the gauge origin $\vec{R} \to \vec{R}'$ is obviously

$$\chi(\vec{r}) = \tfrac{1}{2}\vec{B} \times (\vec{R} - \vec{R}')\cdot\vec{r} \tag{2.2.15}$$

In view of the definition (2.2.1) we may now expect that the current density depends on the choice (gauge) of $\vec{A}$. This is, however, not the case. All physical observables (including the current density) must be gauge-independent (provided that for each gauge the Schrödinger equation is solved exactly, or that one satisfies certain stationarity conditions [8]).

In order to see the gauge independence of $\vec{j}(\vec{r})$ one must note that a change of $\vec{A}$ according to (2.2.14) causes a change of the wave function, that we shall study in the next section. Anyhow, as we shall show, only the total $\vec{j}$ is gauge independent, not the two contributions $\vec{j}_p$ and $\vec{j}_d$ individually. This means that one should avoid a premature physical interpretation of $\vec{j}_p$ and $\vec{j}_d$ separately. Even the names 'paramagnetic' and 'diamagnetic' contributions must be taken with great care.

No gauge problems arise for atoms. We consider, for simplicity's sake, a one-electron atom in an s-state. A 'natural' choice of the gauge origin is, of course, the position of the nucleus. One can show that for this choice the imaginary part of ψ vanishes at least to first order in B. To this order $\vec{j}$ is entirely given by $\vec{j}_d$ and we get

$$\vec{j} = -\frac{e^2}{mc}\vec{A}\psi^*\psi = -\frac{e^2}{2mc}\vec{B}\times\vec{r}\cdot\psi^*\psi \tag{2.2.16}$$

The current density vector $\vec{j}$ is hence orthogonal to the field vector $\vec{B}$ and to the position vector $\vec{r}$, and is proportional to r times the electron density ϱ. For a field in z-direction the $\vec{j}$-vectors lie in planes parallel to the xy- plane. They follow the tangents of concentric circules as illustrated on Fig. 2. This diamagnetic ring current (symbolized as rotating clockwise) has already been postulated in the beginning of this century by Langevin [9].

If one chooses a different (off-center) gauge origin, $\vec{j}_p$ no longer vanishes and $\vec{j}_d$ is changed such that the sum of $\vec{j}_p$ and $\vec{j}_d$ remains invariant. Of course, either contribution becomes very complicated and hard to evaluate, and there is no reason *not* to choose the natural gauge origin.

In molecules, however, there is *no* natural gauge origin and no possible choice of the gauge which makes j_p vanish. On the other hand one will expect that e.g. the inner shell electrons in a molecule behave like in an atom, and that one should choose 'different gauges for different electrons' in order to avoid spurious contributions of opposite sign.

In approximate methods gauge-invariance is not automatically guaranteed. In the Hartree–Fock-scheme it holds e.g. in the limit of a complete basis [8].

2.3 The Schrödinger equation in a magnetic field

The general recipe to get the Schrödinger equation for an atom or a molecule in a magnetic field (often referred to as the principle of 'minimal coupling') consists of replacing the momentum operator $\vec{p} = (\hbar/i)\nabla$ in the following way (in the Gaüssion system of units)

$$\vec{p} \rightarrow \vec{\pi} = \vec{p} + \frac{e}{c}\vec{A} \tag{2.3.1}$$

with $\vec{A}$ the vector potential, related to the magnetic field strength $\vec{B}$ via (2.2.9).

Imposing the 'Coulomb gauge condition' (which is no loss of generality)

$$\operatorname{div}\vec{A} = 0 \tag{2.3.2a}$$

which implies

$$\vec{p}\cdot\vec{A}\psi = (\vec{p}\vec{A})\psi + \vec{A}(\vec{p}\psi) = \vec{A}\vec{p}\psi \tag{2.3.2b}$$

where $(\vec{p}\vec{A})$ or $(\vec{p}\psi)$ means that $\vec{p}$ acts only on $\vec{A}$ or ψ respectively, the Hamiltonian (for a one-electron system with the potential energy V) becomes

$$H = \frac{1}{2m}\pi^2 + V = \frac{1}{2m}p^2 + V + \frac{e}{mc}\vec{A}\cdot\vec{p} + \frac{e^2}{2mc^2}\vec{A}^2 \tag{2.3.3}$$

There is an additional term that describes the interaction of the electron *spin* with the magnetic field. We ignore it since we are only considering systems with zero total spin.

For $\vec{A}$ due to a homogeneous external field we can in view of (2.2.13) rewrite H as

$$H = T + V + \frac{e}{2mc}\vec{B}\cdot\vec{l}_R + \frac{e^2}{8mc^2}\{B^2\cdot(\vec{r}-\vec{R})^2 - [B\cdot(\vec{r}-\vec{R})]^2\} \tag{2.3.4}$$

where T is the kinetic energy and $\vec{l}_R$ the angular momentum operator referring to the origin $\vec{R}$

$$T = \frac{1}{2m}\vec{p}^2; \quad \vec{l}_R = (\vec{r}-\vec{R})\times\vec{p} \tag{2.3.5}$$

In an n-electron system one must sum both T and the $\vec{B}$-dependent terms over all electrons. In going from (2.3.3) to (2.3.4) use has been made of a well-known formula from vector analysis

$$(\vec{a}\times\vec{b})\cdot\vec{c} = \vec{a}\cdot(\vec{b}\times\vec{c}) \tag{2.3.6}$$

We note that H as given by (2.3.4) contains a 'perturbation' linear in the field strength $\vec{B}$ and another one quadratic in $\vec{B}$.

If some of the nuclei have non-zero spin and hence permanent magnetic moments, any of these nuclei gives a contribution

$$\vec{A}_N = \frac{\vec{\mu}\times(\vec{r}-\vec{\varrho})}{|\vec{r}-\vec{\varrho}|^3} \tag{2.3.7}$$

to the vector potential, where $\vec{\mu}$ is the magnetic moment of the nucleus and $\vec{\varrho}$ its position. Other gauges for A_N are possible, but $\vec{\varrho}$ is an obvious natural gauge origin—and we shall never consider a different one. (Note in this respect the diffference to the vector potential (2.2.13) due to an external homogeneous magnetic field.)

Let $\vec{A}_{tot}$ consist of two parts, one like (2.2.13) and the other like (2.3.7), then the Hamiltonian (2.3.4) becomes

$$H = T + V + \frac{e}{mc}(\vec{A}\cdot\vec{p} + \vec{A}_N\cdot\vec{p}) + \frac{e^2}{2mc^2}(\vec{A}^2 + \vec{A}_N^2 + 2\vec{A}\vec{A}_N)$$

$$= H_0 + H_{10} + H_{01} + H_{20} + H_{02} + H_{11} \tag{2.3.8}$$

$$H_0 = T + V \tag{2.3.9a}$$

$$H_{10} = \frac{e}{mc}\vec{A}\cdot\vec{p} = \frac{e}{2mc}\vec{B}\cdot\vec{l}_R \tag{2.3.9b}$$

$$H_{01} = \frac{e}{mc}\vec{A}_N\cdot\vec{p} = \frac{e}{mc}\frac{\vec{\mu}\cdot\vec{l}_\varrho}{|\vec{r}-\vec{\varrho}|^3}; \quad \vec{l}_\varrho = (\vec{r}-\vec{\varrho})\times\vec{p} \tag{2.3.9c}$$

$$H_{20} = \frac{e^2}{2mc^2}\vec{A}^2 = \frac{e^2}{8mc^2}\{B^2(\vec{r}-\vec{R})^2 - [\vec{B}\cdot(\vec{r}-\vec{R})]^2\} \tag{2.3.9d}$$

$$H_{02} = \frac{e^2}{2mc^2}A_N^2 = \frac{e^2}{2mc^2}\frac{[\vec{\mu}\times(\vec{r}-\vec{\varrho})]^2}{|\vec{r}-\vec{\varrho}|^6} \tag{2.3.9e}$$

$$H_{11} = \frac{e}{mc^2}AA_N = \frac{e^2}{2mc^2}\frac{\{\vec{B}\times(\vec{r}-\vec{R})\}\{\vec{\mu}\times(\vec{r}-\varrho)\}}{|\vec{r}-\vec{\varrho}|^3}$$

$$= \frac{e^2}{2mc^2}\frac{(\vec{B}\cdot\vec{\mu})[(\vec{r}-\vec{R})\cdot(\vec{r}-\vec{\varrho})] - [\vec{B}\cdot(\vec{r}-\vec{\varrho})][(\vec{r}-\vec{R})\cdot\vec{\mu}]}{|\vec{r}-\vec{\varrho}|^3} \tag{2.3.9f}$$

There are terms linear and quadratic in $\vec{B}$ or $\vec{\mu}$, but also a term (H_{11}) bilinear in $\vec{B}$ and $\vec{\mu}$.

2.4 Gauge Transformations

A reader less interested in the mathematics may skip this section.

Let us apply to the Hamiltonian (2.3.4) a unitary transformation with the operator U

$$U = e^{-i\Lambda}; \quad U^\dagger = e^{i\Lambda}; \quad \Lambda = \Lambda(\vec{r}); \quad U^\dagger U = 1 \tag{2.4.1}$$

with Λ a real local (and hence hermitean) operator

$$H \to H' = U^\dagger H U \tag{2.4.2}$$

Such a unitary transformation leaves the eigenvalues E of H invariant, but modifies the eigenfunctions, such that the Schrödinger equation is changed in the following way

$$H\psi = E\psi \rightarrow H'\psi' = E\psi'$$
$$\psi \rightarrow \psi' = U^{\dagger}\psi = e^{-i\Lambda}\psi; \quad \psi^* \rightarrow \psi'^* = e^{i\Lambda}\psi^* \tag{2.4.3}$$

By elementary manipulations using the Hausdorff formula

$$e^{-X}He^{X} = H + [H, X] + \frac{1}{2!}[[H, X], X] + \cdots. \tag{2.4.4}$$

the commutation relation between p_x and x

$$[p_x, x] = p_x x - x p_x = \frac{\hbar}{i} \tag{2.4.5}$$

and the fact that two local operators commute, one gets

$$H' = T + V + \frac{e}{mc}(\vec{A} + \frac{\hbar c}{e}\operatorname{grad}\Lambda)\cdot\vec{p} + \frac{e^2}{2m^2}(\vec{A} + \frac{\hbar c}{e}\operatorname{grad}\Lambda)^2 \tag{2.4.6}$$

i.e. the transformed Hamiltonian H′ is just equal to the original Hamiltonian with a different gauge of the vector potential

$$\vec{A} \rightarrow \vec{A}' = A + \frac{\hbar c}{e}\operatorname{grad}\Lambda \tag{2.4.7}$$

A shift of the gauge origin is described by

$$\Lambda = \frac{e}{2\hbar c}\vec{B} \times (R' - \vec{R})\vec{r} \tag{2.4.8}$$

Let us now take the definition (2.2.1) of the current density and insert (2.4.7) and (2.4.3). We get

$$\begin{aligned}
\vec{j}_p \rightarrow \vec{j}'_p &= \frac{ie\hbar}{2m}[e^{i\Lambda}\psi^* \operatorname{grad}(e^{-i\Lambda}\psi) - e^{-i\Lambda}\psi \operatorname{grad}(e^{i\Lambda}\psi^*)] \\
&= \frac{ie\hbar}{2m}\{e^{i\Lambda}\psi^*[-i(\operatorname{grad}\Lambda)\psi + \operatorname{grad}\psi]e^{-i\Lambda} \\
&\quad - e^{-i\Lambda}\psi[i(\operatorname{grad}\Lambda)\psi^* + \operatorname{grad}\psi^*]e^{i\Lambda}\} \\
&= \vec{j}_p + \frac{e\hbar}{m}\psi^*\psi \operatorname{grad}\Lambda
\end{aligned} \tag{2.4.9}$$

$$\vec{j}_d \rightarrow \vec{j}'_d = -\frac{e^2}{mc}\left[\vec{A} + \frac{\hbar c}{e}\operatorname{grad}\Lambda\right]\psi^*\psi = \vec{j}_d - \frac{e\hbar}{m}\psi^*\psi \operatorname{grad}\Lambda \tag{2.4.10}$$

$$\vec{j} \rightarrow \vec{j}' = \vec{j}'_p + \vec{j}'_d = \vec{j}_p + \vec{j}_d = \vec{j} \tag{2.4.11}$$

The current density is hence gauge invariant, but the components $\vec{j}_p$ and $\vec{j}_d$ are not.

2.5 Perturbation Theory

We consider a molecule in a static external magnetic field. Some or all of the nuclei may have non-zero spin and hence permanent magnetic moments $\vec{\mu}(K)$, which also affect the eigenstates of the Hamiltonian. One has hence to deal with several perturbations simultaneously. The energy depends on the vector $\vec{B}$ of the external field and the vector $\vec{\mu}(K)$. The $\alpha\beta$-component of the nuclear magnetic shielding (chemical shift) tensor of the K-th nucleus is formally defined as

$$\sigma_{\alpha\beta}(K) = \frac{\partial^2 E}{\partial \mu_\alpha(K) \partial B_\beta}; \quad \alpha, \beta = x, y, z \tag{2.5.1}$$

In (2.5.1) α, β are any two of the cartesian coordinates. The nine components of σ consitute the shift tensor

$$\boldsymbol{\sigma} = \begin{pmatrix} \sigma_{xx} & \sigma_{xy} & \sigma_{xz} \\ \sigma_{yx} & \sigma_{yy} & \sigma_{yz} \\ \sigma_{zx} & \sigma_{zy} & \sigma_{zz} \end{pmatrix} \tag{2.5.2}$$

$$\chi_{\alpha\beta} = \frac{\partial^2 E}{\partial B_\alpha \partial B_\beta}; \qquad \boldsymbol{\chi} = \begin{pmatrix} \chi_{xx} & \chi_{xy} & \chi_{xz} \\ \chi_{yx} & \chi_{yy} & \chi_{yz} \\ \chi_{zx} & \chi_{zy} & \chi_{zz} \end{pmatrix} \tag{2.5.3}$$

Note that $\boldsymbol{\chi}$ is a symmetric tensor ($\chi_{\alpha\beta} = \chi_{\beta\alpha}$), while $\boldsymbol{\sigma}$ is not ($\sigma_{\alpha\beta} \neq \sigma_{\beta\alpha}$).

$\boldsymbol{\sigma}$ and $\boldsymbol{\chi}$ arise as the coefficients of the expansion of E in powers of μ_α and B_β

$$E = E_0 - \frac{1}{2} \sum_{\alpha,\beta} B_\alpha \chi_{\alpha\beta} B_\beta + \sum_K \sum_{\alpha,\beta} \mu_\alpha(K) \sigma_{\alpha,\beta}(K) B_\beta + \sum_{\alpha,\beta,\gamma} \cdots \tag{2.5.4}$$

The terms linear in B_α or μ_α vanish for a molecule in a non-degenerate state and the terms of higher order in B_α or μ_β are (for the field strengths used in practice) so small that they can safely be neglected.

$\boldsymbol{\sigma}(K)$ is, of course, a property related to the K-th nucleus, while $\boldsymbol{\chi}$ is a property of the entire molecule.

One can understand pictorally the mechanism that leads to $\boldsymbol{\sigma}$ and $\boldsymbol{\chi}$. The external field $\vec{B}$ creates a current density in the molecule. This current density gives rise to another magnetic field. The nuclei 'feel' this induced field in addition to the external field, i.e. a resulting field different from the pure external field. This leads to $\boldsymbol{\sigma}$. The magnetic moment of the induced current density also interacts with the external field, which is the origin of $\boldsymbol{\chi}$.

The standard way to evaluate $\chi_{\alpha\beta}$ or $\sigma_{\alpha\beta}$ is perturbation theory, more precisely, multiple perturbation theory, because there are several perturbations (even two components B_α and B_β of $\vec{B}$ may be regarded as two independent perturbations).

Let us briefly summarize the formalism of double perturbation theory that is needed for our purposes. We start from a Hamiltonian in the form

$$H = H_0 + \kappa V_{10} + \lambda V_{01} + \kappa^2 V_{20} + \kappa\lambda V_{11} + \lambda^2 V_{02} \tag{2.5.5}$$

where κ and λ are parameters that measure the perturbations (in our case κ may e.g. be B_x and λ may represent μ_y or B_y. We shall later come to the explicit expressions for V_{10}, V_{01} etc. Note that the 1st subscript counts orders with respect to the 1st perturbation, the 2nd subscript to the second perturbation. One expands the wave function ψ and the energy in analogy to (2.5.5)

$$\psi = \psi_0 + \kappa\psi_{10} + \lambda\psi_{01} + \kappa^2\psi_{20} + \kappa\lambda\psi_{11} + \cdots \tag{2.5.6a}$$

$$E = E_0 + \kappa E_{10} + \lambda E_{01} + \kappa^2 E_{20} + \kappa\lambda E_{11} + \cdots \tag{2.5.6b}$$

with the only difference that the expansion of H stops after the term in λ^2, while those of ψ and E continue, they are infinite power series expansions. Insertion of (2.5.5, 6) into the Schrödinger equation and collecting terms of the same powers in κ and λ leads to

$$(H_0 - E_0)\psi_0 = 0 \tag{2.5.7a}$$

$$(V_{10} - E_{10})\psi_0 + (H_0 - E_0)\psi_{10} = 0 \tag{2.5.7b}$$

$$(V_{01} - E_{01})\psi_0 + (H_0 - E_0)\psi_{01} = 0 \tag{2.5.7c}$$

$$(V_{20} - E_{20})\psi_0 + (V_{10} - E_{10})\psi_{10} + (H_0 - E_0)\psi_{20} = 0 \tag{2.5.7d}$$

$$(V_{02} - E_{02})\psi_0 + (V_{01} - E_{01})\psi_{01} + (H_0 - E_0)\psi_{02} = 0 \tag{2.5.7e}$$

$$\begin{aligned}(V_{11} - E_{11})\psi_0 &+ (V_{10} - E_{10})\psi_{01} \\ &+ (V_{01} - E_{01})\psi_{10} + (H_0 - E_0)\psi_{11} = 0\end{aligned} \tag{2.5.7f}$$

and analogous terms that involve higher orders in κ or λ. Of course, (2.5.7a) is the ordinary (unperturbed) Schrödinger equation, that has to be solved first. One then gets by some rather elementary manipulations of (2.5.7) that

$$E_{10} = \langle\psi_0|V_{10}|\psi_0\rangle \tag{2.5.8a}$$

$$E_{01} = \langle\psi_0|V_{01}|\psi_0\rangle \tag{2.5.8b}$$

i.e. that the 'first order corrections' E_{10} and E_{01} to the energy are simply expectation values calculated with the unperturbed wave function ψ_0. The 2nd order energies are given as

$$E_{20} = \langle\psi_0|V_{10} - E_{10}|\psi_{10}\rangle + \langle\psi_0|V_{20}|\psi_0\rangle \tag{2.5.9a}$$

$$\begin{aligned}E_{11} &= \langle\psi_0|V_{10} - E_{10}|\psi_{01}\rangle + \langle\psi_0|V_{11}|\psi_0\rangle \\ &= \langle\psi_0|V_{01} - E_{01}|\psi_{10}\rangle + \langle\psi_0|V_{11}|\psi_0\rangle\end{aligned} \tag{2.5.9b}$$

$$E_{02} = \langle\psi_0|V_{01} - E_{01}|\psi_{01}\rangle + \langle\psi_0|V_{02}|\psi_0\rangle \tag{2.5.9c}$$

For their evaluation ψ_{10} or ψ_{01} are needed that have to be obtained from solution of (2.5.7b) or (2.5.7c). One sees two important things from (2.5.9). Firstly, each of the 2nd order energies consists of two terms, one that involves a first order operator (V_{10} or V_{01}) and a first order wave function (ψ_{10} or ψ_{01}) and another term that is the expectation value of a 2nd order operator (V_{20}, V_{11} or V_{02}) with the unperturbed wave function. Secondly, the 'mixed' 2nd order energy E_{11} can be evaluated in two independent ways, either using V_{10} and ψ_{01} or V_{01} and ψ_{10}.

The equivalence of the two alternative expression (2.5.9b) is often referred to as the exchange theorem of double perturbation theory [10]. It can be interpreted pictorially for our example. E_{11}, which is essentially the chemical shift, can be *either* understood as the interaction of the magnetic field due to the current density that has been induced by the external field, with the magnetic moment of the nucleus, *or* as the interaction of the current density induced by the nuclear magnetic moment with the external field.

In the case of magnetic perturbations one will usually solve Eq. (2.5.7b) for the external magnetic field as perturbation V_{10} and get the first order wave function ψ_{10} due to the external magnetic field. The various E_{11} can then be calculated for the perturbations V_{01} due to magnetic fields of all the nuclei, using always the same ψ_{10}.

In other words, we construct first the current density induced by the external magnetic field and then study its interaction with the various nuclei.

One should note that the simple expressions (2.5.8, 9) only hold if the 'unperturbed Schrödinger equation' (2.5.7a) has been solved exactly. This is usually not possible. However, one can show that the expressions (2.5.8, 9) are meaningful even in terms of approximate wave functions, provided that certain stationarity conditions for the energy expection value $\langle \psi_0 | H | \psi_0 \rangle$ are satisfied [5]. These hold e.g. if ψ_0 is obtained from Hartree–Fock theory.

By comparison of (2.3.9) with (2.5.5) we easily find the explicit expressions for V_{10} etc. Let us choose

$$\kappa = B_\alpha, \quad \lambda = \mu_\beta; \qquad \alpha, \beta = x, y, z \tag{2.5.10}$$

then

$$V_{10} = \frac{e}{2mc} l_{R\alpha} \tag{2.5.11a}$$

$$V_{01} = \frac{e}{mc} \frac{l_{\varrho\beta}}{|\vec{r} - \vec{\varrho}|^3} \tag{2.5.11b}$$

$$V_{20} = \frac{e^2}{8mc^2} \{(\vec{r} - \vec{R})^2 - (r_\alpha - R_\alpha)^2\} \tag{2.5.11c}$$

$$V_{11} = \frac{e^2}{2mc^2} \frac{\delta_{\alpha\beta}(\vec{r} - \vec{R})(\vec{r} - \vec{\varrho}) - (r_\alpha - \varrho_\alpha)(r_\beta - R_\beta)}{|\vec{r} - \vec{\varrho}|^3} \tag{2.5.11d}$$

where the subscripts α and β designate the respective components (x, y, z) of the indicated vectors. $l_{\vec{R}\alpha}$ and $l_{\vec{\varrho}\beta}$ are components of the angular moment operator referring to the origins $\vec{R}$ and $\vec{\varrho}$ respectively (see Eq. (2.3.5)). Note that although we take the external field as the 1st perturbation, in the definition (2.5.1) of the shift tensor it is formally the 2nd perturbation.

2.6 The Gauge Origin Problem

The Hamiltonian H is defined in terms of the vector potential $\vec{A}$, this means that H depends on the gauge of $\vec{A}$. Physical observables like the energy E or the current density $\vec{j}$ must be gauge independent. We have already seen in Sect. 2.4 that the total $\vec{j}$ is gauge independent, but that either of the two contributions $\vec{j}_d$ and $\vec{j}_p$ depends on the gauge. One sees easily that of the two contributions to E_{20} according to (2.5.9a) the first corresponds to $\vec{j}_p$ and the second to $\vec{j}_d$. They are hence also referred to as paramagnetic and diamagnetic contributions to E_{20}, respectively (E^d_{20} is always positive, E^p_{20} always negative, which explains the names diamagnetic and paramagnetic respectively)

$$E_{20} = E^d_{20} + E^p_{20} \tag{2.6.1}$$

$$E^d_{20} = \langle \psi_0 | V_{20} | \psi_0 \rangle; \qquad E^p_{20} = \langle \psi_0 | V_{10} | \psi_{10} \rangle \tag{2.6.2}$$

Similarly E_{11} as given by (2.5.9b) can be decomposed as

$$E_{11} = E^d_{11} + E^p_{11} \tag{2.6.3}$$

$$E^d_{11} = \langle \psi_0 | V_{11} | \psi_0 \rangle; \qquad E^p_{11} = \langle \psi_0 | V_{01} | \psi_{10} \rangle \tag{2.6.4}$$

It turns out that only E_{20} and E_{11} are gauge-independent, while E^d_{20}, E^p_{20}, E^d_{11}, E^p_{11} individually are not. Moreover, even E_{20} and E_{11} are gauge independent *only* if ψ_0 and ψ_{10} are determined exactly, which is usually not possible. In coupled Hartree–Fock E_{20} and E_{11} are gauge independent in the limit of a complete basis [8].

Practically E_{20} and E_{11} are hence gauge dependent, which is very inconvenient. We have seen in Sect. 5 that there is a 'natural' gauge origin for atoms, namely the position of the nucleus, and that for this choice the paramagnetic current vanishes. Consistently also E^p_{20} and E^p_{11} vanish for this natural gauge, such that E_{20} and E_{11} are evaluated rather easily. For spherically symmetric atoms the results are for any of the free diagonal elements ($\alpha = \beta$, with $\alpha, \beta = x, y, z$; k counts the electrons)

$$E_{20} = E^d_{20} = \langle \psi_0 | V_{20} | \psi_0 \rangle = \frac{e^2}{12mc^2} \left\langle \psi_0 \left| \sum_{k=1}^{m} r_k^2 \right| \psi_0 \right\rangle \tag{2.6.5}$$

$$E_{11} = E^d_{11} = \langle \psi_0 | V_{11} | \psi_0 \rangle = \frac{e^2}{3mc^2} \left\langle \psi_0 \left| \sum_{k=1}^{n} \frac{1}{r_k} \right| \psi_0 \right\rangle \tag{2.6.6}$$

For a one-electron atom the sum over k reduces to a single term and ψ_0 becomes a one-electron function. Let us now choose an 'unnatural' gauge origin $\vec{R} = (R, 0, 0)$ with $\vec{0} = (0, 0, 0)$ the position of the nucleus. Then we get for the diamagnetic contributions to E_{20} and E_{11}

$$E^{d}_{20} = \frac{e^2}{12mc^2}\langle \psi_0 | r^2 | \psi_0 \rangle + \frac{e^2}{8mc^2} R^2 \tag{2.6.7}$$

$$E^{d}_{11} = \frac{e^2}{3mc^2}\left\langle \psi_0 \left| \frac{1}{r} \right| \psi_0 \right\rangle + \frac{e^2}{2mc^2} R^2 \left\langle \psi_0 \left| \frac{1}{r^3} \right| \psi_0 \right\rangle \tag{2.6.8}$$

The paramagnetic contributions, when evaluated exactly, become

$$E^{p}_{20} = -\frac{e^2}{8mc^2} R^2 \tag{2.6.9}$$

$$E^{p}_{11} = -\frac{e^2}{2mc^2} R^2 \left\langle \psi_0 \left| \frac{1}{r^3} \right| \psi_0 \right\rangle \tag{2.6.10}$$

such that E_{20} and E_{11} remain invariant. However, the original expressions for E_{20} and E_{11} (see 2.5.9, 11) look rather different

$$E^{p}_{20} = \langle \psi_0 | V_{10} - E_{10} | \psi_{10} \rangle = \frac{e}{2mc} \langle \psi_0 | l_{R\alpha} | \psi_{10} \rangle \tag{2.6.11}$$

$$E^{p}_{11} = \langle \psi_0 | V_{01} - E_{01} | \psi_{10} \rangle = \frac{e}{mc} \left\langle \psi_0 \left| \frac{l_\beta}{r^3} \right| \psi_{10} \right\rangle \tag{2.6.12}$$

with ψ_{10} solution of (see 2.5.7b and 2.5.11)

$$\frac{e}{2mc} l_{R\alpha} \psi_0 + (H_0 - E_0)\psi_{10} = 0 \tag{2.6.13}$$

In order to show the equivalence between (2.6.11, 12) and (2.6.9, 10) some manipulations are necessary [3, 5] which hold for the exact ψ_0 and ψ_{10} and which involve commutator relations between $\vec{r}$ and $\vec{p}$. If one does not solve (2.6.13) exactly, the computed E^{p}_{20} and E^{p}_{11} will differ from the exact expressions (2.6.9, 10) and will not compensate (2.6.7, 8)—which are easily evaluated to high accuracy (since they are just expectation values) — to yield the correct expressions (2.6.5, 6).

Use of an 'unnatural' gauge origin hence causes substantial errors, because E^{d}_{20} and E^{d}_{11} are obtained with high accuracy, but E^{p}_{20} and E^{p}_{11} with low accuracy (unless one uses very large basis sets, such that one can come close to the exact solution of (2.6.13)) and the errors do not compensate each other.

As mentioned in Sect. 2.2 there is no natural gauge origin for molecules. However, it is possible, to choose different gauge origins for different orbitals, as is done in the IGLO method (see Sect. 2.7).

2.7 Classification of Methods for the Ab-initio Calculation of Chemical Shifts

All available standard methods are of coupled-Hartree–Fock (CHF) type [11] (see Sect. 2.13 for methods beyond CHF). This means that the wave function is chosen as a single Slater determinant built up from n molecular orbitals (MOs) each doubly occupied (with α and β spin) for a 2n-electron system. The MOs depend on the external magnetic field in such a way that the stationarity condition for the energy

$$\delta\left\{\int \phi^* \mathrm{H}\phi \,\mathrm{d}\tau \Big/ \int \phi^* \phi \,\mathrm{d}\tau\right\} = 0 \tag{2.7.1}$$

holds in the presence of the magnetic field, independently of its strength.

In the conventional CHF approach [12] a common gauge origin is taken for all MOs. Since such a common gauge origin implies a poor description for nearly all atoms in a molecule (with the exception of small hydrides like CH_4 for which the gauge origin at the C atom is accetable), spurious diamagnetic contributions (see Sect. 2.6) have to be cancelled by corresponding paramagnetic contributions. In order to obtain the latter with sufficient accuracy, rather large basis sets are required (exact gauge invariance only holds for a complete one-electron basis). This makes the conventional CHF method very time-consuming and limits it to rather small molecules.

The GIAO method [13] is a variant of CHF. Its main idea goes back to F. London [14]. In this approach one does not expand an MO φ_i in terms of the AOs χ_μ

$$\varphi_i = \sum_\mu c_{i\mu}\chi_\mu \tag{2.7.2}$$

as in conventional CHF, but rather in terms of gauge-transformed AOs

$$\varphi_i = \sum_\mu \tilde{c}_{i\mu}\tilde{\chi}_\mu \tag{2.7.3a}$$

$$\tilde{\chi}_\mu(\vec{r}) = e^{-i\Lambda_\mu(\vec{r})}\chi_\mu(\vec{r}) \tag{2.7.3b}$$

$$\Lambda_\mu = \frac{e}{\hbar c}\vec{B} \times (\vec{R} - \vec{R}_\mu)\cdot\vec{r} \tag{2.7.3c}$$

where $\vec{R}_\mu$ is the origin (position of the nucleus) of the AO χ_μ. Each AO has thus its own (natural) gauge origin and cancellations of spurious diamagnetic and paramagnetic terms are avoided. Much smaller basis sets than with standard CHF are sufficient. However, the evaluation of the necessary integrals (mainly those of the electron–electron interaction) over the gauge-transformed AOs $\tilde{\chi}_\mu$ becomes much more complicated and computer-time consuming, such that again the method is limited to relatively small molecules. Unpublished results by P. Pulay [15] indicate that it is possible to speed up GIAO calculations considerably by using for the integral evaluation the knowhow of gradient programs.

In the IGLO (*i*ndividual *g*auge for *l*ocalized *o*rbitals) method [3, 4] one starts with a Hartree-Fock calculation in the conventional basis $\{\chi_\mu\}$, i.e. one only needs the integrals over the χ_μ. One then transforms the canonical MOs, as they usually result from a Hartree–Fock calculation, to localized MOs (LMOs) [16]

$$\varphi_k \rightarrow \psi_k = \sum_l U_{kl}\varphi_l \tag{2.7.4}$$

where U_{kl} is a unitary matrix. In the case of CH_4 one gets e.g. one LMO that is essentially equal to the 1s-AO of C and four LMOs that are localized in each of the four CH-bonds.

Now for each LMO an individual gauge origin (usually at the centroid of charge of this LMO) is chosen and the CHF calculation is performed for these gauge origins. Like in GIAO the cancellation of large spurious diamagnetic and paramagnetic terms is avoided. The main difference is that in IGLO all AOs within one MO have the same gauge factor (like 2.4.3), they are so to say, in phase, while in GIAO the gauge factor depends on the AO, different AOs within one MO have hence different phases.

If one chooses individual gauges for different MOs, new coupling terms are introduced that were not present in conventional CHF. Some of these, like the 'resonance corrections' (2.8.12g, h) are easily evaluated, others like the 'exchange corrections' (2.8.4, 2.8.12e. f) are difficult to compute exactly, but turn out to be very small, such that its sufficient to construct them in a simple but consistent approximate way (for details see Ref. 3, 4). A similar simplification of the GIAO method does not appear to be possible. This is one of the reasons why IGLO wins over conventional CHF as well as over GIAO. As seen in this review IGLO has no serious difficulties with 'real' chemical molecules, up to the size of the pagodane dication $C_{20}H_{20}^{2+}$.

A method proposed by Okninski and Sadlej [17a], which has never been applied numerically, bears some resemblences to IGLO, although it has been formulated in a different philosophy. The idea of different gauges within a molecule has also been considered by Maestro and Moccia [17b], by Levy and Ridard [17c] and finally by Hansen and Bouman [17d].

These authors [17d] gave the name LORG (*l*ocalized *or*bital *l*ocal origin) to their method and admitted that it has many similarities with IGLO.

An analysis of the difference between IGLO and LORG has only been given recently [5]. This difference is subtle but significant. In IGLO the first order perturbation corrections to the orbitals are obtained from a stationarity condition, while LORG does not satisfy such a condition. This makes the LORG equations simpler but less consistent. While IGLO automatically furnishes the magnetic suceptibilities together with the chemical shifts this is neither the case for LORG nor for GIAO. It also appears that LORG converges somewhat more slowly to the basis set limits than does IGLO. A numerical comparison of IGLO and LORG calculations has recently been performed [38b]. The computer times indicated there refer to a rather old IGLO program and are not representative for the performance of the IGLO program available now.

2.8 The Working Equations of IGLO

The IGLO program [4] starts with a conventional closed shell Hartree Fock-calculation in a gaussian basis (see Sect. 2.10). The MOs are, according to the criterion of Boys [16], transformed to localized MOs $\psi_k \equiv \psi_{k0}$ ($k = 1, n$ if $2n$ is the number of electrons). The 1st order perturbation corrections to the MOs are then $i\psi_{k1}$ with ψ_{k1} determined from

$$\{(1 - P_0)\tilde{F}_{k1} - \tilde{P}_{k1}F_0\}\psi_{k0} + (1 - P_0)F_0\psi_{k1} = 0 \tag{2.8.1}$$

In (2.8.1) F_0 is the conventional closed-shell-Fock operator and P_0 the projector to the unperturbed states ψ_{k0}

$$F_0 = h_0 + \sum_{j=1}^{n} (2J_0^j - K_0^j) \tag{2.8.2}$$

$$P_0 = \sum_{j=1}^{n} |\psi_{j0}\rangle\langle\psi_{j0}| \tag{2.8.3}$$

$$\tilde{F}_{k1} = \tilde{h}_{k1} - \sum_{j=1}^{n} \{\tilde{K}_1^j + [K_0^j, \Lambda_j - \Lambda_k]\} \tag{2.8.4}$$

$$\tilde{P}_{k1} = \sum_{j=1}^{n} \{|\varphi_{j1}\rangle\langle\psi_{j0}| - |\psi_{j0}\rangle\langle\psi_{j1}| + [\psi_{j0}\rangle\langle\psi_{j0}, \Lambda_k - \Lambda_j]\} \tag{2.8.5}$$

Here h_0 is the unperturbed one-electron Hamiltonian, J_0^j and K_0^j are contributions of the jth MO ψ_{j0} to the ordinary Coulomb and exchange operators, $\tilde{K}_1^j$ is the corresponding first-order correction. It is defined by its matrix elements

$$\langle a|K_1^j|b\rangle = \int a(1)\psi_{j1}(1)\frac{1}{r_{12}}\psi_{j0}(2)b(2)d\tau_1 d\tau_2$$
$$- \int a(1)\psi_{j0}(1)\frac{1}{r_{12}}\psi_{j1}(2)b(2)d\tau_1 d\tau_2 \tag{2.8.6}$$

$i\tilde{h}_{k1}$ is the first order one-electron operator due to the external magnetic field at the gauge corresponding to the centroid $\vec{R}_k$ of the charge of ψ_{k0}

$$\tilde{h}_{k1} = \frac{e}{2mci}\vec{B} \times (\vec{r} - \vec{R}_k)\cdot\vec{p}$$
$$= \frac{e}{2mci}\vec{B}\cdot\vec{l}_k; \quad \vec{l}_k = (\vec{r} - \vec{R}_k) \times \vec{p} \tag{2.8.7}$$

and Λ_k is the gauge function

$$\Lambda_k(\vec{r}) = \frac{e}{2c\hbar}(\vec{R}_k \times \vec{B})\cdot\vec{r} \tag{2.8.8}$$

Since $\tilde{F}_{k1}$ and $\tilde{P}_{k1}$ depend via (2.8.4) and (2.8.5) on the ψ_{j1}, Eq. (2.8.1) must be solved iteratively. To speed up the convergence the DIIS method of Pulay [18] is applied.

From ψ_{k1} the chemical shifts are evaluated as

$$\sigma = \sigma^d + \sigma^{p0} + \sigma^{p1} \tag{2.8.9}$$

$$\sigma^d = 2\sum_{k=1}^{n} \langle \psi_{k0}|h'_{k2}|\psi_{k0}\rangle \tag{2.8.10a}$$

$$\sigma^{p0} = -4\sum_{j,k=1}^{n} \langle \psi_{k0}|h_{01}|\psi_{j0}\rangle\langle \psi_{j0}|\Lambda_j - \Lambda_k|\psi_{k0}\rangle \tag{2.8.10b}$$

$$\sigma^{p1} = -4\sum_{k=1}^{n} \langle \psi_{k0}|h_{01}(1-P_0)|\psi_{k1}\rangle \tag{2.8.10c}$$

with

$$h_{01} = \frac{e}{mci}\frac{\vec{\mu}\times(\vec{r}-\vec{\varrho})}{|\vec{r}-\vec{\varrho}|^3}\cdot\vec{p} = \frac{e}{mci}\frac{\vec{\mu}\cdot\vec{l}_\varrho}{|\vec{r}-\vec{\varrho}|^3};\quad \vec{l}_\varrho = (\vec{r}-\vec{\varrho})\times\vec{p} \tag{2.8.10d}$$

$$h'_{k2} = h_{11} - [h'_1, \Lambda_k] = \frac{e^2}{2mc^2}\frac{\{\vec{B}\times(\vec{r}-\vec{R}_k)\}\{\vec{\mu}\times(\vec{r}-\vec{\varrho})\}}{|\vec{r}-\vec{\varrho}|^3} \tag{2.8.10e}$$

In order to get the full σ tensor, Eq. (2.8.1) has to be solved for the non-equivalent cartesian components of $\vec{B}$. The evaluation of the magnetic susceptibility is straightforward as well, but a little more complicated. One gets (again for one tensor component)

$$E_2 = E^d + E^{p0} + E^{p1} + E^{x0} + E^{x1} + E^{r0} + E^{r1} \tag{2.8.12a}$$

$$E^d = 2\sum_k \langle \psi_{k0}|\tilde{h}_{2k}|\psi_{k0}\rangle \tag{2.8.13b}$$

$$E^{p0} = -2\sum_{k,l} \langle \psi_{k0}|\tilde{h}_{k1}|\psi_{l0}\rangle\langle \psi_{l0}|\Lambda_l - \Lambda_k|\psi_{k0}\rangle \tag{2.8.12c}$$

$$E^{p1} = -2\sum \langle \psi_{k0}|\tilde{h}_{k1}(1-P_0)|\psi_{k1}\rangle \tag{2.8.12d}$$

$$E^{x0} = \frac{1}{2}\sum_{k,l} \langle \psi_{k0}|[[K_0^l, \Lambda_k - \Lambda_l], \Lambda_k - \Lambda_l]|\psi_{k0}\rangle + \sum_{k,l,m} \langle \psi_{k0}|2[K_0^m, \Lambda_m - \Lambda_k]|\psi_{l0}\rangle\langle \psi_{l0}|\Lambda_k - \Lambda_l|\psi_{k0}\rangle \tag{2.8.12e}$$

$$E^{x1} = 2\sum_{k,l} \langle \psi_{k0}|[K_0^l, \Lambda_k - \Lambda_l](1-P_0)|\psi_{k1}\rangle - \sum_{k,l,m} \langle \psi_{k0}|K_1^m|\psi_{l0}\rangle\langle \psi_{l0}|\Lambda_k - \Lambda_l|\psi_{k0}\rangle \tag{2.8.12f}$$

$$E^{r0} = \sum_{k,l} \langle \psi_{l0} | F_0 | \psi_{k0} \rangle \langle \psi_{k0} | (\Lambda_l - \Lambda_k)^2 | \psi_{l0} \rangle$$

$$+ 2 \sum_{k,l,m} \langle \psi_{l0} | F_0 | \psi_{k0} \rangle \langle \psi_{k0} | \Lambda_l - \Lambda_k | \psi_{m0} \rangle \langle \psi_{m0} | \Lambda_m - \Lambda_l | \psi_{l0} \rangle \tag{2.8.12g}$$

$$E^{r1} = 2 \sum_{k,l} \langle \psi_{l0} | F_0 | \psi_{k0} \rangle \langle \psi_{k0} | (\Lambda_l - \Lambda_k)(1 - P_0) | \psi_{l1} \rangle \tag{2.8.12h}$$

It should be noted that all matrix elements are evaluated exactly, only the commutators involving $\Lambda_k - \Lambda_j$ like (2.8.4) and (2.8.12e, f) are calculated by means of a completeness insertion. (This introduces small errors into these matrix elements, but they are negligible, except for very small basis sets, e.g. those of DZ type).

It is not always necessary to calculate all 9 tensor components of $\boldsymbol{\sigma}$ or all 6 of $\boldsymbol{\chi}$, since some components vanish and others are identical by symmetry. Rather than to characterize a tensor by its cartesian components, it is often better to indicate instead the principal values (eigenvalues) and the corresponding principal axes. For a nonsymmetric tensor like $\boldsymbol{\sigma}$ there are some problems with its 'diagonalization' i.e. the construction of the principal values and axes and with the correspondence of these to their experimental counterparts. For not too large asymmetries it is justified replacing $\boldsymbol{\sigma}$ by its symmetric part $\tilde{\boldsymbol{\sigma}}$ with $\tilde{\boldsymbol{\sigma}}_{ij} = \frac{1}{2}(\sigma_{ij} + \sigma_{ij})$ and to diagonalize it. The effect of the asymmetry of $\boldsymbol{\sigma}$ on physically measurable quantities has only recently received some interest [19].

For symmetric top (especially linear) molecules (e.g. CO_2 or NH_3) two principal values of $\boldsymbol{\chi}$ are identical, and it is then convenient to indicate the isotropic part χ and the anistropy $\Delta\chi$ instead of the three principal values

$$\chi = \tfrac{1}{3}(\chi_{\parallel} + 2\chi_{\perp}) \tag{2.8.13a}$$

$$\Delta\chi = \chi_{\parallel} - \chi_{\perp} \tag{2.8.13b}$$

The analogous definitions are possible for σ of those atoms that lie on an n-fold symmetry axis ($n > 2$). In the less symmetric case one defines in addition to the isotropic part the 1st and 2nd anisotropies [20]

$$\chi = \tfrac{1}{3}(\chi_{11} + \chi_{22} + \chi_{33}) \tag{2.8.14a}$$

$$\Delta\chi_1 = \chi_{11} - \tfrac{1}{2}(\chi_{22} + \chi_{33}) \tag{2.8.14b}$$

$$\Delta\chi_2 = \chi_{22} - \chi_{33} \tag{2.8.14c}$$

where χ_{11} is the principal component largest in absolute value and χ_{22} the second one. If only the first anisotropy is documented, it is usually the $\Delta\chi$ rather than $\Delta\chi_1$. Analogous definitions are used for σ and $\Delta\sigma$.

2.9 Orbital Contributions to the IGLO Results

One sees from Eqs. (2.8.10) that each tensor component of σ can be written as a sum of orbital contributions

$$\boldsymbol{\sigma} = \sum_{k=1}^{n} \sigma(k) \tag{2.9.1}$$

$$\sigma(k) = \sigma^{d}(k) + \sigma^{p0}(k) + \sigma^{p1}(k) \tag{2.9.2}$$

$$\sigma^{d}(k) = 2\langle \psi_{ko} | h'_{k2} | \psi_{ko} \rangle \tag{2.9.3a}$$

$$\sigma^{p0}(k) = -4 \sum_{j=1}^{n} \langle \psi_{ko} | h_{01} | \psi_{jo} \rangle \langle \psi_{jo} | \Lambda_j - \Lambda_k | \psi_{ko} \rangle \tag{2.9.3b}$$

$$\sigma^{p1}(k) = -4\langle \psi_{ko} | h_{01}(1 - P_0) | \psi_{k1} \rangle$$
$$= \frac{-4e}{mci} \left\langle \psi_{ko} \left| \frac{\vec{\mu} \cdot \vec{l}_{\varrho}}{|\vec{r} - \vec{\varrho}|^3} (1 - P_0) \right| \psi_{k1} \right\rangle \tag{2.9.3c}$$

Each orbital contribution $\sigma(k)$ consists of three parts. Of these $\sigma^d(k)$ can easily be interpreted as a local diamagnetic contribution and $\sigma^{p1}(k)$ as a local paramagnetic contribution. In (2.9.3c) the projection $(1 - P_0)\psi_{k1}$ means that only that part of the first order perturbation function ψ_{k1} is taken, which is orthogonal to all occupied MOs ψ_{ko}. The complementary part of ψ_{k1} which is 'in the space' of the occupied MOs appears in (2.9.3b); it is entirely expressible in terms of the unperturbed LMOs ψ_{ko}. There are hence arguments in favour of regarding this term as either paramagnetic (arising from $P_0\psi_{k1}$ and involving only a 1st order operator h_{01}) or as diamagnetic (since its construction only requires the ψ_{ko}). In most papers on IGLO calculations, in which diamagnetic and paramagnetic contributions are displayed separately, σ^{p0} is taken together with σ^{p1} as the paramagnetic term σ^p. Note that in LORG calculations [17] where similar terms arise, σ^{p0} is usually taken together with σ^d as the diamagnetic term. One might also regard the individual terms in the sum over j (2.9.3b) as the quantities $\sigma^{p0}(k, l)$ to be documented and regard these as 'non-local' terms. In doing so, one would, however, not gain much relevant additional information.

In analogy to (2.9.1–3) also the magnetic susceptibility (2.8.12) can be decomposed [3, 4]. For E^d, E^{p0} and E^{p1} the analogy to σ^d, σ^{p0} and σ^{p1} is perfect. One can also decompose the 'exchange correction' (2.8.12e, f) into LMO contributions, while the resonance corrections (2.8.12g, h) involve off-diagonal elements $\langle \psi_{10} | F | \psi_{ko} \rangle$ of the Fock operator and are hence better regarded as sums of non-local contributions. Fortunately these are usually quite small (of the order of few percent of the susceptibility). Let us note here only the local diamagnetic and paramagnetic contributions

$$E^{d}_{11}(k) = 2\langle \psi_{ko} | \tilde{h}_{k2} | \psi_{ko} \rangle \tag{2.9.4a}$$

$$E^{p1}_{11}(k) = -2\langle \psi_{ko} | \tilde{h}_{k1}(1 - P_0) | \psi_{k1} \rangle$$
$$= -\frac{e}{mci} \langle \psi_{ko} | \vec{B} \cdot \vec{l}_k (1 - P_0) | \psi_{k1} \rangle \tag{2.9.4b}$$

From the expressions (2.9.3) and their counterparts for the susceptibilities one can easily find out how local paramagnetic and diamagnetic contributions arise.

We see e.g. from (2.9.4b) that E^{p1}_{11} vanishes if ψ_{k0} is an eigenfunction of the angular momentum operator $\vec{l}_k$ in direction of the field $\vec{B}$ (independent of the eigenvalue). This is e.g. the case if ψ_{k0} is an LMO of an atomic core. But even LMOs that describe a chemical bond are often rather close to spherically symmetric and are hence almost eigenfunctions of the local angular momentum operator $\vec{l}_k$ with eigenvalue zero. So paramagnetic contributions can only arise if there is a significant deviation of the LMOs from spherical symmetry.

The same argument applies to (2.9.3c) provided that the centroid of charge of ψ_{k0} coincides with the position of the nucleus (i.e. that $\vec{\varrho} = \vec{R}_k$ and hence $\vec{l}_\varrho = \vec{l}_k$). If $\vec{\varrho}$ is sufficiently far from $\vec{R}_k$, the factor $|\vec{r} - \vec{\varrho}|^3$ makes the integrand in (2.9.3c) very small. So significant paramagnetic contributions to σ are expected only from LMOs ψ_{k0} for which $\vec{R}_k$ is rather close to $\vec{\varrho}$.

In order to see what determines the magnitude of paramagnetic contributions it is convenient to have a closer look at Eq. (2.8.1). Let the ψ_a be the eigenfunctions of F_0 corresponding to unoccupied (virtual orbitals).

$$F_0\psi_a = \varepsilon_a\psi_a \tag{2.9.5}$$

Let us then expand $(1 - P_0)\psi_k$ in the basis of the ψ_a

$$(1 - P_0)\psi_{k1} = \sum_a c_{ka}\psi_a \tag{2.9.6}$$

and scalar multiply (2.8.1) from the left by ψ_b. The result is

$$\langle\psi_b|\tilde{F}_{k1}|\psi_{k0}\rangle - \sum_{j(\neq k)} c_{jb}\langle\psi_{j0}|F_0|\psi_{k0}\rangle = c_{kb}(\varepsilon_k - \varepsilon_b) \tag{2.9.7a}$$

$$\varepsilon_k = \langle\psi_{k0}|F_0|\psi_{k0}\rangle \tag{2.9.7b}$$

The dominating contribution to the matrix element $\langle\psi_0|\tilde{F}_{k1}|\psi_{k0}\rangle$ is

$$\langle\psi_b|\tilde{h}_{k1}|\psi_{k0}\rangle = \frac{e}{mci}\langle\psi_b|\vec{B}\cdot\vec{l}_k|\psi_{k0}\rangle \tag{2.9.8}$$

Often the paramagnetic terms are dominated by the contribution of a single virtual MO ψ_a. Let this be the case and let us assume that the off-diagonal elements $\langle\psi_{j0}|F|\psi_{k0}\rangle$ of the Fock operator can be neglected, then the expression (2.9.3) for $\sigma^{p1}(k)$ becomes

$$\sigma^{p1}(k) \approx \frac{4e^2}{m^2c^2}(\varepsilon_k - \varepsilon_b)^{-1}\left\langle\psi_{k0}\left|\frac{\vec{\mu}\cdot\vec{l}_\varrho}{|\vec{r} - \vec{\varrho}|^3}\right|\psi_b\right\rangle\langle\psi_b|\vec{B}\cdot\vec{l}_k|\psi_{k0}\rangle \tag{2.9.9}$$

In order that this contribution has a large absolute value, the energy-denominator $(\varepsilon_b - \varepsilon_k)$ must be small, i.e. ψ_b must be a 'low-lying virtual state', and the two matrix elements must be large. Now the matrix element $\langle\psi_0|\vec{B}\cdot\vec{l}_k|\psi_{k0}\rangle$ vanishes if ψ_{k0} is eigenfunction of the local angular momentum operator $\vec{l}_k$ projected on the direction $\vec{B}$ of the magnetic field, e.g. if ψ_{k0} is invariant with respect to a rotation around an axis parallel to $\vec{B}$ through the gauge origin. The matrix element $\langle\psi_{k0}|\vec{\mu}\cdot\vec{l}_\varrho/|\vec{r} - \vec{\varrho}|^3|\psi_a\rangle$ vanishes if ψ_{k0} is an eigenfunction of the angular momentum operator $\vec{l}_\varrho$ projected to the direction of the magnetic momentum $\vec{\mu}$

of the considered nucleus. So if ψ_{k0} is spherically symmetric around its gauge origin or around the nucleus i.e. either eigenfunction of $\vec{l}_k$ or $\vec{l}_\varrho$ with eigenvalue zero, the expression (2.9.9) will vanish. Only the nonspherical 'contaminations' contribute.

Consider the zz-diagonal element of σ^{p1}. Then the matrix elements in (2.9.9) become appreciable if action of $\vec{l}_k$ (and of $\vec{l}_\varrho$) in z-direction transforms ψ_{k0} to an orbital which has large overlap with the low-lying virtual MO ψ_a.

What is the action of an l_z-operator on an orbital? If ψ is of σ-type, i.e. rotationally symmetric about the z-axis, it is annihilated, if it is of π-type, i.e. of the form xf(r, z) or yf(r, z), it is rotated by 90°

$$l_z xf(r,z) = \frac{\hbar}{i}\left\{x\frac{\partial}{\partial y}(xf) - y\frac{\partial}{\partial x}(xf)\right\} = i\hbar yf(r,z) \tag{2.9.10a}$$

$$l_z yf(r,z) = \frac{\hbar}{i}\left\{x\frac{\partial}{\partial y}(yf) - y\frac{\partial}{\partial x}(yf)\right\} = -i\hbar xf(r,z) \tag{2.9.10b}$$

if ψ is of δ-type, i.e. of the form xyf(r, z) or $(x^2 - y^2)f(r,z)$, it is rotated by 45°

$$l_z xyf(r,z) = \frac{\hbar}{i}\left\{x\frac{\partial}{\partial y}(xyf) - y\frac{\partial}{\partial x}(xyf)\right\} = -i\hbar(x^2 - y^2)f(r,z) \tag{2.9.11}$$

For an arbitrary orbital its σ-component is annihilated, its π-component rotated by 90°, its δ-component rotated by 45° etc.

If ψ is of σ, π, etc type with respect to an axis parallel to l_z, but shifted by a distance R in x-direction one gets for a σ type MO (l'_z is the z-component of $\vec{l}_\varrho$)

$$l'_z\psi = \frac{\hbar}{i}\left\{(x-R)\frac{\partial}{\partial y} - y\frac{\partial}{\partial x}\right\}f(r,z) = i\hbar R\frac{\partial\psi}{\partial y} \tag{2.9.12}$$

if ψ is of π-type

$$l'_z\psi_x = \frac{\hbar}{i}\left\{(x-R)\frac{\partial}{\partial y} - y\frac{\partial}{\partial x}\right\}\{xf(r,z)\} = i\hbar\left\{yf + R\frac{\partial\psi_x}{\partial y}\right\}$$

$$l'_z\psi_y = \frac{\hbar}{i}\left\{(x-R)\frac{\partial}{\partial y} - y\frac{\partial}{\partial x}\right\}\{yf(r,z) = i\hbar\left\{-xf + R\frac{\partial\psi_y}{\partial y}\right\} \tag{2.9.13}$$

Action of an off-center angular momentum operator hence changes a σ to a π-type-MO and a π-type MO to a rotated and deformed π-type MO i.e. a π-MO 'contaminated' by δ-contributions. The larger R, the higher is the δ-type contribution in (2.9.13).

We illustrate the LMO analysis of the IGLO calculation for the simple example of the LiH molecule (Table 1). Here the LMOs are rather close to 1s-AOs of Li^+ and H^- respectively (the latter being somewhat deformed) such that no large paramagnetic contributions to the susceptibility are expected. This is, in fact, confirmed. Note that for the field parallel to the molecular axis paramagnetic contributions vanish for symmetry reasons, since the MOs are eigenfunctions of the angular momentum operator in direction of the molecular axis.

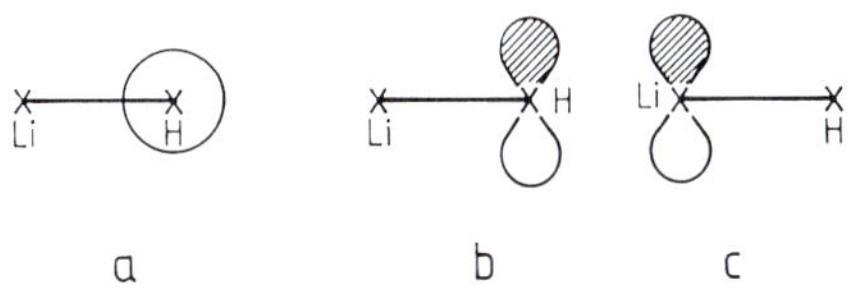

Fig. 3a–c. Illustration of the origin of a paramagnetic contribution to the current density around the Li nucleus in LiH for the field perpendicular to the paper plane. (**a**): LMO of the LiH bond (approximately AO of H^-), (**b**): action of the angular momentum operator l through Li parallel to $\vec{B}$ on (a), (**c**): low-lying unoccupied 2p-AO of Li

For the chemical shifts the situation is a little more complicated. One sees that $\sigma(^1H)$ is dominated by σ^d of the bond LMO b(LiH) which is 26.98 ppm for the perpendicular component. There is a very small $\sigma^p(b)$ of -0.45 ppm and the total contribution of 1s(Li) to $\sigma_\perp(^1H)$ is -0.74 ppm. This is so since Li^+ even in LiH is nearly spherically symmetric. For $\sigma(^{6,7}Li)$ the dominating contribution is again σ^d(1s, Li), but now there is also a substantial paramagnetic contribution of the bond LMO. This is so because on one hand this LMO is not entirely spherically symmetric. On the other hand action of the angular momentum operator $\vec{l}_\varrho$ of Li through the Li nucleus leads to an orbital with has large overlap with a low-lying unoccupied pπ-MO of Li. This is illustrated in Fig. 3.

In Table 1 we have always referred to IGLO results. Let us now compare IGLO and CHF with a common gauge origin for the two MOs (Table 2). We make this comparison for $\sigma(^1H)$ only. The basis sets are large enough that gauge invariance is nearly achieved. We see that the contribution of the 1s(Li)-MO to $\sigma(^1H)$ is very small for either the gauge at H, at Li or in an IGLO calculation,

Table 1. Contributions to IGLO results for the susceptibility χ, and the absolute chemical shifts $\sigma(^1H)$ and $\sigma(^{6,7}Li)$ in LiH

			$-\chi$	$\sigma(^1H)$	$\sigma(^{6,7}Li)$
⊥	1s(Li)	dia	0.71	−0.24	95.54
		para	−0.02	−0.51	−0.08
		sum	0.69	−0.74	95.46
	b(LiH)	dia	8.23	26.98	2.61
		para	−1.67	−0.45	−14.26
		sum	6.56	26.53	−11.65
	Total	dia	8.94	26.75	98.15
		para	−1.69	−0.96	−14.35
	Non local		−0.47	—	—
		sum	6.78	25.79	83.81
∥	1s(Li)	dia	0.70	0.57	95.58
	b(LiH)	dia	8.46	27.62	5.83
	Total	dia	9.16	28.18	101.41
iso			7.58	26.59	89.67

Basis IV was used. Units are ppm cgs for χ and ppm for σ. ⊥ means that the magnetic field is perpendicular to the molecular axis, ∥ that it is parallel

Table 2. Contributions to $\sigma_{\perp}(^1H)$ in LiH (component perpendicular to the bond). Comparison of coupled Hartree–Fock with gauge origin at H or Li and IGLO

MO	Type	CHF(H)	CHF(Li)	IGLO
1s(Li)	σ_d	17.33	−0.25	−0.24
	σ_p	−17.61	−0.15	−0.51
	sum	−0.28	−0.40	−0.74
b (LiH)	σ_d	27.90	19.28	26.98
$\sim H^-$	σ_p	−1.67	6.58	−0.45
	sum	26.23	25.87	26.53
Total		25.95	25.46	25.79

Basis IV was used. Units are ppm

but that for CHF(H) large σ_d and σ_p components arise, that practically cancel. For the gauge at Li there is no such cancellation, but now there is a substantial σ_p of the bond MO, which neither arise for CHF(H) nor for IGLO.

2.10 Basis Sets

In order to make comparisons between different molecules easy, basis sets have been standardized. They are characterized by numbers I, II, III, IV with the size of the basis sets increasing with increasing number. The smallest basis set used (smaller than basis I) is given a different designation, namely DZ (for double zeta). The use of this basis set is justified only in special cases, mainly for hydrocarbons. This basis allows calculations of relatively large molecules, for which basis I to IV would be prohibitive. In this review we shall mainly refer to results obtained with basis II.

All these sets are constructed from Huzinaga bases augmented by polarization functions. The choice of the latter is a result of our own experience.

The DZ basis is a (7s, 3p) Huzinaga [21] basis in the contraction (4, 1, 1, 1; 2, 1) for a first row atom like C and a 3s basis in the contraction (2, 1) for hydrogen.

Basis sets I and II are (9s, 5p) Huzinaga sets augmented by one set of d-functions for the first row atoms B–F, 9s augmented by 3p or 4p1d for Li or Be respectively and 5s augmented by one p set for hydrogen, Basis I and II differ in the contraction scheme. Basis I: $(5, 4 \times 1; 3, 1, 1)$; Basis II $(5, 4 \times 1; 2, 1, 1, 1)$.

For 2nd row atoms basis II is a (11s, 7p) Huzinaga basis in the contraction $(5, 6 \times 1;\ 2, 5 \times 1)$ augmented by two d sets.

Basis sets III and IV start from (11s, 7p) sets in the contraction $(5, 6 \times 1;$ $2, 5 \times 1)$ for first row atoms, (12s, 8p) sets in the contraction $(5, 7 \times 1;\ 2, 6 \times 1)$ for 2nd row atoms, 6s (3, 1, 1, 1) for hydrogen. They are augmented by two p sets for hydrogen, two d sets for 1st row atoms and three d sets for 2nd-row atoms (basis III) or larger polarization sets (basis IV).

Table 3. Exponents of the polarization functions

BASIS II		BASIS III	
H	$\eta_p = 0.65$	H	$\eta_p = 0.33, 1.3$
Li	$\eta_p = 0.19, 0.75, 3.0$	C–F	$\eta_d = 0.35, 1.4$
Be	$\eta_p = = 0.058, 0.18, 0.88, 2.2$; $\eta_d = 0.5$	P, S	$\eta_d = 0.25, 0.8, 2.5$
B	$\eta_d = 0.7$		
C–F	$\eta_d = 1.0$		
Na	$\eta_d = 0.2, 0.8$		
Mg	$\eta_d = 0.25, 1.0$		
Al	$\eta_d = 0.3, 1.2$		
Si, P	$\eta_d = 0.35, 1.4$		
S, Cl	$\eta_d = 0.4, 1.6$		

The exponents of the polarization functions are listed in Table 3.

The dependence of the IGLO results of the chosen basis have been studied in detail in various original papers [4, 22–26]. In this review, we shall, for the sake of consistency, mostly display results obtained with a single type of basis set, namely basis II, which has turned out to be a good compromise between accuracy and computational effort. This implies that the IGLO results referred to here will not always be the best possible ones.

So far mainly basis sets built from gaussian lobes have been used. Cartesian gaussians can be chosen alternatively, and it has turned out [27] that the results with these two types of basis sets usually differ very little (of the order of less than 1 ppm). As far as the necessary computer time is concerned, lobes on the whole are superior for calculations with small to medium quality basis sets (i.e. DZ or basis II) for elements of the 1st row (e.g. hydrocarbons), cartesian gaussians (using the Pitzer version of the HONDO program), win, when AOs with high l-values (d, f, ..) are involved and/or the symmetry is high. These comparisons were made on a CYBER 205 and the advantage of the use of lobes is in part due to a rather effective vectorization of the lobe integral program [28]. For more details on computer times see Sect. 3.13.

2.11 Units and Sign Conventions

The chemical shift is dimensionless. It is usually given in ppm. We have chosen the following sign convention. Absolute chemical shifts, as they result directly from a calculation, are chosen positive if they are energy-raising (shielding, diamagnetic) and negative if they are energy-lowering (deshielding, paramagnetic). For relative shifts we use—in agreement with a now generally accepted convention—the definition

$$\delta = \sigma_{ref} - \sigma \qquad (2.11.1)$$

where σ_{ref} is the absolute shift of a reference compound. Hence $\delta > 0$ means that the shift is to lower field (to higher frequency; paramagnetic, deshielding), while $\delta < 0$ means a shift to higher field (diamagnetic, shielding).

For susceptibilities, in spite of recommendations to use the SI system, the standard unit is still

$$1\,\text{ppm cgs} = 10^{-6}\,\text{erg(gauss)}^{-1}\,\text{mol}^{-1} = 10^{-6}\,\text{cm}^3/\text{mol} \tag{2.11.2}$$

the conversion factor to atomic units is

$$1\,\text{a.u.} = 1(a_0)^3 = 0.0892394\,\text{cm}^3/\text{mol} \tag{2.11.3}$$

With some reservations a conversion to SI is possible (the reservation has to do with the fact that there is no unique unit for the magnetic moment in the SI system)

$$1\,\text{ppm cgs} = 4\pi \cdot 10^{-12}\,\text{m}^3/\text{mol} \tag{2.11.4}$$

The recommended sign convention is that $\chi < 0$ means diamagnetism and $\chi > 0$ paramagnetism, i.e. the sign of χ is opposite to that of E_2, which is somewhat unfortunate.

2.12 Current Density Plots. The Example of Formaldehyde

When a molecule is placed into a homogeneous static magnetic field, a current density is induced. Plots of such current densities are very helpful for an understanding of the magnetic properties of the molecule. This current density induces additional fields at the position of the nuclei. If such an additional field weakens the applied external field one finds shielding (high field shift), if it strengthens the field, deshielding (down field shift) is realized.

Most important are usually currents around the nucleus. A diamagnetic current around a nucleus (drawn as clockwise) causes shielding, a paramagnetic current (counter clockwise) causes deshielding. The effects of off-center ring currents are less easily to realize. For a nucleus close to the center of the ring current a diamagnetic current causes shielding and a paramagnetic current deshielding. With increasing distance between the nucleus and the center the effect decreases and it even changes its sign for large enough distance.

Figure 4 shows a plot of the current density induced in the formaldehyde molecule H_2CO for two orientations of the molecule relative to the field. For didactic reasons we have excluded the contributions of the K-shells of C and O. In these shells diamagnetic currents are induced, which shield the respective nuclei by amounts that are practically the same in all molecules containing C or O.

One sees from Fig. 4 that the total current density is a sum of local currents in various LMOs. One recognizes in Fig. 4a (for the field perpendicular to the molecular plane) diamagnetic currents around the CH bonds and in the lone pairs of O, but a paramagnetic current around C, which is in part responsible for the downfield shift of the ^{13}C resonance in H_2CO (as compared to e.g. CH_4). An even stronger deshielding of the C atom arises (see Sect. 3.4) for the field perpendicular to the plane which bisects the HCH angle (Fig. 4b).

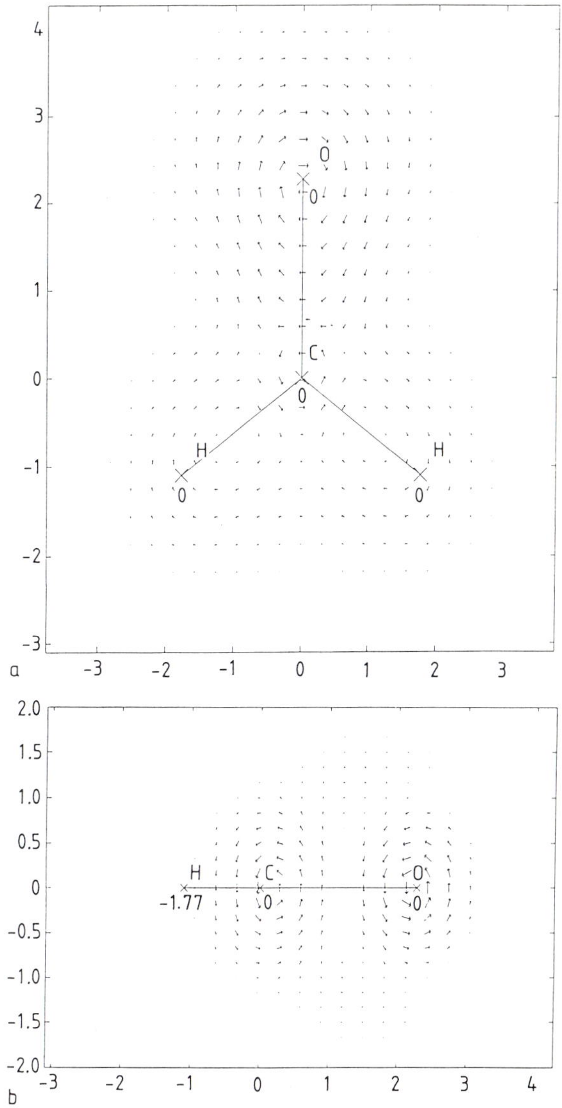

Fig. 4. The current density (excluding the K-shell contribution of C and O) in formaldehyde H_2CO (**a**) in the molecular plane, (**b**) in the plane bisecting the HCH angle for a field perpendicular to the respective plane. Coordinate scales in a_0

The relation between the current density $\vec{j}$ and the chemical shift σ at the position $\vec{\varrho}$ of the nucleus can be formulated in terms of Biot–Savart's law according to which the induced field at $\vec{\varrho}$ is

$$\vec{B}_{ind} = \frac{1}{c}\int \frac{(\vec{r} - \vec{\varrho}) \times \vec{j}(\vec{r})d^3r}{|\vec{r} - \vec{\varrho}|^3} \tag{2.12.1}$$

The integrand of (2.12.1) has been termed shielding density [29]. A plot of the shielding density gives information as to the weights of the various regions in the molecule concerning the shift of a particular nucleus.

2.13 Improved Computational Methods

As mentioned in Sect. 2.7, all available standard methods are of coupled Hartree–Fock type. This approximation (together with the IGLO concept) has turned out to be surprisingly good for the great majority of applications. There are certainly cases, where one has to go beyond CHF, i.e. beyond approximating the wave function as a single Slater determinant.

Although it is in principle understood [30] how one should proceed, the computations of chemical shifts and magnetic susceptibilities published so far [31–34], by means of many-body perturbation theory [31], multiconfigurational SCF [32] or configuration interaction [33] are far from satisfactory, since they are too poor on the level without correlation. These calculations indicate that for molecules like H_2, LiH, HF correlation effects are small [31], while for more critical systems like BH the results differ somewhat [32, 33].

The failure of 'beyond-Hartree–Fock' calculations is not too surprising, since methods beyond CHF suffer from the gauge origin problem at least to the same extent as conventional CHF methods. It appears desirable to combine beyond-CHF methods with the IGLO concept, which would allow reliable studies of correlation effects on magnetic molecular properties.

In addition to the inclusion of correlation effects it will also become important to take care of relativistic effects. Some pilot studies on an otherwise low level of theory have been performed [35]. One must conclude that relativistic effects play a major role for heavy elements (starting from the 3rd or 4th row), but that they are non-negligible even for light elements (say of the 2nd row).

A discussion of both correlation and relativistic effects is outside the scope of this review.

3 Results and Their Interpretation

3.1 General Remarks

In the following systematic study of chemical shifts we put the emphasis on the interpretation of the dependence of a chemical shift on the electronic structure, and on the comparison of experiment with theory for a large number of molecules calculated with the same type of AO-basis. The basis is usually basis II (see Sect. 2.10). This implies that the theoretical values given here *are not always the best available ones*. Those from basis III or IV are usually better, but such calculations are only possible for relatively small molecules. We have decided to exclude a detailed comparison of basis sets from this review. This has, of course, been done in the original literature [1, 4, 22–26]. Our decision implies that reference to non-IGLO results from the literature would not make too much sense, because differences in the results are often due to the use of different basis sets. In fact, with comparable basis sets (and for identical geometries) the results of IGLO, GIAO or LORG calculations are usually rather close, and the closer—as it should be—the larger the basis sets are. Only conventional CHF requires much larger basis sets for comparable quality, except for simple hydrides like CH_4. In view of the practical difficulties with the other methods (see Sect. 2.7) a wealth of data comparable with that available from IGLO does not exist for any other method.

For the analysis of σ in terms of LMO contributions, results from other methods (except possibly LORG) would not be very helpful since these contributions only appear in IGLO (and related ones in LORG) calculations.

Among the pioneering as well as recent CHF calculations one can mention those of Refs. 12 and 36, among GIAO calculations those of Ref. 37 while LORG results are found in Refs. 17 and 38.

The current literature has been reviewed in the NMR series of the specialist's periodical report [39], mainly in articles by Raynes and by Jameson. A very recent review, stressing mainly applications to small molecules, is that by Chesnut [40].

In comparing theoretical with experimental chemical shifts two general problems arise

1. Theoretically one computes absolute shifts σ, experimentally one measures relative shifts δ with respect to some 'standard'.
2. In theory one studies one isolated molecule in its equilibrium geometry, in experiment an ensemble of molecules, usually in the condensed phase.

One can switch from the experimental δ-scale to an 'experimental' σ-scale if one knows σ for a 'reference molecule'—which will usually not be the 'standard' (with respect to which δ is defined), for which σ can be obtained from its rotational paramagnetism and a diamagnetic complement from ab-initio calculations [20]. An experimental absolute scale for ^{13}C has e.g. been constructed [41] with CO as the reference molecule. If such an experimental σ-scale is available, it is best to use this scale for the comparison of theory with experiment.

It is also possible to switch from the theoretical σ-scale to a theoretical δ-scale, e.g. if the standard molecule is accessible to IGLO calculations. For ^{13}C shifts tetramethylsilane (TMS) is used as the standard. IGLO-calculations with the basis sets DZ and II have been performed for TMS, so theoretical δ-values are available. If one wants to use other basis sets one can use a theoretical reference, e.g. CH_4 and convert to the conventional scale by correcting for the experimental $\delta(^{13}C)$ difference between TMS and CH_4. (See, however, the following paragraph). For some nuclei the standards are too poorly defined from a theoretical point of view, e.g. 85% aqueous phosphoric acid for $\delta(^{31}P)$. In such a situation a theoretical δ-scale can be defined for a reference molecule, here PH_3 and the experimental δ of PH_3 is used to adjust the theoretical δ-scale to the experimental one. It may happen that the theoretical reference was not a good choice, the agreement between theory and experiment for this molecule may either be unusually good or unusually bad. In such cases the theoretical δ-values will be subject to spurious systematic errors. This can be avoided if one does not join the two δ-scales by referring to a single molecule, but for a few molecules. One constructs then an optimum regression line between experimental δ and theoretical δ

$$\sigma = a\delta + b$$

Ideally a should be equal to -1 (one may or may not impose this) and b is then taken as σ of the standard.

The experimental counterparts to a calculated σ or δ should be those obtained in the gas phase, preferentially extrapolated to zero density and zero temperature, or even to the equilibrium geometry. Most experimental data have, however, been obtained in condensed phase and the just-mentioned extrapolations are hardly possible. Gas-to-liquid shifts vary for ^{13}C between ~ 0 and ~ 10 ppm (sometimes they are even larger) and one must be aware that poor agreement between theory and experiment may in part be due to neglect of the gas-to-liquid shift. These shifts may also have some influence on the δ-scale. The difference between $\sigma(^{13}C)$ in TMS and CH_4 in the gas phase is ~ 7 ppm, in good agreement with the corresponding difference between IGLO calculations with basis II. So for gas phase shifts and basis-II calculations it does not matter whether one uses TMS as standard with or without CH_4 as reference. In the liquid phase the difference between TMS and CH_4 is only 2.3 ppm, it therefore matters whether or not one uses CH_4 as an auxiliary reference. In DZ calculations the difference TMS to CH_4 is only 0.2 ppm. For other nuclei (like N or O) gas-to-liquid shifts are much larger and they may have either sign. For details see Refs. 25, 42 and 43.

Most calculations reported in this review have been performed for experimental geometries from standard sources [44, 45], provided that such geometries were available. For molecules unknown from experiment, ab-initio geometries obtained on SCF level with a moderate basis (typically 6-31G*) were used. Sometimes geometry optimizations on higher level were necessary (see Sect. 3.2). In critical cases explicit information on the geometry optimization are given. For calculations taken from published work some information on the geometry can be found in the original references. In case of doubt geometry data are available on request.

The theoretical results discussed here do not discriminate between isotopes (see also Sect. 3.11).

3.2 Proton NMR Shifts

From the experimental point of view [46] ^1H-NMR is very popular, mainly due to the omnipresence of hydrogen in all kinds of molecules, to the high sensitivity of ^1H (as compared to other nuclei), and the fact that the naturally most abundant isotope of H is 'NMR active' and not disturbed by quadrupole or other effects.

For theoretical calculations ^1H is much less suited (than many other nuclei like ^{13}C), mainly for the following reasons

(a) ^1H shifts are rather small, in particular the range of absolute isotropic ^1H shifts varies only between ~ 10 and ~ 32 ppm with the majority of values between 26 and 32 ppm. A very high accuracy of the theoretical calculations would be required for reliable comparisons or predictions. For other nuclei the range of σ is several hundred ppm and less absolute accuracy of the calculations is needed to make them useful.
(b) Hydrogen atoms are usually at the periphery of a molecule and hence much affected by inter-molecular forces, e.g. H-bonds. In fact gas-to-liquid shifts and solvent effects are relatively very large for ^1H shifts. Calculated values that refer to isolated molecules, may bear little relation to experiments in the pure liquid or in solution.
(c) Strong changes in the electronic structure may have little effect on $\sigma(^1H)$, because various contributions to $\sigma(^1H)$ may compensate each other, e.g. in the sequence NaH, $MgH_2 \cdots$, HCl.

The chemical shift of an isolated H-atom with respect to a bare proton can be calculated exactly, it is

$$\sigma(^1H) = \tfrac{1}{3}\alpha^2 = 17.75\,\text{ppm}$$

where $\alpha = 1/137.03602$ is the fine-structure constant.

For the 2-electron systems H^- and H_2 at least the coupled Hartree value can easily be calculated to any desired accuracy. In the case of H^- it is 24.34 ppm, while on a CI level, i.e. with inclusion of electron correlation one obtains 24.24 ppm [47], the correlation effect is hence extremely small. Even smaller are relativistic corrections. The basis-II-IGLO-value for H^- is 26.57 ppm, but for negative ions like H^- basis II is not good enough, inclusion of a single 'diffuse' s-function reduces the shift to 24.41 ppm, not far from the CHF-limit.

For H_2 basis II yields $\sigma_{iso} = 26.45$ ppm (at r_e) which is not changed on increasing the basis. The contribution of electron correlation is supposed [31] to be less than 0.2 ppm (increasing the shielding). An 'experimental' value of σ_{iso} has been estimated long ago by Ramsey et al. [48] from the rotational paramagnetism of HD and a theoretical value for σ_d with respect to the center of mass. The result $\sigma_{iso} = 26.2$ ppm has been confirmed by more recent studies [49]. To compare theory with experiment one must at least note that theoretical values refer to the equilibrium bond length $r_e = 1.401a_0$, while experiment implies an average over the zero point motion, which to 1st-order can be simulated by choosing the equilibrium distance $r_0 = 1.449a_0$ of H_2 where one gets 25.92 ppm (with correlation probably ~ 26.1 ppm). For a study of the effect of vibrational averaging on $\sigma(^1H)$ in H_2 see Ref. 49.

From inspection of $\sigma(^1H)$ of H^+, H and H^- one may expect that $\sigma(^1H)$ in arbitrary hydrides depends essentially on the density of electrons near the

Table 4. LMO contributions to the absolute isotropic shift $\sigma(^1H)$ in ppm in various 1st row hydrides

	K	XH	XH′	lp	σ	$\Delta\sigma$
LiH	−0.07	26.70	—	—	26.64	2.2
BeH_2	0.00	28.58	−0.10	—	28.48	4.5
BH_3	0.00	29.55	−2.47	—	24.62	7.5
BH	0.49	37.34	—	−13.10	24.74	13.1
CH_3^+	0.00	26.65	−5.12	—	16.41	14.1
CH_4	0.00	25.09	1.99	—	31.07	10.8
NH_3	0.00	20.92	3.02	4.27	30.92	16.4
H_2O	0.00	17.05	3.98	4.75	30.54	20.8
HF	0.01	12.70	—	5.15	28.15	24.2

K means the K-shell of X;
XH the LMO of the XH bond involving the considered proton;
XH′ another XH bond;
lp a lone pair.
Any contribution refers to a single XH bond or a single lone pair.
$\Delta\sigma$ is the anisotropy defined by Eqs. (2.8.13b) and (2.8.14b).
$\sigma(^1H)$ for TMS in the same basis (II) is 31.03

proton, i.e. the electronic charge (or in XH_n-molecules on the electronegativity of X). One sees from Table 4 that this is not so. In the series of the saturated neutral hydrides XH_n from LiH via CH_4 to HF the chemical shift rises little from LiH to BeH_2 and decreases strongly to BH_3, then jumps up to CH_4, remains almost constant from CH_4 via NH_3 to H_2O and decreases slightly to HF. If σ were dominated by the charge density at the proton a monotonic decrease of $\sigma(^1H)$ from LiH to HF should be found.

If one considers the various LMO contributions to $\sigma(^1H)$ in Table 4 one sees that the one of the XH-bond has nearly the expected behaviour i.e. it decreases with the electron density at H—at least from BH_3 via CH_4 to HF. However the contributions of the other XH bonds (in which the considered H is not involved) and of the lone pairs are far from being negligible. In the sequence CH_4 to HF the decrease of the XH-bond contributions (less shielding) is nearly compensated by the other contributions (increased shielding). On the 'left' side of CH_4 these 'other' contributions are deshielding rather than shielding. This difference causes the big 'jump' from BH_3 to CH_4.

Obviously in the 'saturated' molecules CH_4 to HF all local contributions to the current density are mainly diamagnetic and create a magnetic field opposite to the external one. The effect of lone pairs is always larger than that of XH bonds although they are more distant to the protons.

In BeH_2, BH_3, CH_3^+, and BH there are low-lying unoccupied MOs which cause significant paramagnetic current densities, provided that action of the angular momentum operator in field direction results in an orbital that has strong overlap with a low-lying unoccupied one.

BH is a prototype of a van-Vleck paramagnetic molecule, although our knowledge about this paramagnetism comes from theory only [4, 7, 31, 50]. The isotropic susceptibility χ_{iso} of BH from IGLO calculations is $\sim +19$ ppm cgs, the perpendicular component $\chi_{\perp}$ is even $+34$ ppm cgs, while parallel to the bond the molecule is diamagnetic ($\chi_{\parallel} = -12$ ppm cgs). The current density in BH for the field perpendicular to the bond is plotted in Fig. 5a. The arrows in counterclockwise direction clearly indicate that this current is paramagnetic. For comparison the typically diamagnetic current density in HF is shown in Fig. 5b.

The mechanism which leads to the paramagnetic current is easily understood. Let z be the molecular axis. Action of the angular momentum operator l_x transforms both the MO of the BH bond and the lone pair to MOs that have large overlap with the low lying unoccupied π_y-MO (see Fig. 6).

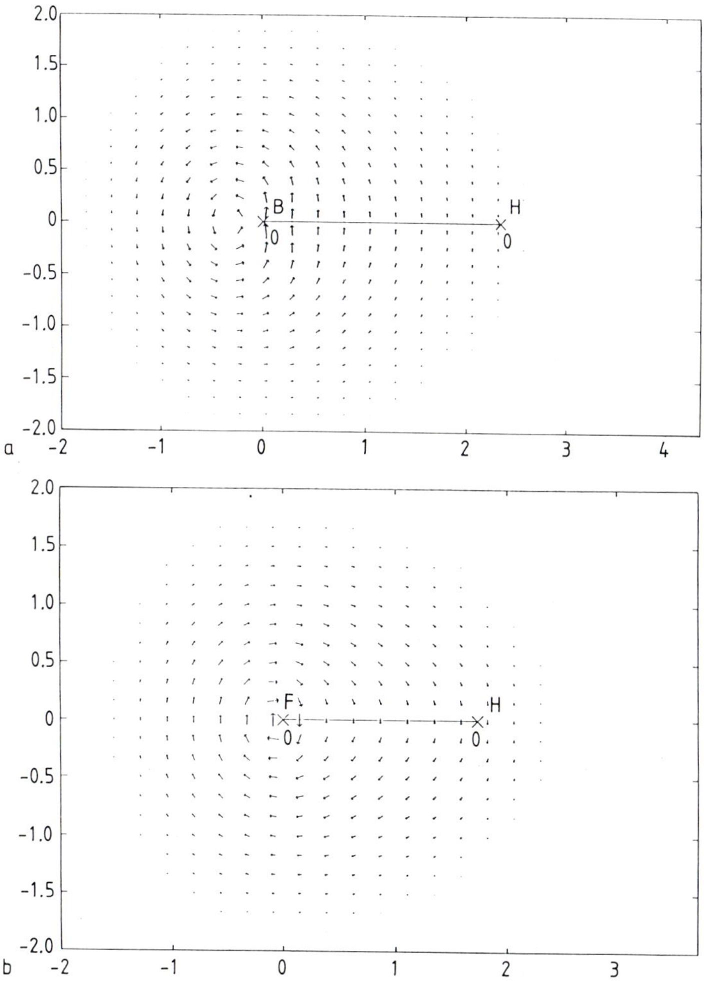

Fig. 5 a, b. The current density (**a**) in BH, (**b**) in HF in a plane containing the molecular axis for the field perpendicular to this plane. Coordinate scales in a_0

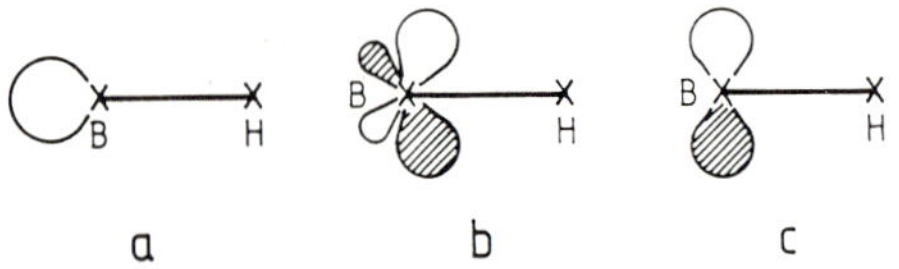

Fig. 6 a–c. Explanation of the occurrence of a paramagnetic current in BH. (**a**): Lone pair on B, (**b**): result of 'rotation' of (a), (**c**): low-lying unoccupied MO

For the molecules discussed in this section it is rather straightforward to understand the physical origin of the current density. It is also not very difficult to understand how these current densities are responsible for the susceptibility and the shift at the 'heavy' atoms. For proton shifts the situation is difficult, because the position of the protons is very off-center with respect to the center of the current density. The relation between current density $\vec{j}$ and shift σ is given by Biot–Savart's law (2.12.1). The factor $|\vec{r} - \vec{\varrho}|^{-3}$ gives the largest weight to regions close to $\vec{\varrho}$, while the dominantly exponential decay of $\vec{j}$ with respect to its 'natural' center favours regions close to this center. It is hard to realize this balance from qualitative inspection of the current density.

Both the BH bond and the lone pair MO give a paramagnetic contribution to the susceptibility and the boron shift. However, what is somewhat puzzling, only the lone pair has a (very large) deshielding effect on the proton shift (-21.5 ppm), while the BH bond MO is shielding (by ~ 12 ppm more than for the field parallel to the bond). We shall encounter more of such puzzling situations.

BH_3 has in common with BH the existence of a low-lying unoccupied π-MO, into which the BH bond MOs can be 'rotated' by the angular momentum operator in the molecular plane perpendicular to a bond. The paramagnetic current is not strong enough to make BH_3 paramagnetic [4], however the deshielding effect on $\sigma(^1H)$ is obvious (see Table 5), each BH bond contributes ~ -4.6 ppm to $\sigma(^1H)$.

The isoelectronic CH_3^+ resembles BH_3, but the effects are much stronger (probably because the positive charge stabilizes the unoccupied π-MO). As far as χ is concerned, CH_3^+ is at least van-Vleck paramagnetic for the field parallel to the molecular plane [4], though the isotropic part of χ is slightly diamagnetic. We see from Table 5 that $\sigma(^1H)$ parallel to the molecular plane has the unusually small value of 13.2 ppm.

If we go from BH_3 with its low-lying uncoccupied π-MO, to B_2H_6 where this MO is involved in bonding, the paramagnetic contributions are strongly reduced [4]. One finds, however, a striking difference between $\sigma(^1H)$ for terminal (27.6 ppm) and bridge (32.3 ppm) H-atoms. The higher shielding for H_b appears plausible in view of the higher electronic charge density. If one looks at the orbital contributions [4] one sees, however, that the contributions of the MOs of the respective BH or BHB bonds differ little and that the contributions of the 'other' bonds are responsible for the higher shielding of the bridge hydrogen. The 'other' 3-center bond has a shielding effect of 2.9 ppm on $\sigma(^1H_b)$.

All proton shifts in Tables 4 and 6 are significantly anisotropiec. The anisotropies actually cover a wider range of values than the isotropic shifts.

Table 5. As Table 4, but separately for parallel and perpendicular field directions

		K	XH	XH′	lp	σ
$\parallel$	LiH	0.58	27.55	—	—	28.12
	BeH_2	0.53	29.61	1.33	—	31.47
	BH_3	0.13	31.13	−4.56	—	22.18
	BH	0.40	29.42	—	3.64	33.46
	CH_3^+	0.10	30.78	−8.86	—	13.16
	H_2O	0.08	18.32	3.66	4.84	34.48
	HF	0.28	23.34	—	6.55	44.27
$\perp$	LiH	−0.73	26.28	—	—	25.89
	BeH_2	−0.26	28.07	−0.72	—	26.99
	BH_3	−0.24	26.40	1.71	—	29.58
	BH	0.54	41.30	—	−21.47	20.31
	CH_3^+	−0.20	18.38	2.38	—	23.93
	H_2O	−0.15	13.52	4.62	2.34	22.67
	HF	−0.13	6.88	—	4.45	20.09

In linear molecules $\parallel$ means parallel to the molecular axis, in planar molecules parallel to the plane, averaged over the two in-plane directions,
$\perp$ is perpendicular to the axis in linear molecules and to the plane in planar molecules.
(Note that in symmetric top molecules the symbols $\parallel$ and $\perp$ are often used differently)

Table 6. As Table 4 for 2nd row hydrides

	K	L	XH	XH′	lp	σ	$\Delta\sigma$
NaH	−0.01	0.05	26.14	—	—	26.20	4.4
MgH_2	0.00	0.04	27.45	0.38	—	27.10	6.2
AlH_3	0.00	0.01	28.52	−0.78	—	26.97	3.9
SiH_3^+	0.00	0.01	27.25	−1.40	—	24.45	2.9
SiH_4	0.00	0.00	27.86	0.13	—	28.24	8.3
PH_3	0.00	−0.04	25.32	1.25	1.86	29.66	10.9
H_2S	0.00	−0.01	21.98	2.32	3.23	30.76	16.3
HCl	0.00	−0.02	18.29	—	4.33	31.25	22.6

L means the L-shell of X

The smallest anisotropy ($\Delta\sigma = 2.1$ ppm) arises for LiH, where the bonding MO is essentially a 1s-AO of H^-, the largest anistropy is found for HF ($\Delta\sigma = 24.2$ ppm).

In all hydrides both the respective XH bond and the 'other' terms contribute to the anisotropy. In the case of BH the contribution of the BH bond goes in the 'wrong' direction i.e. its anisotropy is opposite to that of the total $\sigma(^1H)$. Very strong anisotropies are often caused by paramagnetic currents. However, even in 'good diamagnetic' molecules like HF a strong anisotropy may arise. In the case of HF it is understandable that the diamagnetic current induced by a field in bond direction creates a magnetic field that is stronger at the position of the proton than the corresponding field created by the current density induced by a field perpendicular to the bond. This field is, so to say, off-center with respect to the proton (Fig. 7). Note that Fig. 7 is a schematic plot of the induced magnetic field, and not of the current density.

The hydrides of second row atoms (Table 6) show the same pattern as far as the individual contributions are concerned, but the overall results differ somewhat from their first-row counterparts. There is no 'jump' between AlH_3 and SiH_4, almost no change from NaH to AlH_3 and a continuous slight increase of $\sigma(^1H)$ from AlH_3 to HCl. Note that the individual contributions vary much more than their sum, i.e. that there is a significant cancellation of opposing effects. The slight variation of $\sigma(^1H)$ does not at all reflect the drastic change in the electronic structure on going from NaH to HCl.

Of great practical interest are the shifts of protons in CH bonds [46]. The orbital contributions to σ (^1H) of various representative molecules are analyzed in terms of the LMO contributions in Table 7.

If one wants to understand the difference between $\sigma(^1H)$ in CH_4 and C_2H_6 from the data in Tables 4 and 7 one can conclude that it is mainly due to the difference of the contributions of a neighbouring CC or CH-bond, while

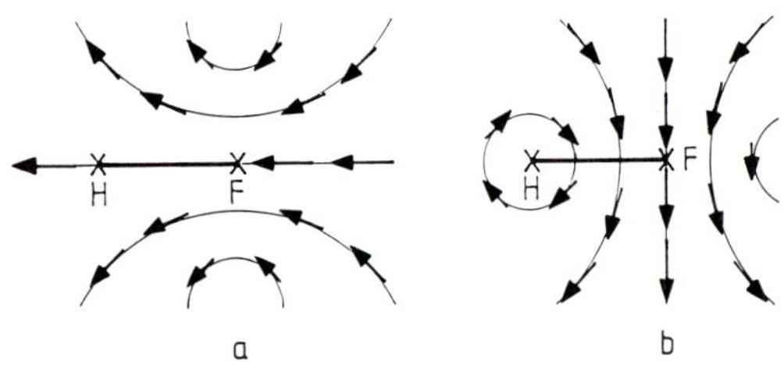

Fig. 7 a, b. Illustration of the different shielding of the proton in HF for a field applied parallel and perpendicular to the molecule axis. The field lines of the induced field are shown schematically for (**a**) the applied field parallel to the HF axis, (**b**) perpendicular to it

Table 7. LMO contributions to the absolute isotropic proton shift $\sigma(^1H)$ for some small organic molecules

	K	CH	CH′	CX	Other	σ	$\Delta\sigma$
C_2H_6	0.00	25.60	1.88	1.12	−0.04	30.45	10.3
c-C_3H_6	−0.03	25.21	1.88	1.69	1.11	31.55	16.4
Tetrahedrane	−0.14	22.45	—	1.02	2.12	27.49	14.4
C_2H_4	−0.02	26.12	−0.17	0.28	−0.49	25.73	5.5
H_2CO	−0.03	29.28	−3.60	−1.00	−2.30	22.39	1.7
H_2CS	−0.01	27.97	−3.27	−2.20	−2.26	20.23	8.6
C_2H_2	−0.11	23.97	—	5.04	0.27	29.18	18.4
HCN	−0.08	23.92	—	3.90	0.16	27.91	17.3
HCP	−0.03	25.29	—	3.12	−0.23	28.16	20.0
C_6H_6	−0.11	25.95	—	−2.65	0.85	24.03	3.4

K refers to the K-shell (1s) of C;
CH is the CH bond in which the considered H is involved;
CH′ another CH bond attached to the same C;
in CX σ and π contributions are summed; in the case of benzene CX represents the sum of the two CC-bond σ and π contributions;
'other' is the sum of all other LMO contributions; for $\Delta\sigma$ see Table 4

non-neighbour effects are negligible. The strong deshielding in ethene (relative to ethane) comes mainly from the contributions of the neighbour bonds CH and CC, which are much less shielding than in C_2H_6. This reduced shielding is due to the paramagnetic currents which are related to the existence of a low-lying antibonding π^*-MO, into which some of the occupied MOs can be 'rotated'. Even non-neighbour effects are non-negligible in C_2H_4. Formaldehyde H_2CO resembles ethene, but the paramagnetic effects are stronger. This is in part due to the smaller energy-denominator for a $n \rightarrow \pi^*$ (compared to a $\sigma \rightarrow \pi^*$) excitation, in part to the fact that the π^*-MO in a heteropolar C=X bond is more concentrated at the C atom, and finally to better overlap of $l_z n$ with π^* as compared to $l_z(XH)$. The current density in H_2CO for various field directions is plotted on Fig. 4.

Acetylene lies between ethane and ethylene and closer to the former. Triple bonds 'behave' somehow as being 'less unsaturated' than double bonds. The gap between the highest occupied and the lowest virtual MO is larger in acetylene than in ethylene, paramagnetic effects are hence less pronounced in the former. The comparison between C_2H_2 and C_2H_4 is not easy because almost all contributions change between C_2H_2 and C_2H_4 or C_2H_6. On the other hand C_2H_2 resembles very much the isoelectronic HCN. Benzene, C_6H_6, is also included in Table 7. The smaller proton shielding, as compared to C_2H_4 comes from a significant paramagnetic effect of the neighboring CC-bonds. The total contribution of the π-system to $\sigma(^1H)$ in benzene is (not documented in Table 7) + 1.2 ppm. The corresponding value in C_2H_4 is 2.4 ppm. If one assumes that in a hypothetical cyclohexatriene with one localized π-bond attached to each carbon, the contribution of this bond should be the same as in ethene, one notes a diminished shielding of the π-system by 1.2 ppm. One may regard this as due to London's ring current [14], however, one should be very careful with statements of this kind. It is curious, by the way, how controversially the existence or nonexistence [51–53] of this ring current in the π-system has been discussed in the literature.

A rationalization of ring strain effects on proton shifts does not appear to be easy, especially if one compares cyclopropane with tetrahedrane in Table 7.

A comparison of various hydrocarbons, both saturated and with isolated double bonds has revealed that many LMO contributions are approximately transferable. This allowed the construction of an ab-initio increment system, which is applicable to molecules which are too large for an IGLO calculation [26]. The results turned out to be satisfactory, although, of course, such a system, which contains no adjustable parameters, cannot compete with an empirical system, as far as the accuracy of the predictions is concerned. However, it gives more physical insight.

It is important to note at this point that an increment system can only be valid if the various LMO contributions are well 'transferable'. Transferability is usually guaranteed if the LMO contributions are dominated by local diamagnetic terms (as is the case in saturated hydrocarbons). Local paramagnetic contributions are usually not transferable, because they involve

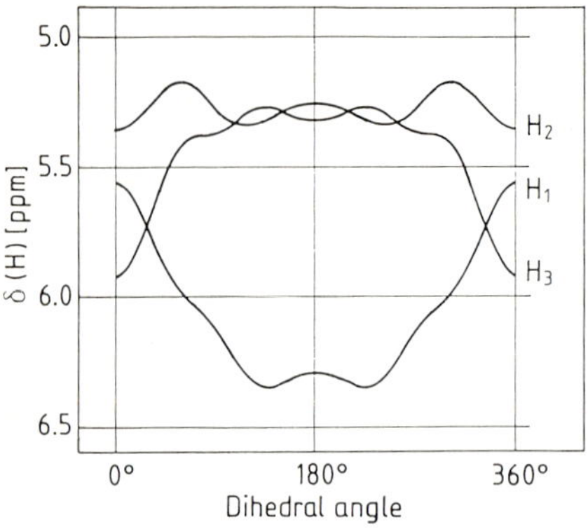

Fig. 8. Dependence of the proton shifts in 1,3-butadiene on the dihedral angle ϑ of a rotation around the CC-single bond. H_1 refers to the proton at C_2, H_2 and H_3 to the E-resp. Z-position at C_1

small energy denominators, and excitation energies are properties of the entire molecule, hence not determined by local effects.

The dependence of the chemical shifts on 'internal rotation' is an indication that an increment system with transferable contributions can only hold as a first approximation. In Fig. 8 we demonstrate the angular dependence of the chemical shift of the three non-equivalent protons in 1,3-butadiene on the dihedral angle θ of a rotation around the C–C single bond. Near the s-*trans* conformation, i.e. at $\theta = 180°$, the CH_2 protons, H_2 and H_3 change their relative positions twice, while the CH proton H_1 resonates approximately 1 ppm downfield. Near the s-*cis* conformation it is H_3 which becomes the most deshielded H atom, and $\sigma(H_1)$ moves upfield. As the variations of the proton resonances with θ cannot be traced back to changes in a few orbital contributions, it appears to be a nontrivial task to incorporate these angular dependencies into an increment system, which would be simple enough to be of practical value. Similar conclusions can be drawn from an analogous discussion of the rotational dependence of ^{13}C NMR shifts (Fig. 13; Sect. 3.4).

Both the results in Fig. 9 and those in Table 8 reveal the average accuracy of IGLO calculations (with basis II) for chemical shifts. The difference between experiment and theory is usually a few tenth of a ppm, provided that one compares the IGLO results with experimental gas phase data. Extrapolation of the gas phase data to zero pressure and temperature is likely to improve the agreement. The gas-to-liquid shifts are usually larger than the discrepancies between IGLO and experimental values, especially if in the liquid phase strong hydrogen bonds are realized [43].

Of course, the IGLO values depend to some extent on the chosen basis set and on the geometry. As far as the basis dependence of $\sigma(^1H)$ is concerned, our general experience has been that on going from basis II to basis IV changes of the order of 0.1 ppm arise, while between DZ and basis II changes of more than 1 ppm are possible. We conclude that basis II is usually sufficiently close to saturation as far as $\sigma(^1H)$ is concerned (except for negative ions like H^-, where 'diffuse' functions have to be added).

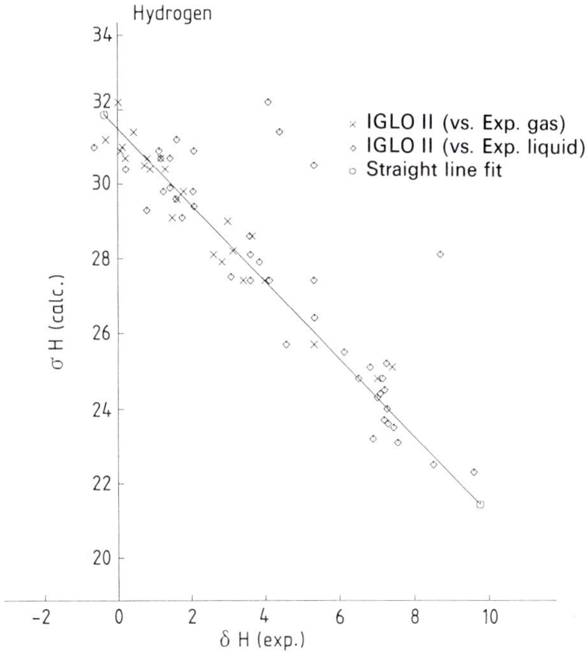

Fig. 9. Theoretical vs experimental proton shifts. See also Table 8. The straight line fit (experimental gas phase data only) yields a slope of -0.92 ± 0.05 and an intercept of 31.2 ± 0.2 with 1.0 as maximum and 0.5 as mean deviation

The variation of $\sigma(^1H)$ with the geometry is relatively large as compared to the overall range. We find e.g. for an increase (decrease) of r(OH) in H_2O by 0.007 Å a change of $\sigma(^1H)$ by -0.24 ppm and $+0.29$ ppm respectively [4]. Similar effects are found on varying the bond angle.

One should expect that use of r_0—rather than r_e—geometries for the IGLO calculation similates to some extent the average over the zero-point vibrations, and should lead to better agreement of theory and experiment. In practice this need not be the case since larger internuclear distances are likely to increase the errors due to neglect of electron correlation.

It is not yet fully understood whether Hartree–Fock theory, on which IGLO is based, is sufficiently accurate and whether effects of electron correlation need to be taken care of. Correlation effects are probably important in cases where low-lying excited states play a role, i.e. where strong paramagnetic effects arise as probably for diazenes or disphosphenes [23, 42]. BH is certainly a case where correlation effects should play a significant role [31, 32, 50], though the published results don't allow quantitative conclusions.

We close this section by summarizing some unpublished results on the effects of hydrogen bonding of $\sigma(^1H)$. We mean here 'cationic' hydrogen bonds (anionic

Table 8. Relative proton shifts $\delta(^1H)$[a]

		IGLO(DZ)	IGLO(II)	Exp(g)	Exp(l)
CH_4		0.2	−0.0	0.13[b]	−0.63[b]
C_2H_6		0.3	0.6	0.88[b]	0.22[b]
C_3H_8	C_1H	0.1	0.3	0.81[c]	
	C_2H	0.3	0.6	1.29[c]	
c-C_3H_6		−0.4	−0.5	0.20[d]	
neo-C_5H_{12}		1.1	1.0	0.97[f]; 0.95[d]	0.92[g]
c-C_6H_{12}		0.4	1.1		1.42[g]; 1.54[c]
C_2H_4		6.4	5.3	5.31[b]	4.57[b]
C_2H_2		4.2	1.9	1.48[b]	1.75[b]
c-C_3H_4	C_1H	10.1	7.8		7.06[e]
	C_3H	0.5	1.7		0.93[e]
Norbornadiene	C_1H	3.1			3.53[e]
	C_2H	8.6			6.75[e]
	C_7H	0.7			2.00[e]
HCN			3.1	2.83[b]	3.85[b]
CH_3CHO	C_1H	11.1	8.7		9.69[g]; 9.72[e]
	C_2H	1.8	1.2	1.92[h]	2.12[g]
CH_3OH	CH	2.4	3.6	3.41[i]	3.34[g]
	OH	−3.8	−1.2	0.02[i]	4.07[g]; 4.11[g]
C_2H_5OH	C_1H	0.9	0.3	1.17[i]	1.17[g]
	C_2H	2.3	2.4	3.65[i]	3.58[g]
	OH	−2.2	−0.4	0.44[i]	4.40[g]
CH_3CO_2H	C_2H		1.6		2.06[g]
HCO_2CH_3	C_1H		7.2		7.98[g]
	C_2H		2.8		3.69[g]
CH_3SH	CH		1.2		1.24[g]
	SH		0.1		2.07[g]
Oxirane			1.7		2.34[e]
CH_3F		3.3	3.6	4.13[h]	4.25[k]
CH_2F_2		4.6	4.6		5.45[k]
CHF_3		5.7	5.5		6.25[k]
CH_3Cl		1.9	2.0		2.99[l]
CH_2Cl_2		3.5	3.6		5.31[l]
$CHCl_3$		5.8	5.8		7.21[m]; 7.27[l]
C_6H_6		8.9	7.0		7.26[m]; 7.27[e]
C_6H_5CCH	o-H		7.5		7.46[m]; 7.42[m]
	m-H		7.3		7.12[m]; 7.23[m]
	p-H		6.5		7.14[m]; 7.24[m]
	CCH		3.5		3.08[g]
C_6H_5F	o-H		6.7		7.05[m]; 6.95[m]
	m-H		7.4		7.34[m]; 7.21[m]
	p-H		6.6		7.13[m]; 7.01[m]
Pyridine	o-H		8.5		8.51[m]; 8.50[e]
	m-H		6.2		7.14[m]; 7.06[e]
	p-H		7.9		7.55[m]; 7.46[e]
Thiophene	α-H		7.2		7.19[e]
	β-H		6.6		7.04[e]
NH_3		−0.9	0.1	0.05, 0.10[b]	1.12[b]
H_2O		−2.3	0.5	0.73[b]	5.32[b]
HF			2.9	2.6[b]	8.70[b]
SiH_4		2.5	2.8	3.13[b]	
PH_3		0.4	1.4	1.61[b]	1.58[b]
H_2S			0.3	0.21[b]	1.42[b]
HCl			−0.2	−0.32[b]	1.60[b]

Table 8. (*Continued*)

		IGLO(DZ)	IGLO(II)	Exp(g)	Exp(1)
CH_3SiH_3	CH	−0.6	−0.6		0.27[n]
	SiH	3.3	3.1		3.71[n]
HPF_4			5.9	7.4[o]	6.8[o]
H_2PF_3			6.2	7.0[o]	6.5[o]
CF_3PH_2			2.9		3.6[p]
HOF			12.4	12.0[q]	

[a]Referenced to TMS; the calculated shielding $\sigma(^1H)$ of TMS is 32.79 and 31.03 ppm for DZ and basis II;
[b]Ref. 90; the gas phase data, which were referenced originally to CH_4 were converted using $\delta(CH_4) = 0.13$ (Ref. 91); the liquid phase data, which were referenced originally to cyclopentane were converted using $\delta(c\text{-}C_5H_{10}) = 1.50$ (Ref. 94);
[c]Ref. 92; [d]Ref. 93; [e]Ref. 46; [f]Ref. 94; [g]Ref. 95; [h]Ref. 96; [i]Ref. 96; [k]Ref. 97; [l]Ref. 98; [m]Ref. 99; [n]Ref. 100; [o]Ref. 101; [p]Ref. 102; [q]Ref. 103

ones like in B_2H_6 have already been discussed at the beginning of this section). The results are found in Table 9.

The hydrogen bond effect is most spectacular in $[FHF]^-$. With basis II we find $\sigma(^1H)$ as low as 11.35 ppm. However for a field parallel to the molecular axis a $\sigma_\parallel$-value of 49.12 ppm is calculated that is even larger than $\sigma_\parallel$ in HF (44.27 ppm), i.e. there is a very strong shielding for a parallel field, either F^- ion is more distant from the proton than in HF, but the shielding effects of two F^- ions accumulate. For the field perpendicular to the FHF^- axis a deshielding of − 7.00 ppm relative to a bare proton is calculated. It is obviously so that at

Table 9. LMO contributions to $\sigma(^1H)$ in hydrogen bonded systems

		XH	X′H	XH′	lp X[a]	lp X′[b]	σ
HF	∥	23.34	—	—	6.55	—	44.27
	⊥	6.88	—	—	4.45	—	20.03
	iso	12.70	—	—	5.15	—	28.15
$[FHF]^-$[c]	∥	11.86	11.86	—	4.19	12.57	49.12
	⊥	−8.91	−8.91	—	1.83	5.49	−7.00
	iso	−1.99	−1.99	—	2.61	7.83	11.35
HF..*H*F[d]		12.11	0.54	—	4.81	−1.55	26.43
*H*F..HF[d]		12.55	—	—	4.84	−0.27	26.82
H_2O		17.05	—	3.93	4.75	—	30.54
H_2O..*H*OH		15.67	0.49	3.87	4.56	−1.22	28.42
H_2O..HO*H*		17.37	—	4.13	5.01	—	31.51
H$_2$O..HOH		16.91	—	3.98	4.76	−0.06	30.35

XH means the bond which the considered proton is directly involved in; X′H means the hydrogen-bond; XH′ the bond to the ‘other’ atom; lp a lone pair.
[a]for one pair;
[b]for one lone pair;
[b]sum of the lone pairs except that involved in the H-bond;
[c]a set of diffuse functions added to basis II for F;
[d]basis I

the FH-distance realized in FHF^- the magnetic field created by the diamagnetic currents around F^- is small but parallel to the applied field (see Fig. 7). The analogous effect in another H bond has been discussed [54].

Similar, but much smaller effects are found for the hydrogen-bonded dimers. A proton involved in an H bond is deshielded by 1–2 ppm relative to isolated HF or H_2O. This reduced shielding has its origin in part in the direct XH contribution and in the contribution of the lone pairs of the other O or F. The observed results for these hydrogen-bonded dimers are at least in the right way toward the observed gas-to-liquid shifts (Table 8). To account quantitatively for these shifts larger aggregates than dimers have to be considered.

3.3 *NMR Shifts of Li, Na, Be, Mg, B, and Al*

The chemical shifts of Li and Na have so far not received very much attention, because Li and Na exist mainly as Li^+ or Na^+ and little insight into the bonding situation is gained from $\sigma(^{6,7}Li)$ or $\sigma(^{23}Na)$. For some recent work on Li shifts see however Ref. 55. The Hartree–Fock limit values for isolated Li^+ and Na^+ are 95.34 ppm and 623.6 ppm respectively [56], the IGLO values for Li^+ and Na^+ with basis II are 95.5 and 623.6 ppm. Some results for $\sigma(^{6,7}Li)$ together with the LMO contributions are displayed in Table 10.

$\sigma(^{6,7}Li)$ in most Li compounds studied so far, is slightly deshielded with respect to Li^+, this deshielding is rather sensitive to the chosen basis, but it varies little between different molecules. LiH, LiF or $LiCH_3$ have within one ppm the same $\sigma(^{6,7}Li)$. The deshielding with respect to Li^+ that occurs in LiH or LiF has been explained in Sect. 2.9.

Table 10. LMO contributions to $\sigma(^{6,7}Li)$[a]

	K	LiX	Other	σ	$\Delta\sigma$[f]
Li_2	95.3	3.6	−0.0	98.8	7.3
LiH[b]	95.2	−5.6	—	89.6	17.7
LiF[b]	94.9	−2.8	−2.3	89.8	23.3
CH_3Li	95.2	−8.1	1.5	88.5	24.1
$(CH_3Li)_2$	95.1	−9.7	1.7	87.1	18.9
C_2H_5Li	95.2	−6.7	1.0	89.4	22.6
C_2H_3Li	95.2	−5.0	−0.5	89.7	17.5
C_2HLi	95.2	−5.2	1.0	91.0	20.9
C_3H_5Li	95.1	−12.5[d]	3.7	86.3	9.9
C_5H_5Li	95.1	−1.2[e]	7.3	101.3	29.2
C_6H_5Li	95.2	−5.1	−1.1	88.9	19.5
SiH_3Li[c]-n-	95.2	−4.2	0.0	91.0	18.5
SiH_3Li[c]-i-	95.2	−2.5	−1.6	91.1	16.6

[a]abbreviations of the LMOs as in Table 4 and 7;
[b]a set of diffuse functions added to basis II for H or F;
[c]-n- refers to a "normal" and -i- to an inverted structure;
[d]sum of the two "π"-bonds localized "at the same" side as the Li atom;
[e]sum of the three "π"-bonds localized "at the same" side as the Li atom
[f]$\Delta\sigma$ is the anisotropy of 2 defined in analogy to (2.8.13, 14)

Since this deshielding is related to the availability of the 2p-AOs of Li, it can only be accounted for correctly if the 2p-AOs are described appropriately by the basis. P-type basis functions, which are regarded as 'polarization functions' in ordinary SCF calculations, become now much more important. In fact a basis of three Huzinaga p-functions (basis II) is necessary to get a rather converged value of the deshielding in LiH with respect to Li^+, namely ~ -6 ppm. With a DZ basis one does not account for this deshielding, here LiH appears even shielded with respect to Li^+.

With basis II (like with smaller sets) almost no difference between $\sigma(^{6,7}Li)$ in the LiH, LiF or $LiCH_3$ molecules is found, and in other Li organic molecules there is little variation of $\sigma(^{6,7}Li)$. Li is a little more deshielded in lithium-allyl and a little less in lithium acetylene. The Li_2 molecule apparently behaves differently because here the Li nucleus is significantly shielded relative to Li^+. Another example of shielding relative to isolated Li^+ is found in lithium cyclopentadienid $Li(C_5H_5)$ ($\sigma = 101.3$ ppm with basis II, $\sigma = 102.7$ ppm with DZ basis) and more so the 'Lithiocene' $Li(C_5H_5)_2^-$ ($\sigma = 106.2$ ppm, DZ basis). It does not appear that any conclusion from $\sigma(^{6,7}Li)$ about 'covalent contributions' to the bond are possible. It would also be premature to make ring currents in $C_5H_5^-$ responsible for the strong shielding of the Li^+ in the compounds involving this ion.

Sodium behaves very much like Lithium. Since we have only studied Na^+, NaH and NaF, we can just comment that $\sigma(^{23}Na)$ is deshielded in NaH or NaF by ~ -60 ppm with respect to 'isolated' Na^+. This deshielding only arises perpendicular to the NaX bond and gives rise to a substantial anisotropy of ~ 80 ppm. Both the deshielding and the anisotropy and much stronger than in Li-compounds. A deshielding of the same order of magnitude has been found experimentally in sodium salts in aqueous solutions, where it is regarded as due to hydration of Na^+ [57].

For Na a problem arises for the first time that is encountered with all 2nd-row atoms. While for 1st row atoms there is no significant difference whether or not the K-shell MOs are included in the localization, this is no longer so for 2nd row atoms. If one localizes the L-shell MOs one obtains essentially sp^3 hybrid like AOs. In a standard IGLO calculation the centroids of charge are then taken as gauge origins. The resulting L-shell contributions are the same as for 2s and 2p-like MOs only if the basis is sufficiently large ('steep' d-type polarization functions for the L-shell are required). One can exclude the K- and L-shell from the localization, but if there are more than one 2nd row atoms of the same kind, the K-shell-AOs will be delocalized, and it is at least necessary to perform a separate localization for K- and L-shell-MOs and of the valence MOs as is done in the default option of the IGLO program.

On Be and Mg very few results are available so far. One can compare the basis II results for $\sigma(^9Be)$ in Be^{2+} (130.7 ppm), Be (149.3 ppm) with that in BeH_2 (95.3 ppm). Similarly one compares $\sigma(^{25}Mg)$ in Mg^{2+} (694.0 ppm), Mg (708.8 ppm) and MgH_2 (450 ppm). One also notes the strong anisotropy in BeH_2 (80.6 ppm) or MgH_2 (360.9 ppm). The deshielding has similar origins as in Li or Na compounds.

Following a suggestion of R. Benn we have calculated 'magnesocene' $Mg(C_5H_5)_2$. Due to the size of this molecule only DZ calculations were possible. The difference of $\sigma(^{25}Mg)$ in $Mg(C_5H_5)_2$ and MgH_2 from a DZ basis (182 ppm) agrees rather well with the experimental difference (184.6 ppm) [58a] between $Mg(C_5H_5)_2$ and $Mg(C_2H_5)_2$. With respect to Mg^{2+} (714.6 ppm in DZ) a slight deshielding of ~21 ppm in $Mg(C_5H_5)_2$ is found, while MgH_2 is deshielded by ~200 ppm. In the sandwich complex all 2p AOs of Mg are obviously used somehow for bonding and are less available as low-lying unoccupied MOs to cause paramagnetic currents. Like for lithiocene it would be premature to invoke a ring current of the π-system to explain the $\sigma(^{25}Mg)$ in magnesocene. DZ-basis calculations are certainly not adequate for far-reaching conclusions.

Boron shifts have been studied to some extent both experimentally (via either of the two nuclei ^{10}B and ^{11}B) and theoretically. IGLO calculations with basis II [59] are compared in Table 11 and Fig. 10 with experimental values for a variety of boron compounds. The agreement between theory and experiment is usually satisfactory. The chemical shifts of carboranes are sensitive to geometry changes, like those of carbocations.

Preliminary results for B_4H_{10} based on a geometry from the literature were rather disappointing, but use of an ab-initio optimized geometry [60] improved

Table 11. Relative boron shifts $\delta(^{10,11}B)$

		IGLO (DZ)	IGLO (II)	Experiment[b]
$OCBH_3$		−41.1	−46.8	−52.0
H_3PBH_3		−33.9	−39.2	−42.5; −43.0
BH_4^-		−33.5	−39.1	−36.0; −40.5
BH_3		79.7	83.7	—
$B(CH_3)_3$		74.9	83.3	84.0–86.8
BF_3		14.1	12.2	9.4[c]
B_2H_6				16.0–18.0; 16.6(g)[c]
B_4H_{10}	$B_{1,3}$	−42.7	−40.4	−40.0[c]
	$B_{2,4}$	−2.9	−2.8	−6.5[c]
B_5H_9	B_1	−52.8	−53.3	−51.8[c]
	B_{2-5}	−8.8	−10.5	−12.5[c]
1,5-$C_2B_3H_5$		12.0	11.4	1.4[d]
1,6-$C_2B_4H_6$		−17.4	−18.3	−19.4[e]
1,2-$C_2B_4H_6$	B_2	−14.7	−15.9	−15.3[e]
	B_6	5.9	2.8	−1.6[e]
2,4-$C_2B_5H_7$	$B_{1,7}$	−17.9	−21.6	−23.5[e]
	B_3	7.5	4.9	5.0[e]
	$B_{5,6}$	6.9	7.1	2.0[e]
1,2-$C_2B_{10}H_{12}$	$B_{9,12}$	1.9		−3.1[f]
	$B_{8,10}$	−10.6		−9.6[f]
	$B_{3,6}$	−16.3		−15.0[f]
	$B_{4,5,7,11}$	−14.3		−13.9[f]
Borazene		27.2	32.6	29.1–30.5

[a]Referenced to $BF_3{\cdot}OEt_2$; B_2H_6 was used as a primary reference for the calculated data, a value of $\delta(B_2H_6, g) = 16.6$ ppm was used for conversion, the calculated shielding $\sigma(^{10,11}B)$ of B_2H_6 is 114.9 and 102.2 ppm for DZ and basis II;
[b]Ref. 104; [c]Ref. 105; [d]Ref. 106; [e]Ref. 107; [f]Ref. 108

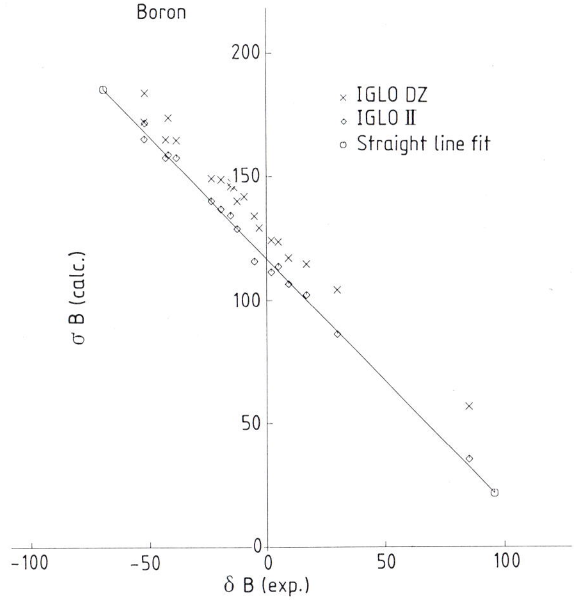

Fig. 10. Theoretical vs experimental boron shifts. See also Table 11. The straight line fit yields a slope of -0.99 ± 0.02 and an intercept of 116.9 ± 2.7 with 6.1 as maximum and 2.6 as mean deviation

the situation considerably. For one isomer of $C_2B_3H_5$ the agreement between theory and experiment is still unsatisfactory even with an optimized geometry.

A decomposition of $\sigma(^{10,11}B)$ into LMO contributions can be found in Table 12. One sees that $\sigma(^{10,11}B)$ of BH_4^- is almost entirely determined by the K-shell contribution, one BH bond only deshields by ~2 ppm. A similar situation is found in other molecules with a filled electron octet around B, like H_3PBH_3 or $OCBH_3$.

Very different is BH_3 with its low-lying unoccupied π-MO, which is responsible for the strongly deshielding contributions of the BH bonds.

Table 12. LMO contributions to $\sigma(^{10,11}B)$[a]

	K	BH	BX	lp	Other	σ	$\Delta\sigma$
BH_4^-	165.7	−1.9	—	—	—	157.9	0.00
BH_3F^-	165.7	−12.7	−8.6	—	2.0	121.1	46.3
$OCBH_3$	165.6	−0.3	1.0	—	−0.4	165.2	22.3
H_3PBH_3	165.7	−3.2	2.7	—	−0.9	158.8	3.3
BH_3	165.7	−43.5	—	—	—	35.1	178.9
$B(CH_3)_3$	165.7	—	−43.7	—	1.0	35.5	141.3
BH_2F	165.7	−39.6	−9.7	—	−0.3	66.7	64.1
BHF_2	165.7	−28.8	−28.8	—	—	95.7	7.8
BF_3	165.7	—	−20.6	—	0.9	106.7	−19.1
Borazene	165.7	−23.1	−26.7	—	−3.0	86.2	51.9
B_2H_6	165.8	−21.2	−8.8[b]	—	−3.0	102.2	54.7
BH	165.8	−177.0	—	−251.6	—	−262.8	692.1

[a]Abbreviations of the LMOs as in Table 4 and 7;
[b]Contribution of one BHB three-center bond

Replacement of H by CH_3 changes extremely little. More interesting is the effect of fluorine substitution. The deshielding by a BF bond in BF_3 is roughly only half that of a BH bond in BH_3. This is obvioulsy related to the π-donating effect of F which allows the $2p_z$-AO on B to participate in bonding and reduces its availability for paramagnetic effects. Note that the shielding effect of fluorine substitutions is limited to those cases where the π-donor effect of F dominates, and that is rather exceptional. In methanes, silanes etc. fluorine substitution always leads to deshielding (see Sect. 3.4.6). Molecules with a filled octet around B behave in this case like CH_4, i.e. fluorine substitution on BH_4^- has a deshielding effect.

In diborane B_2H_6 the unoccupied MO of BH_3 is also used to some extent for bonding and the deshielding is considerably reduced with respect to BH_3, but not as much as in BH_4^-. In borazene the situation is somewhat similar.

An extreme case is (the experimentally unknown) BH, which is a van-Vleck paramagnetic molecule (see Sect. 3.2).

For $\sigma(^{27}Al)$ there are again very few theoretical data available, but Al resembles B to a large extent. Most Al compounds are strongly deshielded with respect to Al^{3+}. Some results for $\sigma(^{27}Al)$ with basis II are AlH_3: 332.6 ppm, $(H_3C)AlH_2$: 316.5 ppm, $HOAlH_2$: 405.6 ppm, $FAlH_2$: 411.4 ppm, AlF_3: 521.6 ppm. In the same basis one gets Al^{3+}: 750.3 ppm.

The calculated difference (basis II) between AlH_3 and Al_2H_6 of 120 ppm (112 ppm for DZ) compares well with the experimental difference of 100–120 ppm for mono- and dimeric alkylubstituted alanes [58b].

3.4 Carbon NMR Shifts in Neutral Molecules

^{13}C is probably, next to ^{1}H, the best-studied nucleus, both from the experimental [61] and the theoretical side [26]. To demonstrate the quality of IGLO calculations on carbon shifts, we have prepared two Tables (13 and 14) on the comparison of theoretical and experimental results. In Table 13 absolute experimental shifts $\sigma(^{13}C)$ in the gas phase at 300 K extrapolated to zero pressure [41] are compared with IGLO values of $\sigma(^{13}C)$ for three basis sets. The experimental values in Table 13 are as close as one can get experimentally to $\sigma(^{13}C)$ of isolated molecules. These values are based on measured (and extrapolated) relative shifts and the absolute $\sigma(^{13}C)$ of one reference molecule (here CO) obtained from microwave spectra together with a theoretical diamagnetic part of σ. All molecules of Ref. 41 are included in Table 13.

The agreement between experiment and IGLO with basis II and III is quite good, as can also be seen from Fig. 11. The deviations are usually of the order of a few ppm. In some cases like CF_4, CS_2, CO the 'errors' are larger. Perfect agreement between theory and experiment cannot be expected, mainly for the following reasons.

(a) The basis is never complete. (In fact, extension of the basis, which is e.g. possible for CH_4 often improves the agreement between experiment and

Table 13. Absolute shifts $\sigma(^{13}C)$

		IGLO(DZ)	IGLO(II)	IGLO(III)	Exp.[a]
CH_4		218.3	198.5	193.8	195.1
TMS		218.1	192.7		188.1
CH_3CN	C_2	211.2	192.2	189.1	187.7
C_2H_6		212.3	185.3	180.9	180.9
CH_3NH_2		191.5	166.1	162.2	158.3
$(CH_3)_2CO$	C_2	192.6	169.2		158.0
CH_3CHO	C_2		167.5		157.2
CH_3OH			144.6		136.6
C_2H_2		139.2	121.7	116.1[b]	117.2
CH_3F		162.7	126.5	122.1	116.8
CH_2CCH_2	C_2		118.4		115.2
HCN			77.2	72.1	82.1
CH_3CN	C_2	84.8	70.4	64.4	73.8
C_2H_4		85.8	62.6	58.3	64.5
CF_4		105.3	83.2	76.5	64.5
CO_2			61.7	50.7[b]	58.8
C_6H_6		86.5	60.4		57.2
OCS			18.9	10.1	30.0
CO[c]			−21.5	−23.7	1.0
CH_3CHO	C_1		−7.2		−6.7
CS_2			−45.7	−56.1	−8.0
$(CH_3)_2CO$	C_1	6.2	−17.0		−13.1
CH_2CCH_2	C_2		−36.1		−29.3

[a]Gas phase data extrapolated to zero density at T = 300K from Ref. 41;
[b]Exponents of the polarization function differ from those given in Table 3, cf. Ref. 22;
[c]The value of 1.0 ppm was used to define the experimental absolute scale

theory. On the other hand comparsion of the results with basis II and III suggests that with these basis sets one is not too far from saturation).

(b) Effects of electron correlation have been neglected. These are likely to play a role for some molecules, e.g. CO.

(c) In experiments at 300 K a Boltzmann average over vibrationally excited states is involved. Experiment and theory differ at least in the averaging over the zero point motion, while theory is done for a fixed geometry.

One sees from Table 13 that the absolute shifts $\sigma(^{13}C)$ obtained with the DZ basis are not very good. For relative shifts $\delta(^{13}C)$ the DZ results are, on the whole, much better (Table 14), due to a rather systematic error in the direction of a more diamagnetic value for all molecules. The error varies, however, if one compares primary, secondary, tertiary or quaternary carbons. It increases in the order primary, secondary, tertiary, quaternary.

One should note that the IGLO values depend somewhat on the chosen localization procedure. The values given here are based on the actual default option of the IGLO program, according to which core and valence MOs are *not* mixed in the localization. In previous papers sometimes 'full' localization was used. The difference between the two localizations vanishes in the limit of a complete basis. For basis II differences of 1 to 2 ppm in $\sigma(^{13}C)$ arise, for the

Table 14. Relative shifts $\delta(^{13}C)$[a]

Molecule		IGLO(DZ)	IGLO(II)	IGLO(III)	Exp.(g)	Exp.(1)[b]
CH_4		−0.1	−5.8		−13.2[g]; −7.0[d]	−2.3
C_2H_6		5.8	7.4	7.1	1.1[g]; 7.2[d]	6.5
CH_3NH_2		26.7	20.6	25.9	29.8[d]	28.3
CH_3OH		42.8	48.1		51.5[d]	49.9
CH_3F		55.5	66.7	66.2	71.9[e]; 71.3[d]	71.6
CH_3SiH_3		−15.0	−14.3			
CH_3PH_2		−5.0	−5.5			−4.4
CH_3SH			6.2			6.5
CH_3Cl		19.3	24.4	25.0[c]		25.6
CH_2F_2		89.4	99.1	99.0	110.6[e]	111.7
CHF_3		103.4	106.3	106.9	119.7[e]	118.8
CF_4		112.8	109.5	111.6	123.7[e]; 123.6[d]	119.9
CH_2Cl_2		43.7	53.2			54.4
$CHCl_3$		69.9	80.6			77.7
CCl_4		102.4	110.1			96.7
$CFCl_3$		121.1	121.4	122.6		117.6[k]
C_2H_5OH	C_1		54.7			57.0
	C_2		13.4			17.6
C_3H_8	C_1	12.5	13.8	13.5	17.2[f]	16.1
	C_2	7.5	14.9	14.8	18.8[f]	16.3
n-C_4H_{10}	C_1	15.7	16.5		11.0[i]	13.1
	C_2	17.2	24.6		24.4[i]	24.9
i-C_4H_{10}	C_1	19.3	19.9		22.2[i]	24.6
	C_2	9.8	22.8		22.7[i]	23.3
n-C_5H_{12}	C_1	17.5	19.2		9.9[h]	13.7
	C_2	18.6	26.8		20.6[h]	22.6
	C_3	28.0	36.1		32.7[h]	34.6
i-C_5H_{12}	C_1		16.0			21.9
	C_2		23.5			29.7
	C_3		21.8			31.7
	C_4		2.3			11.4
Neo-C_5H_{12}	C_1	33.8	33.1			31.6[l]
	C_2	10.8	25.0			28.0[l]
c-C_3H_6		−1.4	−7.3			−2.8
c-C_6H_{12}		14.0	18.8			27.6
C_2H_4		132.4	130.1	129.7	117.4[g]; 128.8[f]; 125.5[d]	123.5
C_2H_2		79.0	71.1		70.9[d]	71.9
c-C_3H_4	C_1	139.3	120.0			108.7
	C_2	13.6	3.6			2.3
Benzene		130.5	132.3		130.9[d]	128.5
Toluene	C_1	136.8				137.8
	C_2	139.6				129.7
	C_3	135.9				128.4
	C_4	133.5				125.5
	C_α	13.9				21.3
Nonbornadiene	C_1	42.8				50.9
	C_2	150.3				143.5
	C_7	67.4				75.5
Fluorbenzene	C_1		166.4			163.3
	C_2		117.9			115.5
	C_3		135.9			130.1
	C_4		127.0			124.1

Table 14. (*Continued*)

Molecule		IGLO(DZ)	IGLO(II)	IGLO(III)	Exp.(g)	Exp.(l)[b]
Phenylacetylene	C_1		119.5			122.3
	C_2		140.2			132.1
	C_3		131.5			128.1
	C_4		132.8			128.2
	C_α		76.7			83.5
	C_β		77.2			71.7
Oxirane		29.7	33.1	33.9		40.6
Flurane	C_2	151.7	148.7			143.6
	C_3	120.9	112.4			110.4
Thiophene	C_2		134.7			125.6
	C_3		128.5			127.3
Pyridine	C_2	159.0	156.3			149.9
	C_3	124.1	123.5			123.8
	C_4	140.3	142.6			136.0
Phosphabenzene	C_2		165.9			154.1
	C_3		143.8			133.6
	C_4		130.1			128.8

[a]Referenced to TMS; the calculated shielding $\sigma(^{13}C)$ of TMS is 218.1 and 192.7 ppm for DZ and basis II; for basis III CH_4 was used as primary reference for the calculated data, a value of $\delta(CH_4) = -5.8$ ppm was used for conversion, the calculated shielding of CH_4 using basis III is 193.8 ppm;
[b]Ref. 61; [c]Basis II used for hydrogen; [d]Ref. 41; [e]Ref. 109; [f]Ref. 110; [g]Ref. 111; [h]Ref. 112; [i]Calculated using the gas to liquid shift of Ref. 113; [k]Ref. 114; [l]Ref. 115

DZ basis differences of the order of 6 ppm. These differences usually cancel in the relative shifts $\delta(^{13}C)$.

The relative shifts $\delta(^{13}C)$ in Table 14 cover a much wider range of molecules than the $\sigma(^{13}C)$ in Table 13. In many cases experimental values both in the gas phase and the condensed phase are given, and one sees that the gas-to-liquid shifts are often of the same order of magnitude as the differences between experiment and theory.

The theoretical $\delta(^{13}C)$ values in Table 14 were obtained with the aid of $\sigma(^{13}C)$ for TMS calculated in the same basis, which is possible for DZ and basis II. For basis III the difference to CH_4 was calculated and the basis-II-difference between CH_4 an TMS was added. This difference (~ 6 ppm) is rather close to the experimental one between gaseous CH_4 and TMS. In the liquid state the experimental difference is only 2.3 ppm, so one must be careful in order to be consistent. On DZ level the same difference is practically zero.

For an understanding of the factors that influence the ^{13}C shifts, an LMO analysis of various typical molecules is useful. The relevant data are in Table 15.

What one realizes from Table 15 is that the contribution of the K-shell of C is almost invariably 200.9 ppm (the exact HF value for the K-shell in a C atom is 201.1 ppm). This value is, moreover, practically isotropic. The contribution of the CH-bond LMO is always negative (deshielding), it is nearly zero in saturated hydrocarbons, but appreciably deshielding (~ -30 ppm) if the C is

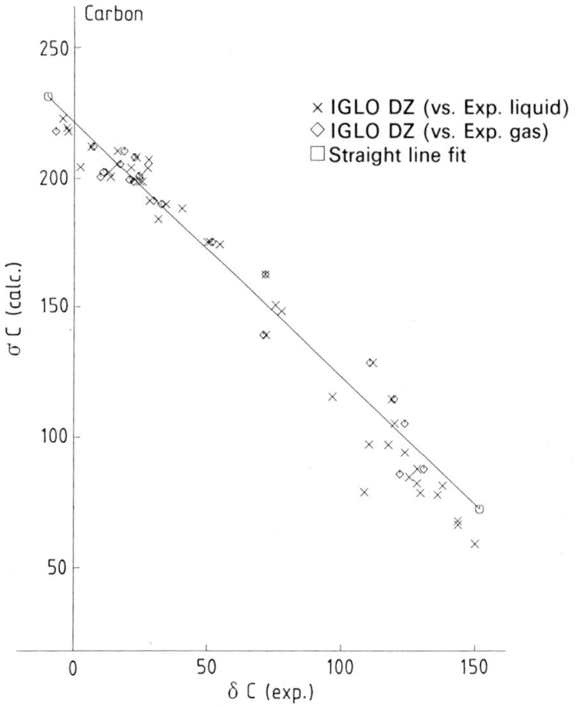
Carbon
× IGLO DZ (vs. Exp. liquid)
◇ IGLO DZ (vs. Exp. gas)
□ Straight line fit
σ C (calc.)
δ C (exp.)
250
200
150
100
50
0
50
100
150

a

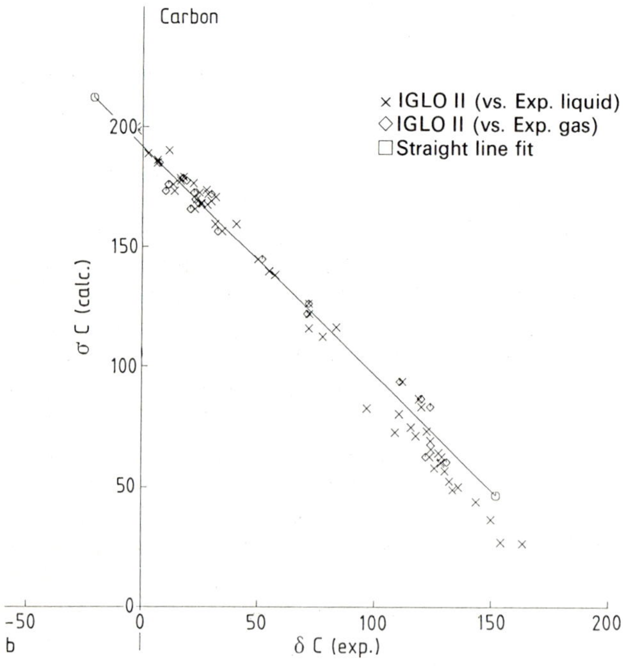
Carbon
× IGLO II (vs. Exp. liquid)
◇ IGLO II (vs. Exp. gas)
□ Straight line fit
σ C (calc.)
δ C (exp.)
200
150
100
50
0
-50
0
50
100
150
200

b

Table 15. LMO contributions to $\sigma(^{13}C)$

	K	CH	CX	Other	σ	$\Delta\sigma$
CH_4	200.9	−0.6	—	—	198.5	0
C_2H_6	200.9	−2.3	−7.8	−0.8	185.3	14
c-C_3H_6	200.9	−1.9	1.8	−0.7	200.0	57
Tetrahedrane	200.9	6.8	4.8	4.7	226.7	−52
C_2H_4	200.9	−31.1	−70.5	−5.6	62.6	180
H_2CO	200.9	−69.5	−70.2	9.5	1.5	180
H_2CS	200.9	−79.2	−107.1	−2.0	−66.7	362
C_2H_2	200.9	−31.5	−38.0	−9.7	121.7	245
HCN	200.8	−36.8	−57.5	−29.4	77.2	305
HCP	200.9	−60.9	−94.8	−27.3	18.0	368
CO_2	200.9	—	−96.4	−42.8	61.7	331
CS_2	201.0	—	−191.3	−55.4	−45.7	494
CO	201.0	—	−184.9	−37.5	−21.5	437
C_6H_6	200.9	−23.8	−114.7[b]	−2.0	60.4	197
CH_3^+	201.0	−125.0	—	—	−175.6	545

[a]Abbreviations of the LMOs as in Tables 4 and 7;
[b]Sum of the two CC bonds ($\sigma + \pi$)

involved in a double or triple bond, the largest effect being calculated for H_2CS, the smallest one in benzene. Tetrahedrane is one of the rare cases where the CH bond (and the CC-bond) contribution is positive (shielding). The effect of the CX (i.e. CC, CN or CO) bond in saturated molecules without ring strain is slightly deshielding while CX multiple bonds deshield strongly. Non-neighbour effects are practically negligible in C_2H_6 and C_2H_4 (these are those of the CH-bonds attached to the other C-atom), but they are rather strong in molecule like HCN, CO, CO_2 where they come mainly from the lone pairs at N or O. It is generally so that non-neighbour lone-pair effects are larger by an order of magnitude than other non-neighbour effects.

The downfield shift of C_2H_4 as compared to C_2H_6 is, of course, due to the availability of the low-lying antibonding π^*-MO and the stronger effect in H_2CO is related to the lower $n \rightarrow \pi^*$-excitation energy (as compared to $\sigma \rightarrow \pi^*$). This has already been discussed in the context of proton shifts. It is noteworthy that the downfield shift of C_2H_4 results to about the same extent from the CH bonds and the CC double bond. Like for $\sigma(^1H)$ there is no simple explanation why the downfield shift is less pronounced in C_2H_2 than in C_2H_4. From the LMO increments in Table 15 one may conclude that one CH-bond has about the same effect in C_2H_4 and C_2H_2, such that C_2H_4 is more deshielded because there are two CH bonds attached to one C in C_2H_4.

Fig. 11 a, b. Theoretical vs experimental carbon shifts. See also Tables 13 and 14. (**a**) DZ data. The straight line fit (experimental gas phase data only) yields a slope of -0.95 ± 0.04 and an intercept of 219.7 ± 3.0 with 18.6 as maximum and 8.4 as mean deviation. (**b**) Basis II data. The straight line fit (experimental gas phase data only) yields a slope of -0.94 ± 0.03 and an intercept of 191.1 ± 2.1 with 14.1 as maximum and 5.9 as mean deviation

From Table 15 one does not get the impression that any LMO contributions are transferable, except those of the K-shell MO. Transferability is found, however, if one compares molecules with similar bonding situation, e.g. aliphatic hydrocarbons, as is done in Table 16. One sees (using basis II), that after the constant 1s contributions, the CH and CC bonds attached to the considered C-atom dominate with increments that are nearly transferable, while non-neighbour CC or CH bonds have practically negligible contributions. The information of Table 16 on the CC and CH contributions to $\sigma(^{13}C)$ in saturated hydrocarbons is extended in Fig. 12. There one can see, that although direct non-neighbour effects are very small, a CC bond contribution depends on the groups that are attached to the 'other' C (i.e. not that, for which $\sigma(^{13}C)$ is studied). For $\sigma(^{13}C)$ in hydrocarbons (including even those with isolated double bonds) an ab-initio increment system has been proposed in Ref. 26, that could e.g. be used for molecules that are too large for an explicit IGLO calculation. The possibilities of such a system have not yet been fully explored. See also the general comments on increment systems for ^1H shifts in Sect. 3.2.

When setting up a ^{13}C NMR increment system we encounter similar problems in the case of ^1H NMR. Again we face the difficulty of incorporating anisotropy effects that show up in the variation of the ^{13}C NMR shifts with the relative orientation of their atoms. As an example, the dependence of the ^{13}C NMR resonances in 1,3 butadiene on the dihedral angle θ describing the rotation of the C–C single bond is shown in Fig. 13. Although we do not find as dramatical changes as for the ^1H NMR data, where the relative positions of the different H atoms changed when rotating s-*trans*-butadiene to s-*cis* butadiene, the variations of $\sigma(^{13}C)$ with θ for the two different C atoms are not at all parallel to each other.

Table 16. LMO contributions to $\sigma(^{13}C)$ for the central C atom[a]

	CH	CX	Other	σ
CH_4	−0.6	—	—	198.5
H_3CCH_3	−2.3	−7.8	−0.8	185.3
$H_2C(CH_3)_2$	−2.4	−8.7	−0.8	177.8
$HC(CH_3)_3$	−1.8	−9.7	−0.3	169.9
$C(CH_3)_4$	—	−8.3	0.0	167.7
H_3CF	−20.8	−12.7	0.8	126.5
H_2CF_2	−34.1	−20.9	2.5	93.6
HCF_3	−37.4	−27.3	4.5	86.4
CF_4	—	−30.5	4.2	83.2
H_3CCl	−10.0	−3.1	0.6	168.3
H_2CCl_2	−18.0	−12.3	−1.0	139.5
$HCCl_3$	−23.2	−20.8	−3.3	112.2
CCl_4	—	−27.8	−7.4	82.6

[a] Abbreviations of the LMOs as in Tables 4 and 7

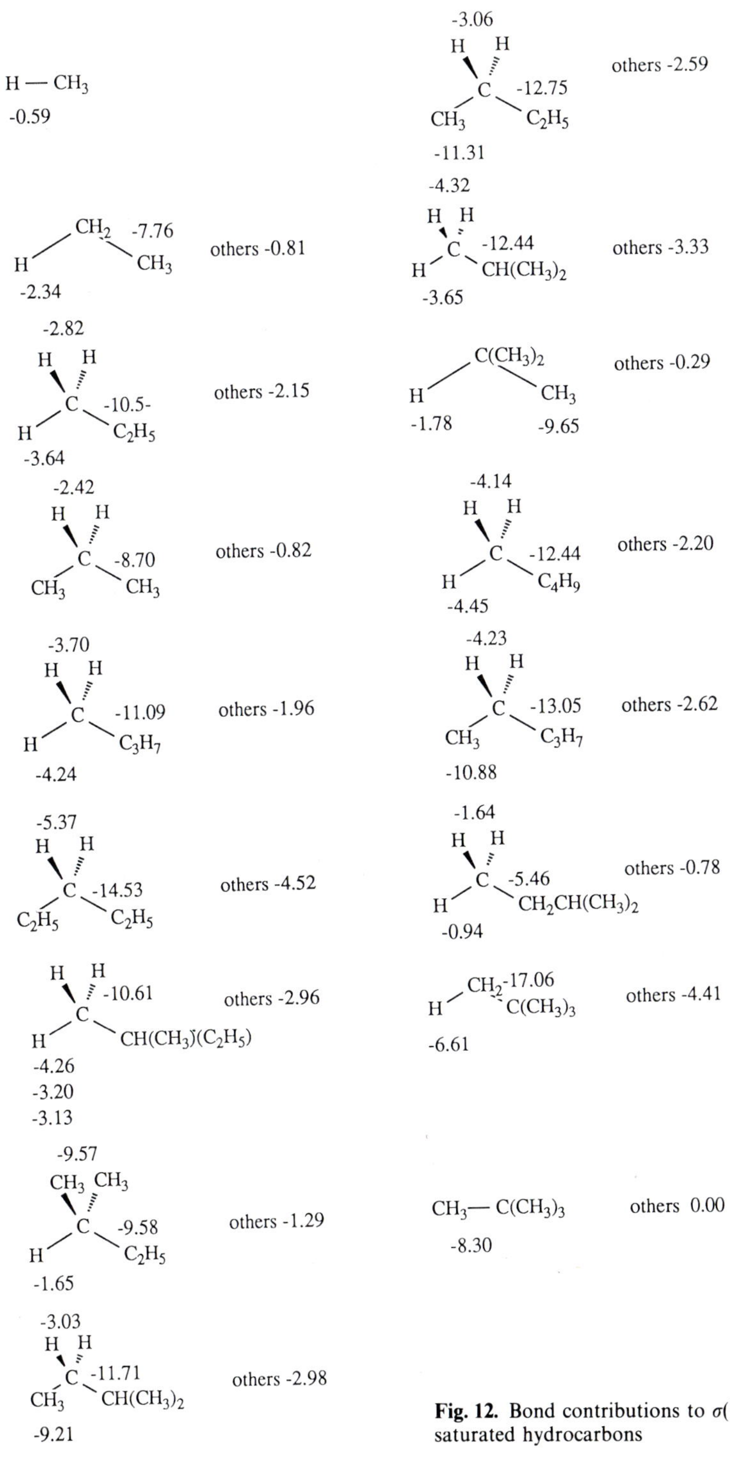

Fig. 12. Bond contributions to $\sigma(^{13}C)$ in saturated hydrocarbons

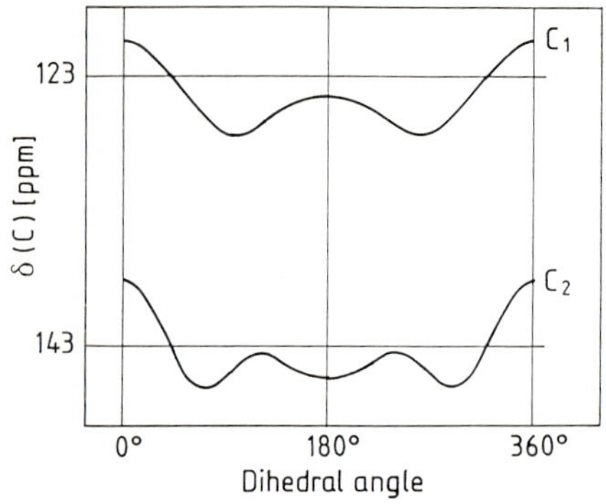

Fig. 13. Dependence of the carbon shifts in 1,3-butadiene on the dihedral angle ϑ of a rotation around the CC-single bond

Often one is faced with problems that, by definition, cannot be understood within an increment system. Consider e.g. the two isomeric 3,4-dimethylhexanes

1 **2**

with **1** the meso-form and **2** the racemic dl-form. One sees from Table 17 that a comparison of the experimental and theoretical shift differences allows an unambigous assignment of the observed spectra of the two isomers. A calculation with a DZ basis is sufficient in this case, provided that one remembers that DZ results are not satisfactory for tertiary carbon atoms, and that hence the results for the C_3 atoms should be disregarded. An assignment is also possible by conventional chemical means [62], but this is certainly more time-consuming.

Going back to Table 16 we realize that transferable bond increments do not exist for halogenated methanes. From CH_4 to FCH_3 the deshielding effect of one CH bond is drastically increased and again so from FCH_3 to F_2CH_2. The contribution of a CF bond in any of these molecules is smaller in absolute value

Table 17. Calculated and experimental (relative) chemical shifts $\delta(^{13}C)$ for the two isomeric 3,4-dimethylhexanes

IGLO(DZ)				Exp.[a]		
C Atom	Rac.	Meso	Δ	Rac.	Meso	Δ
1	14.4	14.6	−0.2	12.1; 11.8	12.1; 11.8	0.0
2	24.4	21.2	3.2	27.5; 27.6	25.6; 25.8	1.9; 1.8
3	29.1	28.2	0.9	38.3; 38.5	39.3; 39.5	−1.0
7	14.0	17.0	−3.0	14.0; 15.8	16.0; 15.8	−2.0

$\Delta = \delta_{rac} - \delta_{meso}$; [a] Ref. 62; Ref. 115

than that of a CH bond. The origin of the deshielding effect of fluorine substitution is the existence of low-lying antibonding σ^* MOs of the CF bonds, which in view of the polarity of the bonds are mainly localized at the C atoms. LMOs of other bonds can be 'rotated' into linear combinations of the σ^* MOs, which explains that the contributions of the other bonds are more deshielding. Chlorine substitution is qualitatively similar to fluorine substitution, but quantitatively different.

The cooperation between experimental and theoretical ^{13}C NMR has recently been particularly successful for studies of the full $\boldsymbol{\sigma}$ tensors, which can be observed in low temperature solid state measurements. In the case of a few polycyclic systems including cyclopropane, bicyclobutane and propellane [63]

Table 18. Principal and average (isotropic) values of the relative shift tensors $\boldsymbol{\delta}$ for some small cyclic and polycyclic systems[a]. Experimental data and IGLO results are given in consecutive lines

		δ_{11}	δ_{22}	δ_{33}	δ_{av}
Cyclopropane	CH_2	22	2	−36	−4
		21	3	−34	−7
Bicyclobutane	CH_2	41	35	22	32
		37	32	4	24
	CH	23	−19	−21	−6
		26	−30	−32	−12
[1.1.1] Propellane	CH_2	138	57	43	79
		132	47	37	72
	C	35	−11	−11	4
		24	−17	−17	9

[a]Referenced to TMS; the calculated shielding $\sigma(^{13}C)$ of TMS is 192.7 ppm (note that basis II was called basis IV in Ref. 63). $\delta_{kh} = \sigma_{ref} - \sigma_{kh}$ where σ_{kh} are the principal values of $\boldsymbol{\sigma}$.

Table 19. Comparison of the principal values of the tensor $\boldsymbol{\delta}(^{13}C)$ of the relative chemical shifts for three types of C-atoms in *n*-eicosane with IGLO calculations for the corresponding C atoms in smaller n-alkanes

	δ_{11}	δ_{22}	δ_{33}	δ_{av}
C_1 in *n*-$C_{20}H_{42}$ exp.[a]	26	22	3	17
C_1 in *n*-C_4H_{10} IGLO	27.3	21.7	0.5	16.5
C_1 in *n*-C_5H_{12} IGLO	31.4	24.3	1.7	19.2
C_2 in *n*-$C_{20}H_{42}$ exp.[a]	37	30	15	27
C_2 in *n*-C_4H_{10} IGLO	36.1	34.7	3.1	24.6
C_2 in *n*-C_5H_{12}IGLO	38.7	27.1	15	26.8
C_{3-18} in *n*-$C_{20}H_{42}$ exp.[a]	50	38	17	35
C_3 in *n*-C_5H_{12} IGLO	55.7	36.8	15.9	36.1

[a]Ref. 116, for δ_{kh} see footnote to Table 18

a unique assignment of the principal values of the $\underline{\underline{\sigma}}$ tensors has e.g. only been possible with the aid of IGLO calculations. Some results are collected in Table 18.

A rather nice so far unpublished example is that of n-eicosane (Table 19). For this molecule a full IGLO calculation is not possible, but in view of the remarks made in connection with increment systems, it should be possible to simulate the CH_3 group of eicosane by that in n-butane or n-pentane in the appropriate configuration, similarly the first and the subsequent CH_2 groups. The agreement between experimental and theoretical principal values of the σ tensors is surprisingly good.

The full σ-tensors have also been studied for various molecules with double and triple bonds involving carbon [64]. The agreement between IGLO and experiment (Table 20) is usually quite good, although the errors for the individual tensor components are somewhat larger than for the isotropic parts, especially for ethylene. The 'basis set II' used in Ref. 64 was not that given this name in the present review. However, calculations repeated with the 'correct' basis II did not change the results very much. The most likely explanation of the discrepancies between theory and experiment is the neglect of correlation effects in the IGLO calculations.

Extensive data on the full σ-tensors are available for carbonic acids. For a few of these experimental and theoretical results are compared in Table 21. Again the agreement is quite good.

Stimulated by the recent investigations of Christl [65], Gleiter [66] and coworkers on unusual ^{13}C NMR shifts of the bridgehead carbons in strained bi- and tricyclic hydrocarbons we studied some simple prototypes for this class of molecules, namely benzvalene, bicyclo[2.1.1]hexene and their dihydro

Table 20. Tensor components for the relative shift $\delta(^{13}C)$ of a few molecules with CC double or triple bonds. Experimental data and IGLO results are given in consecutive line

		δ_{11}	δ_{22}	δ_{33}	δ_{av}
C_2H_2		150	150	−90	70
	a	154	154	−92	71
	b	153	153	−92	71
C_2H_4		234	120	24	126
	a	273	113	13	133
	b	270	105	10	130
H_2CCO	CH_2	39	4	−27	3
	a	47	3	−30	7
	CO	265	239	77	194
	a	282	249	65	199
H_2CCCH_2	CH_2	158	54	23	78
	a	173	52	12	79
	C	233	233	175	214
	a	248	248	193	230

[a] Basis I; [b] Basis II

Table 21. Tensor components of the relative shift $\delta(^{13}C)$[a]

Molecule		IGLO δ_{iso}	δ_{11}	δ_{22}	δ_{23}	Experiment δ_{iso}	δ_{11}	δ_{22}	δ_{33}	
HCOOH		171	94.3	125	294	166				g
$(HCOOH)_2$		170	92.5	135	281					
CH_3COOH	CO_2	177	98.7	138	296	188	110	184	269	h
						176				g
	CH_3	18.1	−8.1	26.4	36.0	33	11	46	46	h
						20.3				g
$(COOH)_2$	b	165	98.4	103	273					
	c	162	97.2	103	287					
						168	112	123	250	h
						160				g
$HCOO^-$		172	109	135	271					
	d	173	103	144	273					
						164	99	168	225	i
CH_3COO^-	CO_2 e	175	98.4	163	265					
	f	175	98.6	163	265					
						180	108	200	230	i
	CH_3 e	25.0	−3.2	19.8	58.5					
	f	25.0	−2.6	18.8	59.0					
						22.9	−3.5	26.5	45.5	i
$(COO^-)_2$	d	175	100	158	266	171	109	158	245	h
CO_3^{2-}		171	110	201	201					
	d	178	115	210	210					
$HCOOCH_3$	CO_2	165	97.8	121	276					
						165	94	136	253	h
							99.5	142	252	h
						161				g
	CH_3	47.2	6.0	64.1	71.4					
						49.1				g

[a]Referenced to TMS; the calculated shielding $\sigma(^{13}C)$ of TMS is 192.7 ppm; [b]*trans* intramolecular bridged; [c]*trans*; [d]A set of diffuse functions added to basis II for O; [e]Angle (OCCH) 0°; [f]Angle (OCCH) 90°; [g]Ref. 61; [h]Ref. 116; [i]Ammonium salt, Ref. 116

Table 22. Relative shifts $\delta(^{13}C)$ for some bi- and tricyclic hydrocarbons[a]

		IGLO	Experiment[b]
Bicyclo[2.1.1]hexane	C_1	37.3	39.9
	C_2	15.8	26.5
	C_5	34.0	39.3
Bicyclo[2.1.1]hexene	C_1	37.7	42.9
	C_2	156.8	143.0
	C_5	66.5	68.0
Tricyclo[3.1.0.0^{2,6}]hexane	C_1	3.3	2.4
	C_2	35.2	34.0
	C_3	26.6	26.1
Benzvalene	C_1	48.7	48.3
	C_2	34.0	36.6
	C_3	146.5	133.7

[a]Referenced to TMS; the calculated shielding $\sigma(^{13}C)$ of TMS is 192.7 ppm; [b]Refs. 61, 65, 66

derivatives. The IGLO results [67] are compared with experimental data in Table 22.

The dramatic effects of the chemical shifts of the homallylic carbons on removal of the C_3–C_6 bond or on saturating the double bond are accurately described by the calculations. A detailed discussion will be given elsewhere [67].

As to the largest molecule so far studied by means of IGLO see the end of Sect. 3.5.

Recently it could be shown that an NMR spectrum regarded as evidence for the synthesis of a dioxirane [68a] was inconsistent with the IGLO prediction for this kind of species. The alternative that a ketoneoxide was produced could be excluded as well. A reinterpretation [68b] of the spectrum of the reaction products of Ref. 68a established that dioxirane was in fact present, but with a different NMR spectrum, that was in good agreement with the IGLO predictions.

3.5 *Carbon NMR Shifts in Carbocations*

Theoretical [1] and experimental results for ^{13}C shifts of carbocations are compared in Table 23.

Carbocations can be divided into two classes, the classical and the non-classical ones. The prototype of classical carbocations is CH_3^+, whereas CH_5^+ or bridged $C_2H_5^+$ and even the allyl- or cyclopropenylcations can be regarded as simple representatives of non-classical carbocations.

The LMO contributions to $\sigma(^{13}C)$ in CH_3^+ have been shown in Table 15. Each of the CH bonds gives (in basis II) a deshielding contribution of -125 ppm. These are related to the existence of a very low-lying unoccupied MO, namely a p_z-MO of C (perpendicular to the molecular plane), into which the bond MOs can be 'rotated'. The absolute chemical shift $\sigma(^{13}C)$ in CH_3^+ is -175.6 ppm from IGLO calculations (basis II) i.e. the carbon atom in CH_3^+ is strongly deshielded even with respect to a bare nucleus. Its shift relative to CH_4 is 370.7 ppm, about at the low-field end of the carbon shift scale.

In classical carbocations like CH_3^+ the positive charge is to a large extent localized at one C atom. There are strong paramagnetic currents for the field direction in the plane that contains C^+ and its three substituents. The deshielding effect depends on the $\sigma \rightarrow \pi$ excitation energy and the latter is sensitive to changes in the geometry. One sees, in fact, from Table 24 that rather small changes of the molecular geometry of CH_3^+ affect $\sigma(^{13}C)$ drastically. In pyramidalizing it (which obviously reduces the $\sigma \rightarrow \pi$ energy gap) a decrease of $\sigma(^{13}C)$ by ~ 100 ppm or more is possible.

In nonclassical cations the electronic charge is delocalized and the p_z-AO of C^+, which is empty in classical cations, becomes involved in bonding and is less available for paramagnetic currents. As a consequence nonclassical cations do not show the strong deshielding of classical ones and behave more like neutral molecules. So CH_5^+, a 'protonated methane', has $\sigma(^{13}C)$ close to that of methane, $C_2H_5^+$, a π-protonated ethene, close to that of ethene etc.

Table 23. Relative shifts $\delta(^{13}C)$ for carbocations[a]

		IGLO(DZ)	IGLO(II)	IGLO(III)	Experiment
CH_3^+		361.0	367.4	373.4	—
$CH_5^+(C_s)$		−10.7	−13.1	−12.4	—
$(H_3C)_2CH^+$	C_1	50.3	44.5		51.5[b]
	C_2	325.4	324.6		320.6[b]
$(H_3C)_3C^+$	C_1	47.0	41.5		48.3[c]
	C_2	346.7	347.7		335.7[c]
Allyl	C_1	233.2	236.3	239.9	—
	C_2	139.4	142.0	142.8	—
1-Methylallyl	CH_2	216.4	216.5		201.5[d]
	CH	142.1	143.4		149.8[d]
	CH	269.1	272.9		255.1[d]
	CH_3	34.8	31.0		36.3[d]
$c\text{-}C_5H_9^+$	C_1	331.9			320.0[e]
	C_2	62.0			71.0[e]
	C_3	19.4			28.0[e]
$c\text{-}C_4(CH_3)_4^{2+}$	C^+	209.2			209.7[c]
	CH_3	18.7			18.8[c]
2-Norbornyl	C_1, C_2	127.4	123.1*		124.5[f]
	C_3	34.4	30.8*		36.3[f]
	C_4	30.5	30.8*		37.7[f]
	C_5	19.4	16.5*		20.4[f]
	C_6	11.4	10.8*		21.2[f]
7-Norbornadienyl	C_7	36.0	27.4*		36.2[g]
	$C_{1,4}$	55.5	49.3*		63.0[g]
	$C_{2,3}$	132.0	129.6*		122.0[g]
	$C_{5,6}$	120.2	114.1*		114.9[g]
Dehydroadamantyl-dication	C^+	7.0			6.6[h]
	CH_2	33.0			35.6[h]
Bicyclohexyl	$C_{1,2\text{ av.}}$	157.8			158.5[i]
	$C_{3,5,6\text{ av.}}$	49.1			45.7[i]
	C_4	43.4			37.5[i]
Guanidinium		165.4	159.4		158.3[c]

[a]Referenced to TMS; the calculated shielding $\sigma(^{13}C)$ of TMS is 218.1, 192.1 and 192.7 ppm for DZ, II (DZ on hydrogen) and basis II; for basis III CH_4 was used as primary reference for the calculated data, a value of $\delta(CH_4) = -5.8$ ppm was used for conversion, the calculated shielding of CH_4 using basis III is 193.8 ppm; [b]Ref. 118; [c]Ref. 61; [d]Ref. 119; [e]Ref. 120; [f]Ref. 121; [g]Ref. 70b; [h]Ref. 73; [i]Ref. 122. *DZ basis for hydrogens

Table 24. Geometry dependence of $\sigma(^{13}C)$ in CH_3^+

∠ HCH	r(Å)	$\sigma(^{13}C)$[a]
120.0°	1.05	−164.1
120.0°	1.07	−169.4
120.0°	1.08	−172.2
120.0°	1.09	−175.1
120.0°	1.11	−181.0
117.0°	1.08	−192.3
109.5°	1.08	−259.1
90.0°	1.08	−721.4

[a]Absolute shift in ppm

In the first paper on IGLO calculations of carbocations [1] it was found that theoretical results for the positively charged C atoms obtained with basis II (or even with a DZ basis) agree with experiment to within a few ppm for non-classical ions, while differences of the order of 20–30 ppm arose for classical ones. It was argued that in view of the localized charge, counterion effects may be more pronounced in classical cations and that it may not be fully justified to compare calculations for free ions with experiment, in which free ions are certainly not realized. It has turned out, however [70] that the discrepancies between theory and experiment for classical carbocations were mainly due to a too naive choice of the geometries for the IGLO calculations. If one uses very carefully optimized structures (based on ab-initio calculations on a sufficiently high level, often including correlation effects) the agreement between theory and experiment becomes almost perfect. That this observation is most pronounced for classical ions (i.e. that for typical nonclassical ions geometries are less critical) is related to the strong sensitivity of $\sigma(^{13}C)$ especially in classical ions on changes in the geometry. This very sensitivity makes IGLO calculations, however, an extremely good test of the geometry realized under the conditions of the NMR experiment.

The geometry optimizations [70] have shown that formally classical ions like the isopropyl or the *t*-butyl cation have their lowest energy not for structures of highest possible symmetry, but rather for distorted geometries, e.g. of C_2 type for the isopropyl cation. These distortions are such that an optimum amount of hyperconjugation is realized and that a genuine classical structure is avoided. Without the challenge from a comparison of IGLO calculations with the measured NMR spectra these new structures might not have been found.

In the case of the isopropyl and isobutyl cations the genuinely classical structures led to $\delta = 364.7$ ppm and $\delta = 369.3$ ppm (basis II) for the positively charged C atom, values close to $\delta = 372.2$ for the experimentally unknown CH_3^+ (where a stabilization by hyperconjugation is not possible). Structure optimization reduced these shifts by ~ 40 or ~ 20 ppm respectively and brought them close to the experimental values (see Table 23).

Often the only available information concerning the structure of a given carbocation is its NMR spectrum. A final answer concerning this structure can often be given if one computes the chemical shifts for various possible structure and compares the results with experiment.

A nice example is that of the 2-norbornyl cation which was the object of a long historical dispute. Comparison of the chemical shifts measured in 'superacid' at low temperature with the values predicted by IGLO for a classical and a non-classical structure leaves no doubt that under the experimental conditions a nonclassical cation is realized [1]. Thus confirming previous evidence by Saunders et al. [71] based on the isotopic perturbation effect.

The situation is less clearcut for the 2-methyl-2-norbornylcation. Its spectrum certainly does not agree with that predicted for a non-classical cation, but it is neither in very good agreement with a classical structure. It appears

[60a] that the equilibrium structure of this ion is actually neither purely classical nor nonclassical and that with the correct structure the agreement between IGLO and experiment is quite good.

Another carbocation the structure of which was disputed, has the formula $C_4H_7^+$, and there were arguments either in favour of a cyclopropylcarbinyl structure **3** or a bicyclobutonium structure **4**

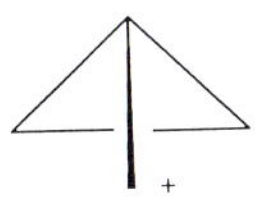

3

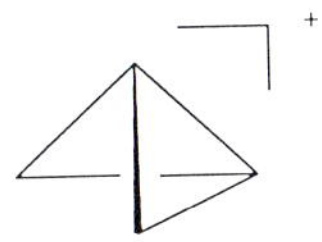

4

Here obviously only structure **4** was compatible [1] with the observed spectrum, although in ab-initio calculations (done on a rather high level), structure **3** had a lower energy. Even for structure **4** the agreement was not perfect. Saunders et al. [72] first repeated the IGLO calculation with an improved geometry of type **4** and then took care of a dynamical equilibrium between the two structures **3** and **4** determining the relative weight such that the experimental NMR spectrum was reproduced. The result was an averaged weight of $\sim 80\%$ for structure **4**, that should lie ~ 0.5 kcal/mol below structure **3**. For methyl substituted homologues of both **3** and **4** which can be isolated and which do no interconvert, the agreement between theory and experiment is nearly perfect, which confirms that one can rely on IGLO calculations for this type of problem.

The example of the bicyclobutonium cation indicates that the structure realized under experimental conditions must not always be the one which has the lowest energy for an isolated molecule. In such cases IGLO calculations allow to check the experimentally most stable structure.

There has recently been some interest in dications. One of these is the 1,3-dehydro-5,7-adamantyl dication synthesized by Schleyer and coworkers [73]. It has unexpectedly small down-field shifts ($\delta = 7$ ppm for bridge C-atoms and $\delta = 33$ ppm for methylene C-atoms). The IGLO (DZ) results calculated without knowing the experimental values, were 6.6 ppm and 35.6 ppm respectively. For a singly charged classical 1-adamantyl cation the δ(C) values are (IGLO/DZ in parenthesis): 337.7 (300.6), 57.7 (66.6). 67.8 (87.6), 37.2 (34.6). One sees again that theory works better for nonclassical cations (even if they are doubly charged). Possibly the geometry of this monocation has to be revised.

Another interesting dication is the cyclobutadienyl dication $C_4H_4^{2+}$, which according to Hückel's rule should be aromatic. Ab initio structure calculations have, however, predicted a nonplanar structure [74]. At least for the tetramethyl derivative of this dication comparison of the experimental NMR spectrum with IGLO calculations for planar and nonplanar geometry were possible and confirmed that the latter is realized.

The largest molecule for which so far IGLO calculations have been performed is the dication $C_{20}H_{20}^{2+}$ of Pagodane. IGLO calculations were

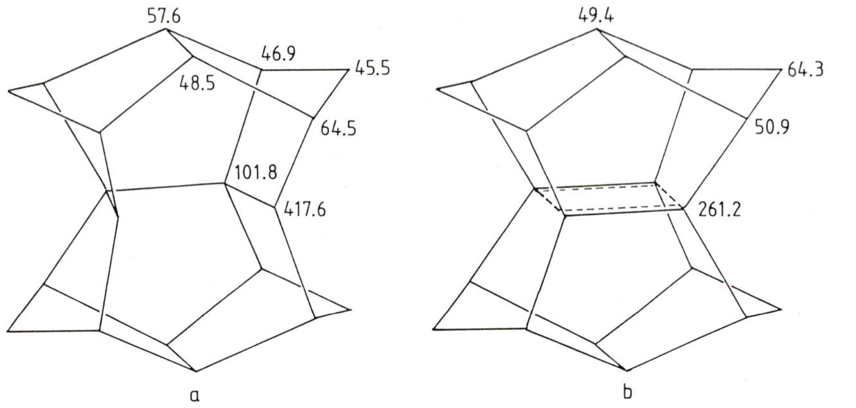

Fig. 14. Computed ^{13}C chemical shifts (IGLO, DZ basis) for classical (**a**) and nonclassical (**b**) pagodane dication

performed in order to answer the question which of two possible structures is realized under experimental conditions, either a classical dication with C_{2v} geometry (a) or non-classical one with D_{2h} symmetry and a two-electron-four-center bond (b) [75]. The corresponding geometries were first optimized on the MNDO level [76]. IGLO calculations with a DZ basis predicted the ^{13}C NMR shifts shown in Fig. 14.

In this case it is so far not possible to conclude definitly which of the two structures is realized. The properly averaged shifts of the classical structure (259.7, 53.1, 55.7, 45.5 ppm) differ rather little from the shifts of the non-classical geometry (261.2, 49.4, 50.9, 64.3 ppm), except for the CH_2 groups (45.5 vs 64.3 ppm). For these the experimental value (65.3 ppm) agrees nicely with that of the non-classical (D_{2h}) structure. One cannot exclude that this agreement is fortuitous, since the IGLO values change on going from the MNDO-optimized geometry to a 3–21G* geometry. Such a geometry has so far only been obtained for the D_{2h} structure with the results 241.6, 47.2, 49.1, 54.0 ppm for the chemical shifts. The experimental values [77] are: 251.0, 52.3, 57.2, 65.3 ppm.

3.6 Silicon NMR Shifts

The basis dependence of $\sigma(^{29}Si)$ has been studied in detail [23] and it has turned out that the smallest acceptable basis, i.e. basis II (see Sect. 2.10) must contain two sets of d-functions on Si. Contrary to $\sigma(^{13}C)$ a DZ basis is not good enough (although it is not as bad as for phosphorus compounds, see Sect. 3.8). For SiH_4, $\sigma(^{29}Si)$ obtained with a DZ basis is e.g. off by ~ 100 ppm (compared to ~ 30 ppm for $\sigma(^{13}C)$ in CH_4). It is, of course, not surprising that d-type polarization functions are important for 2nd row atoms. It has not yet been explored whether for a limited class of molecules DZ calculations of $\sigma(^{29}Si)$ might be acceptable.

The agreement between IGLO calculations and experiment for $\sigma(^{29}Si)$ is usually very good, as is seen from Fig. 15 and Table 25. When Jameson and Jameson recently published their absolute scale of $\sigma(^{29}Si)$ and compared theory with experiment [78] they compared their results with the conventional CHF calculations of Tossel et al. [79] which are not so satisfactory, except for SiH_4 and, to some extent, Si_2H_6—which are small enough for conventional CHF.

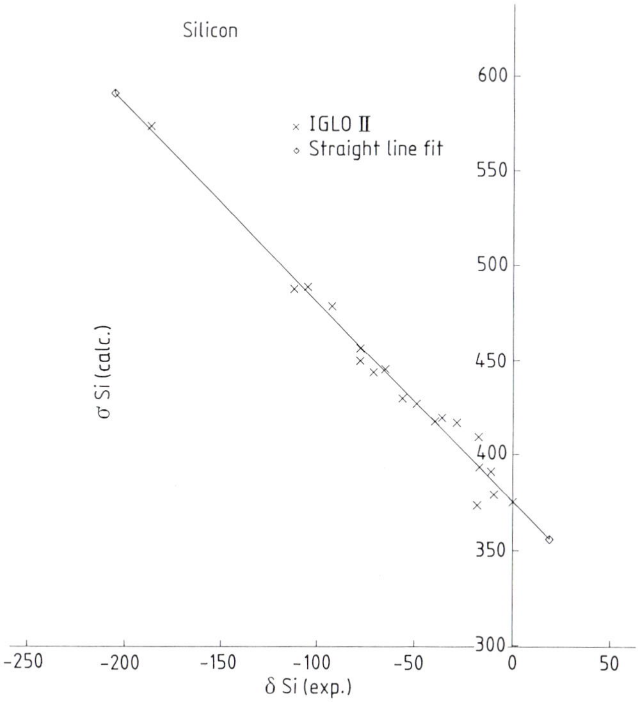

Fig. 15. Theoretical vs experimental silicon shifts. See also Table 25. The straight line fit yields a slope of -1.05 ± 0.04 and an intercept of 375.3 ± 3.0 with 20.9 as maximum and 7.6 as mean deviation

Table 25. Relative shifts $\delta(^{29}Si)$[a]

	IGLO	Experiment[b]		IGLO	Experiment[b]
SiH_4	−103.3	−93.1; −91.9	SiF_4	−111.8	−109.9−−113.7
H_3CSiH_3	−70.0	−65.2; −64.9	SiF_6^{2-}	−198.5	−185.3; −187.7
$(H_3C)_2SiH_2$	−42.4	−37.7; −41.5	H_3SiCl	−44.2	−36.1
$(H_3C)_3SiH$	−18.9	−15.5−−18.5	H_2SiCl_2	−15.8	−11.0
$(H_3C)_4Si$	0.0	0.0	$HSiCl_3$	−3.9	−9.3−−9.8
Si_2H_6	−113.8	−104.8	$SiCl_4$	1.3	−16.6−−20.0
H_3SiF	−16.9	−17.4	Si_2F_6	−74.6	−77.9
H_2SiF_2	−41.6	−28.5	H_3CSiF_3	−54.5	−56.2
$HSiF_3$	−81.3	−77.8	H_3SiCF_3	−68.3	−71.1[c]

[a]Referenced to TMS; the calculated shielding $\sigma(^{29}Si)$ of TMS is 375.1 ppm; [b]Ref. 123; [c]Ref. 124

More recently Jameson (Ref. 39, Vol. 18) noted the good agreement of the absolute Si shifts with the results of IGLO calculations.

An analysis of $\sigma(^{29}Si)$ in terms of LMO contributions is found in Table 26. The bond increments of an SiH or SiC bond are much more deshielding than their CH or CC counterparts. This is certainly related to the lower excitation energies for 2nd row atoms, in particular the $p \rightarrow d$ excitation energies.

In comparing methylated silanes with the parent compounds one finds approximate transferability of bond increments. This is not the case for fluorinated silanes, as it has not been for fluorinated methanes.

The change from Si_2H_6 to Si_2H_4 is qualitatively similar to that from C_2H_6 to C_2H_4, but the downfield shift is now almost entirely due to the contribution of the SiSi double bond. In C_2H_4 the CC double bond was only responsible for about half the observed effect.

The principal values of the shift tensors in Si_2H_4 show the same pattern as in C_2H_4, but the anisotropy is much larger. For Si_2H_4 no experimental results are available, but the experimental results for a peraryl substituted Si_2R_4 [80] agree at least qualitatively with our prediction for Si_2H_4.

One sees from Table 26 that even the L shell contributions are far from transferable.

The hypervalent ions SiF_5^- and SiF_6^{2-} do not show any unexpected behaviour. The deshielding of an SiF bond contribution decreases from SiF_4 to SiF_6^{2-}. This is understable since in going from SiF_4 to SiF_6^{2-} d-AOs on Si become to some extent involved in bonding and are hence less available for

Table 26. LMO contributions to $\sigma(^{29}Si)$[a]

	L	SiH	SiX	Other	σ	$\Delta\sigma$
SiH_4	237.2	−62.6	—	—	478.4	0.0
H_3CSiH_3	231.2	−64.5	−87.9	−0.7	445.1	68.5
$(H_3C)_2SiH_2$	225.5	−64.4	−86.7	−2.1	417.4	56.6
$(H_3C)_3SiH$	227.7	−63.9	−87.9	−2.2	393.9	−45.1
$(H_3C)_4Si$	225.9	—	−85.9	−3.2	375.1	0.0
Si_2H_6	231.1	−56.1	−67.0	−3.0	488.8	−22.7
Si_2H_4[b]	207.5	−80.0	−206.0	−11.2	326.3	193.9
H_3SiF	235.5	−89.3	−63.0	8.4	409.1	94.2
H_2SiF_2	244.9	−98.5	−71.3	15.4	416.7	43.0
$HSiF_3$	262.3	−92.0	−76.1	18.2	456.4	17.0
SiF_4	279.6	—	−78.3	24.2	486.9	0.0
SiF_5^-	284.5	—	−56.6/−45.3[c]	7.3	527.5	−80.8
SiF_6^{2-}	286.6	—	−36.1	7.5	573.5	0.0
H_3SiCl	230.3	−81.4	−66.4	3.6	419.3	75.4
H_2SiCl_2	229.4	−91.1	−79.2	6.0	390.9	34.0
$HSiCl_3$	235.4	−95.3	−88.1	7.1	379.0	−29.8
$SiCl_4$	241.3	—	−92.6	6.6	373.7	0.0
H_3SiOH	236.2	−81.2	−75.3	8.4	421.9	96.3
H_3SiOH_2	227.6	−62.2	−77.8	−4.5	451.9	55.8
Si_2F_6	260.2	—	−77.3/−93.2[d]	18.3	449.7	29.5

[a]The K-shell contribution is always 496.1 ppm; [b]Planar geometry assumed; [c]Eq./ax.; [d]SiF/SiSi

paramagnetic currents. Small paramagnetic effects are an indication of 'saturation' and of a rather large gap between occupied and unoccupied MOs.

In contrast to the amount of data on carbocations only few is known about silecenium ions either experimentally or theoretically. Preliminary calculations on SiH_3^+ and $Si(CH_3)_3^+$ indicate that they are as much deshielded as carbocations relative to the saturated compounds ($\delta(SiH_3^+) = 265$ ppm, $\delta(Si(CH_3)_3^+) = 351$ ppm). The effect of methyl substitution is about as large as in the saturated silicon compounds and thus larger than for hydrocarbons.

3.7 Nitrogen NMR Shifts

Both ^{14}N and ^{15}N are NMR-active nuclei, but with either of them there are experimental problems, some of which have been surmounted recently [81], such that in the past years the knowledge about experimental nitrogen shifts has increased considerably [82]. For IGLO calculations there are, of course, no special difficulties with $\sigma(^{14,15}N)$ [42, 83, 84].

What makes nitrogen a little more difficult than carbon is the presence of a lone pair which has a large, and even long-range effect. The lone pair is also responsible for strong solvent and temperature dependences of nitrogen chemical shifts.

The original papers on IGLO calculations on nitrogen shifts [42, 83, 84] are very recent and up to date and we can refer to them for details.

An analysis of $\sigma(^{14,15}N)$ in terms of LMO contributions is given in Table 27. We can start from NH_3 as a molecule with rather small (shielding) contributions of an NH bond or the lone pair. Replacing H by CH_3 changes rather little, while NH_2, OH or F have (in this order) an increasing deshielding effect—similar to what has been observed for other nuclei, but in more pronounced form. The contribution of the lone pair on N is more strongly affected by this substitution. Substituents from the 2nd row like Cl have much smaller effects, and even phenyl behaves as a weakly electronegative substituent. In aziridine C_2H_4N an increased shielding of $\sigma(^{14,15}N)$ relative to NH_3 is found, similar to $\sigma(^{13}C)$ in cylcopropane.

Very strong deshielding is found when N is involved in a double bond. The down-field shift in N_2H_2 is spectacular, although it is expected that IGLO calculations exaggerate the paramagnetic effects. $\sigma(^{14,15}N)$ in N_2H_2 is also strongly geometry dependent. The difference between the *cis* and *trans* isomers is ~ 100 ppm. The difference between N_2H_4 and N_2H_2 is much larger than that between C_2H_6 and C_2H_4, probably because lone pairs are more effective (together with low-lying unoccupied MOs) to generate paramagnetic currents (but the relevant transition energy may also be smaller).

Other double bonds involving nitrogen behave similarly. As an example we have included FNO (with an NO double bond) in Table 26. The formal CN double bonds in heteroaromatics shows paramagnetic effects in somewhat attenuated form, like the CC bonds in C_6H_6 as compared to those in C_2H_4.

Table 27. LMO contributions to $\sigma(^{14,15}N)$[a]

	NH	NX	lp	Other	σ	$\Delta\sigma$
NH_3	4.3	—	0.3	—	249.1	−31.8
NH_4^+	2.8	—	—	—	247.4	0.0
H_3CNH_2	5.9	0.8	−6.8	−1.3	240.6	37.7
H_2NNH_2	−3.0[b]	−1.6	−22.5	−1.0	205.1	73.8
$HONH_2$	−18.2	−5.9	−50.3	−1.4	142.3	165.4
FNH_2	−38.2	−7.2	−90.2	−0.6	61.9	283.9
H_2PNH_2	8.6[b]	4.4	20.2	−1.0	276.7	55.4
$ClNH_2$	−6.8	11.9	−18.3	4.8	220.8	58.3
$(H_3C)_2NH$	5.1	0.7	−8.8	−0.8	233.0	70.0
$(H_3C)_3N$	—	5.5	10.5	2.1	265.0	−64.4
Aniline	−0.1	−15.7	−5.8	−10.4	204.0	89.1
Aziridine	0.1	12.8	−1.6	−0.3	259.9	120.8
cis-HNNH	−155.8	−183.6	−258.8	−30.4	−392.5	840.4
trans-HNNH	−132.8	−197.7	−341.2	−46.4	−482.0	1002.4
cis-FNNF	−67.4[c]	−185.3	−158.7	−21.9	−197.1	275.3
H_2CNH	−62.7	−137.1	−150.4	−18.8	−132.8	585.5
HPNH	−59.7	−223.2	−259.9	−54.4	−361.1	983.2
FNO	−118.3[c]	−169.3	−272.2	10.4	−313.2	570.6
N_2	—	−146.0	−151.1	−55.1	−116.1	691.8
HCN	—	−126.7	−138.6	−19.1	−48.2	590.6
H_3CCN	—	−125.6	−123.8	−22.7	−36.0	532.0
H_3CNC	−55.2[d]	−47.4	—	−37.1	96.3	426.0
FCN	—	−159.2	−71.3	−8.5	97.1	383.6
PN	—	−337.2	−323.2	−71.5	−495.7	1206.5
ClCN	—	−117.0	−113.9	−22.2	−16.9	557.6
H_3NO	−6.2	−27.2	—	−4.1	186.2	56.5
NO_3^-	—	−133.1	—	−16.3	−179.3	366.7

[a]The K-shell contribution is always 236.1, abbreviations of the LMOs as in Tables 4 and 7; [b]Average of the two symmetry non equivalent NH bonds; [c]NF; [d]NC(CH_3)

A very detailed IGLO study on nitrogen heteroaromatics has recently been published [83].

The deshielding (with respect to NH_3) in pyrrole is relatively small ($\sigma \sim 100$ ppm), pyridine ($\sigma \sim -100$ ppm) is more deshielded and the largest deshielding is found for molecules with three neighbouring N atoms. The difference between pyrrole and pyridine is obviously related to the integration of the lone pair in pyrrole into the π-system.

Triple bonds like that in N_2 display similar but weaker paramagnetic effects. HCN resembles very much the isoelectronic N_2.

Very different from double bonds are semipolar bonds like the NO bond in the hypothetical H_3NO. Here the deshielding effect is not very large; there is, of course, no lone pair on N.

In Table 28 and Fig. 16 experimental and theoretical results for $\sigma(^{14,15}N)$ are compared. The agreement is in part quite good, but in some cases rather unsatisfactory. It is particularly poor for the terminal N in H_2CN_2 (difference 160 ppm, for the central atom 'only' 40 ppm with basis II). In nitriles (H_3CCN or C_6H_3CN) the difference between experiment and IGLO (basis II) is ~ 50 ppm,

Table 28. Relative shifts $\delta(^{14,15}N)$[a]

		IGLO(II)	IGLO(III)	Experiment
NH_3		0.0	0.0	0.0
H_3CNH_2		8.5	9.0	1.3[b]
H_2NNH_2		49.9	49.8	45.4[b]
HNF_2		308.8	323.0	342[h]
$(H_3C)_2NH$		16.1	18.0	6.7[b]
$(H_3C)_3N$		−15.9		13.0[b]
Aziridine		−11.9	−16.0	−11.4; −8.5[b]
HCN		297.3	306.7	290.5[k]
H_3CCN		285.1	295.2	242.5[k]; 239.5[b]
C_6H_5CN		298.8		258.7[b]
F_3CCN		309.0		
H_3CNC		152.8	168.1	162.0[f]
$H_2C\mathit{N}N$		247.0	267.7	286[f]
$H_2CN\mathit{N}$		559.5	557.9	397[f]
Aniline		45.1		56.5[b]
N_2		365.1		364.5[k]
NP		729.5		612.5[k]
NSF		1016.8	986.9	264.7[l]
NSCl		1398.8	1357.1	
NSF_3		147.8	153.8	
NF_3		346.2	361.8	371[i]
NF_4^+		265.8	286.4	288[p]
N_2F_4		319.2		339[p]
cis-N_2F_2		441.9	457.4	384[p]
trans-N_2F_2		558.7	558.3	446[p]
Pyrrole		154.9		148.8[c]
Pyridine		353.4		324[d]; 317.2[c]; 316.5[k]
Pyridazine		488.2		400.4[f]
Pyrimidine		319.7		294.8[c]; 295.5[e]
Pyrazine		386.8		366.7[c]; 332.0[g]
1, 3, 5-Triazine		289.9		282.9[c]; 281.7[f]
1, 2, 4, 5-Tetrazine		470.3		385[f]
Borazene		104.1		98[f]
Cytosine	N_1	127.3		151.8[f]; 152.1[m]; 123 ± 4[n]
	N_3	215.4		207.9[f]; 206.7[m]; 189 ± 4[n]
	NH_2	69.8		93.0[f]; 93.0[m]; 74 ± 4[n]
Adenine	N_1	243.4		235.8[o]; 224.7[o]; 237.4[f]; 234.4[c]
	N_3	230.6		230.8[o]; 223.8[o]; 224.5[f]; 221.3[c]
	N_7	266.2		226.6[o]; 215.2[o]; 242.1[f]; 239.3[c]
	N_9	139.6		170.8[o]; 172.1[o]; 170.5[f]; 168.4[c]
	NH_2	61.6		79.8[o]; 78.0[o]; 82.0[f]; 80.2[c]

[a]Referenced to NH_3; the calculated shielding $\sigma(^{14,15}N)$ of NH_3 is 249.1 and 255.6 ppm for basis II and III resp.; [b]Ref. 82; [c]Ref. 125; [d]Ref. 126; [e]Ref. 127; [f]Ref. 128; [g]Ref. 129; [h]Ref. 130; [i]Ref. 131; [k]Gas phase data, Ref. 132; [l]Ref. 133; [m]Ref. 134; [n]Ref. 135; [o]Ref. 136; [p]Ref. 137

while isonitriles look better (~ 10 ppm). Basis III results are somewhat better, but the big discrepancies don't disappear. In heteraromatic systems the overall agreement is fair, but differences between theory and experiment of ~ 100 ppm arise in cases where two nitrogen atoms are next neighbours. Again it does not appear that these discrepancies vanish if one uses very large basis sets. It is not known to which extent better geometries might improve the situation. It is

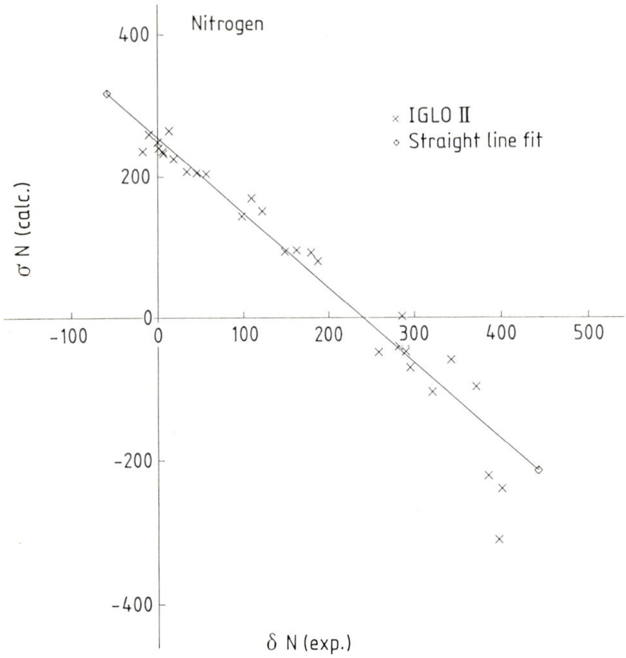

Fig. 16. Theoretical vs experimental nitrogen shifts. See also Table 28. The straight line fit (FSN and CH_2NN excluded) yields a slope of -1.06 ± 0.04 and an intercept of 255.0 ± 8.5 with 69.4 as maximum and 28.7 as mean deviation

likely that in these cases the limits of the one-determinant approximation (on which IGLO is based) are reached and that one ought to include electron correlation effects [83].

That IGLO calculations can be applied to problems involving 'biomolecules' can be seen from the publication [84] on purine and pyrimidine bases, in particular to the question at which position adenine is first protonated.

3.8 Phosphorus NMR Shifts

^{31}P is probably, together with ^{19}F and after ^{1}H and ^{13}C one of the most important nuclei in NMR. While many features of ^{13}C NMR have been rationalized by means of empirical rules long before systematic and accurate calculations and their interpretation became possible, a qualitative understanding of the rules that govern ^{31}P shifts has so far appeared to be a hopeless task.

The first observation in IGLO calculations of phosphorus shifts is a rather large dependence on the choice of the basis set [23]. Basis II to which we refer in this review, is acceptable for molecules with PX single bonds, including

Table 29. LMO contributions to $\sigma(^{31}P)$[a]

	L	PH	PX	lp	Other	σ	$\Delta\sigma$
PH_3	271.0	−51.8	—	−73.3	—	575.5	−26.2
PH_5	282.4	−46.5/−36.5[b]	—	—	—	603.2	243.3
H_3CPH_2	261.4	−54.0	−79.6	−102.6	−5.1	499.4	92.8
$(H_3C)_2PH$	257.9	−51.1	−78.6	−112.9	−7.8	462.1	76.1
$(H_3C)_3P$	253.0	—	−76.0	−116.6	−11.6	429.8	−14.9
H_2NPH_2	260.5	−75.2	−74.6	−132.1	−8.3	428.3	204.6
$HOPH_2$	255.9	−100.1	−67.8	−169.4	−7.0	351.5	307.3
FPH_2	246.3	−124.6	−60.6	−206.1	0.7	264.3	430.7
F_2PH	242.0	−181.0	−84.4	−268.2	1.3	158.5	356.9
PF_3	265.9	—	−103.8	−230.2	−5.4	252.2	311.9
PF_4^+	302.4	—	−116.3	—	21.9	392.2	0.0
PF_4^-	280.7	—	−85.4/−51.9[b]	−185.2	−22.6	331.7	323.4
PF_5	300.4	—	−84.1/−67.9[b]	—	−3.0	442.6	−150.0
PF_6^-	311.3	—	−52.3	—	−5.4	524.8	0.0
OPF_3	287.8	—	−90.0	−199.4[c]	−47.1	384.6	280.3
SPF_3	284.7	—	−94.5	−250.3[d]	24.3	308.6	377.8
OPH_3	260.7	−91.6	—	−75.4[c]	−42.3	401.5	272.5
H_3PO_4	281.9	—	−104.4	−102.5[c]	−45.5	353.9	182.7
PH_4^+	256.2	−79.6	—	—	—	470.9	0.0
P_4	308.0	—	−2.7	56.6	−8.6	881.4	−432.7
H_2PPH_2	267.9	−50.1	−63.8	−74.5	−7.7	554.8	60.9
HPPH	163.9	−148.1	−512.9	−382.1	−89.2	−435.1	1438.8
H_3CPPCH_3	171.0	−173.0[e]	−483.1	−337.9	−89.6	−383.0	1246.0
FPPF	172.5	−81.2[f]	−653.5	−315.3	−86.4	−430.7	837.6
H_2CPH	221.7	−134.3	−320.3	−213.2	−20.9	66.1	344.8
F_2CPH	260.4	−84.8	−205.6	−90.2	−15.8	397.2	197.1
P_2	176.4	—	−581.5	−354.9	−159.4	−386.0	2030.7
HSiP	259.5	—	−319.6	−99.5	−39.6	334.0	963.0
HCP	264.0	—	−258.6	−138.8	−42.9	356.9	994.1
H_3CCP	266.0	—	−259.8	−118.9	−47.3	373.2	841.8
H_2NCP	279.0	—	−239.9	−68.6	−32.5	471.2	461.4
HOCP	294.9	—	−148.8	−47.9	−23.5	607.8	529.7
FCP	302.6	—	−101.7	−38.7	−17.7	679.7	528.6
NP	220.8	—	−438.6	−232.1	−158.1	−74.8	1628.0

[a]The K-shell contribution is always 533.2, abbreviations of the LMOs as in Tables 4 and 7; [b]Eq./ax.; [c]PO (three LMOs); [d]PS (one LMO); [e]PC; [f]PF

semipolar and hypervalent bonds, but less satisfactory for molecules with genuine PP double or triple bonds. For these molecules larger basis sets, with at least three d-sets should be used. The double zeta (DZ) basis is of no use for phosphorus compounds. $\sigma(^{31}P)$ obtained from a DZ basis is in error with respect to basis II by ~60 ppm for PH_3, by ~300 ppm for P_4 and by almost 1000 ppm for P_2.

Again (see Sect. 3.6), the L-shell contributions to $\sigma(^{31}P)$ are far from being invariant. Their variation is much larger for P than for Si. While this contribution varies between ~240 and 310 ppm in the majority of P compounds, values as low as 160–180 ppm are found in molecules with PP double or triple bonds. (See Table 29).

The next observation (in the sence P resembles N, but the effects are more pronounced) is that the contribution of the phosphorus lone pair varies much with the bonding situation.

Relatively transparent is the change of $\sigma(^{31}P)$ on successive methyl substituion of PH_3 (Table 29). Here the L-shell contribution varies little and the PH or PC increments are nearly transferable. There is further a monotonic increase (in absolute value) of the lone pair contribution, which is responsible for $\sim 1/4$ of the total change in σ. The bulk of the change in $\sigma(^{31}P)$ results here from the difference between the PH and PC increments.

In the series of the fluorophosphanes the situation is (like for tbe fluorosilanes, see Sect. 3.6) more complicated. On the way from PH_3 to PF_3 there is a small non-monotonic variation of the L-shell contribution and a very strong, again non-monotonic, variation of the contribution of the lone pair. The increment of one PF bond increases somewhat with the degree of F substitution. Rather remarkable, finally, is the strong increase of the contribution of one PH bond with fluorine substitution (from -52 ppm in PH_3 to -181 ppm in PHF_2). Both the total σ and its anisotropy $\Delta\sigma$ have a non-monotonic dependence on the degree of F substitution, with the extremum of σ at PHF_2 and that of $\Delta\sigma$ at PH_2F. It is hard to see how one can give a simple explanation of these trends.

Somewhat more straightforward is the comparison in the series PF_3, PF_4^+, PF_4^-, PF_5 and PF_6^-. The change from PF_3 to PF_4^+ is mainly due to the disappearence of the lone pair, while the L-shell and PF increments don't vary dramatically. In PF_4^- there is again a lone pair, which is responsible for most of the difference to PF_4^+, the contributions of axial and equatorial bonds differ somewhat, and the smaller average of the PF increments compensates to some extent the lone-pair effect. The difference between PF_4^- and PF_5 is due to the fact that a lone pair is replaced by a PF bond. In PF_6^- there is an additional PF bond, but the average bond increment has become smaller. The reduction of the PF bond increments from PF_4^+ to PF_6^- is like that from SiF_4 to SiF_6^{2-}, and has to do with the reduced 'availability' of the 3d-AOs of P on increased F substitution.

A rather puzzling aspect of phosphorous shifts is that non-nearest neighbours can have a large effect. This is e.g. seen in the series X–C≡P on Table 29. The increment of the X–C bond varies astonishingly little between X = H, CH_3, NH_2, OH or F and it goes even in the 'wrong' direction. The increase of the shielding of P by as much as ~ 300 ppm from HC≡P to FC≡P comes to $\sim 2/3$ from a change in the contribution of the CP triple bond and to $\sim 1/3$ from the lone pair on P. The substituent X has obviously a strong effect on the electronic structure of the CP triple bond and even on the lone pair at the P atom.

If one compares experimental and theoretical $\sigma(^{31}P)$ as is done in Table 30 and Fig. 17 the agreement is usually satisfactory, with two exceptions, one is PN (which is isovalence-electronic to P_2, for which no experimental shift is known), the other FCP. In the case of PN, which is at the paramagnetic (low-field) end, this is not too surprising. The strong paramagnetism is related

Table 30. Relative shifts $\delta(^{31}P)$[a]

	IGLO(II)	Exp.		IGLO(II)	Experiment
PH_3		−238 – −241[b]	SPF_3	−27	−32[k]
PH_4^+	−135	−101[c]	$(H_3C)_2PH$	−127	−99[b]
H_3CPH_2	−164	−164[b]	$(H_3C)_3P$	−94	−61 – −63[b]
F_3CPH_2	−119	−129[d]	H_3SiPH_2	−276	−274[l]
H_2PPH_2	−219	−204[b]	P_4	−545	−553[m]
PCl_3	186	215–220[b]	H_3PO_4	−18	
PF_3	83	97[b]	HCP	−21	−32[n]
H_3CPF_2	197	257[e]	*i*-C_3H_7CP	−29	−64[o]
H_2ClCPF_2	172	202[f]	*t*-C_4H_9CP	−31	−69[n]
F_3CPF_2	142	160[g]	neo-$C_5H_{11}CP$	−21	−51[o]
F_2PPF_2	189	226[h]	C_6H_5CP	−33	−32[n]
HPF_4	−80	−54[i]	F_2CPH	−43	−62[d]
H_2PF_3	−57	−24[i]	FCP	−344	−207[d]
PF_5	−106	−80[c]	PN	410	302[p]
PF_6^-	−189	−144 – −148[b]	Phosphabenzene	227	211[q]
OPF_3	−50	−36[b]			

[a]Referenced to 85% H_3PO_4; PH_3 was used as primary reference for the calculated data, a value of $\delta(PH_3) = -240$ ppm was used for conversion, the calculated shielding $\sigma(^{31}P)$ of PH_3 is 575.5 ppm; [b]Ref. 138; [c]Ref. 132; [d]Ref. 102; [e]Ref. 139; [f]Ref. 140; [g]Ref. 141; [h]Ref. 142; [i]Ref. 102; [k]Ref. 143; [l]Ref. 144; [m]Ref. 145; [n]Ref. 146; [o]Ref. 147; [p]Ref. 132; [q]Ref. 148

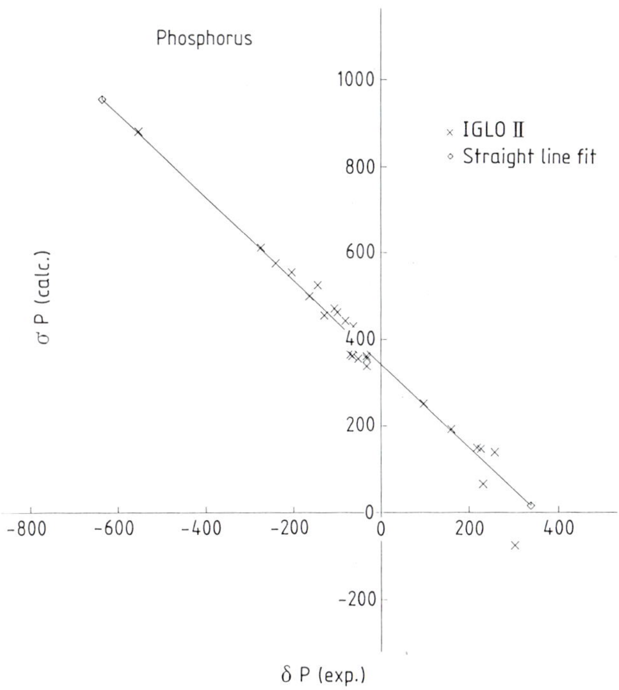

Fig. 17. Theoretical vs experimental phosphorus shifts. See also Table 30. The straight line fit (PN excluded) yields a slope of -0.97 ± 0.03 and an intercept of 343.1 ± 5.6 with 53.5 as maximum and 26.3 as mean deviation

to a low-lying excited state, and in such cases the Hartree–Fock approximation (on which IGLO is based) becomes rather poor. In such situations one overestimates the paramagnetism, as it happens here. The other case FCP, which is very much in the high field region, is harder to understand. For P_4, at the high field end, a molecule almost entirely determined by local diamagnetic contributions, the agreement between theory and experiment is nearly perfect, provided one takes the gas phase value. The ^{31}P shift of P_4 is very sensitive to solvent effects. Solvents probably interact with the lone pair.

Of course, FCP is not an ideal diamagnetic molecule in the same sense as P_4, but it is not easy to understand why of all XCP molecules FCP is the only one for which there is a significant discrepancy between experiment and theory. There is also no problem for the related molecule F_2CPH with a double instead of a triple bond. In the experiments on FCP this molecule has not been isolated, but has been characterized otherwise [85] and for the ^{19}F shift theory and experiment agree well. One probably has to wait on one hand for theoretical calculations that go beyond Hartree–Fock, on the other hand possibly for new experiments.

3.9 Oxygen and Sulfur NMR Shifts

Oxygen is a difficult element for NMR studies, since the only NMR active nucleus ^{17}O has small natural abundance and a large quadrupole moment. Therefore less experimental results are available and some of them are rather uncertain. For sulfur the situation is even worse. For IGLO calculations oxygen is not more difficult than nitrogen or fluorine, but in view of the lack of experimental data a systematic study of oxygen has not yet been performed. Nevertheless a lot of theoretical information on ^{17}O shifts exists, since in IGLO calculations of oxygen-containing compounds the oxygen shifts are calculated automatically, though they were rarely published. Some of these results are presented in Table 31 together with the pertinent LMO contributions. Like for N or P the lone-pair contributions vary much and play a dominant role. Comparing $\sigma(^{17}O)$ in various molecules one sees that H_2O is near the diamagnetic (high field) end of the scale.

More shielded is $\sigma(^{17}O)$ in H_3COH, and particularly in ClOH and H_2POH. This is noteworthy since $\sigma(^{17}O)$ in FOH, HOOH or H_2NOH is significantly deshielded. The pair ClOH/FOH is, in this sense, very similar to the pair ClF, F_2 to be discussed in Sect. 3.10.

Strong deshielding of $\sigma(^{17}O)$ arises either from OF bonds or if O is involved in a double bond, like in formaldehyde, H_2CO, or other aldehydes as well as ketones. Oxygen in heteroaromatics such as furane is also appreciably deshielded. COS and CO_2 differ significantly. While COS behaves as expected for O=X systems, $\sigma(^{17}O)$ in CO_2 is much less deshielded. Obviously CO_2 is a rather stable molecule with two 3-center-4-electron bonds and a rather large gap between occupied and unoccupied MOs.

Table 31. LMO contributions to $\sigma(^{17}O)$[a]

		OH	OX	lp	Other	σ	$\Delta\sigma$
H_2O		16.1	—	1.2	—	305.7	50.9
H_3O^+		11.4	—	−9.1	—	296.4	−51.0
H_3O^+ (planar)		8.9	—	15.5	—	313.2	−55.4
H_3COH		19.3	18.8	8.7	2.1	328.8	94.0
H_3SiOH		18.7	1.6	13.4	−1.8	315.1	38.3
H_2NOH		7.6	16.0	−22.5	4.2	254.0	118.3
H_2POH		22.2	16.5	19.9	3.5	353.2	57.4
HOOH		−17.5	13.2	−75.6	8.1	123.8	319.8
HSOH		31.4	32.1	14.7[b]	14.1	378.1	91.0
FOH		−64.6	−1.5	−183.8	4.2	−158.3	692.8
F_2O		—	−54.3	−282.6	−23.5	−426.0	174.5
ClOH		12.5	41.4	2.3	23.9	353.6	101.1
H_2BOH		−2.4	−63.0[c]	−28.6[d]	−16.6	145.1	268.0
H_2AlOH		12.4	−17.3[c]	6.6[d]	−9.4	263.5	70.9
F_2CO		—	−96.4	−64.8	−21.4	23.8	391.8
H_2CO		—	−207.0	−239.0	−47.1	−460.9	1301.8
CO		—	−153.4	−144.8	−60.4	−87.2	765.6
CO_2		—	−33.0	−10.9	−7.2	220.1	299.6
COS		—	−99.9	−57.5	−35.0	78.8	518.9
O_3	O_1	—	−1026.5	−978.5[b]	−312.1	−3051.0	4381.4
	O_2	—	−849.8	—	−154.7	−2945.6	4644.9
ONH_3		—	−197.3	26.8[d]	3.0	103.7	322.3
OPH_3		—	8.7	37.4[d]	−0.1	317.2	−192.8
OPF_3		—	−35.5	13.5[d]	−15.1	234.0	95.3
OSF_2		—	−177.1[e]	—	−38.3	155.8	228.2
$OSCl_2$		—	−181.4[e]	—	−23.9	65.9	383.5
$OS(CH_3)_2$		—	46.5[e]	—	0.2	318.0	292.4
O_2SF_2		—	−88.7[e]	—	−30.3	152.2	107.7
Furane		—	−103.3	−51.1[f]	22.1	43.7	340.5
SO_2		—	−213.4	−140.6	−28.4	−307.8	646.6
NO_3^-		—	−131.9	−90.4	−46.0	−88.2	295.6
CO_3^{2-}		—	−34.7	−23.3	−24.0	165.9	116.1
SO_3		—	−78.6	−51.3	−41.1	48.9	138.0
c-C_2H_4O		—	32.9	−0.9	7.7	342.8	194.4
HCO_2H	OH	−10.6	51.0	−22.7	−24.5	139.7	205.1
	CO	—	−147.2	−112.0[b]	−33.6	−133.6	624.8

[a]The K-shell contribution is always 271.2, abbreviations of the LMOs as in Tables 4 and 7; [b]Average; [c]One of two OX 'banana' bonds; [d]One lone pair; [e]Sum of OX and lp; [f]Sum of σ and π-type lone pairs

Stronger deshielding than in F_2O is only found for O_3. The theoretical values for $\sigma(^{17}O)$ of the central and terminal O atoms in ozone are − 2946 and − 3051 ppm respectively (these are basis II values, in Ref. 22 various basis sets were compared, but the results were not very basis-set dependent), while experimental counterparts of − 743 and − 1309 ppm have been reported [86]. This is by far the worst case encountered so far in comparing IGLO calculations with experiment. It is to be expected that the IGLO results in this case are not very reliable, because O_3 is not well described by a single Slater determinant. The very large calculated shifts indicate the presence of extremely small energy denominators, the smallness of which is certainly exaggerated. On the other hand nothing is known on the reliability of the experimental values. Ozone is certainly one of the examples that should be recalculated taking care of correlation effects. In SO_2, which is isovalence-electronic with O_3, the difference between IGLO and experiment is 'only' 100 ppm, for other oxygen compounds for which experimental values are available, the agreement is usually not too bad (see Table 32 and Fig. 18).

$\sigma(^{17}O)$ in semipolar bonds varies strongly, between ~ 317 ppm in H_3PO or $(CH_3)_2$ SO and 56 ppm in F_2SO, but the deshielding is certainly less than in genuine XO double bonds.

For ^{33}S shifts a detailed theoretical study has recently been published [24] from which the following data are taken (see Table 33 and Fig. 19). Agreement between IGLO (basis II) and experiment is usually satisfactory. In some cases (e.g. SF_6 or thiophene) the agreement is significantly improved if basis III is used.

Of the molecules studied, H_2S is nearly at the diamagnetic (high field) end, only in thiirane C_2H_4S and COS a stronger shielding of the S atom is found (this is in agreement with experiment). The strongest deshielding (even stronger than in SO_2—where the agreement between theory and experiment is not so

Table 32. Relative shifts $\delta(^{17}O)$

		IGLO(II)	Experiment
H_2O		0	$-36.1(g)^b$; 0
H_3O^+		9	9.13^c
CH_3OH		-23	-38^h; -37^d
C_2H_5OH		22	8^h; 6^d
H_2O_2		182	174^e
F_2O		732	$781(g)^b$; $799(1)^b$
Oxirane		-37	-49^h; -56^e
Furane		262	240^h; 228^e
CH_3CHO		697	592^d
$(CH_3)_2CO$		673	569^d
HCO_2H^f		303	253^e
$CH_3CO_2H^f$		289	251^e
$HCOOCH_3$	OCH_3	139	143^h; 137^e
	CO	413	364^h; 359^e
HCO_2^-		299	300^e
CO_3^{2-g}		160	192^e
CO		393	$350(g)^b$
CO_2		86	$-64.5(g)^b$
COS		227	$200(g)^b$; $205(1)^b$
$(CH_3)_2SO$		-12	$13\text{–}20^{d,e}$
Cl_2SO		240	292^d
F_2SO_2		154	148^e
SO_2		614	513^e
SO_3		49	188^e
OPF_3		72	66^e
NO_3^-		393	410^e

[a]Referenced to $H_2O(1)$; the calculated shielding $\sigma(^{17}O)$ of H_2O is 305.7 ppm; [b]Ref. 43d; [c]Ref. 149; [d]Ref. 132; [e]Ref. 150; [f]Average of CO and OH; [g]A set of diffuse functions added to basis II for O; [h]Ref. 151

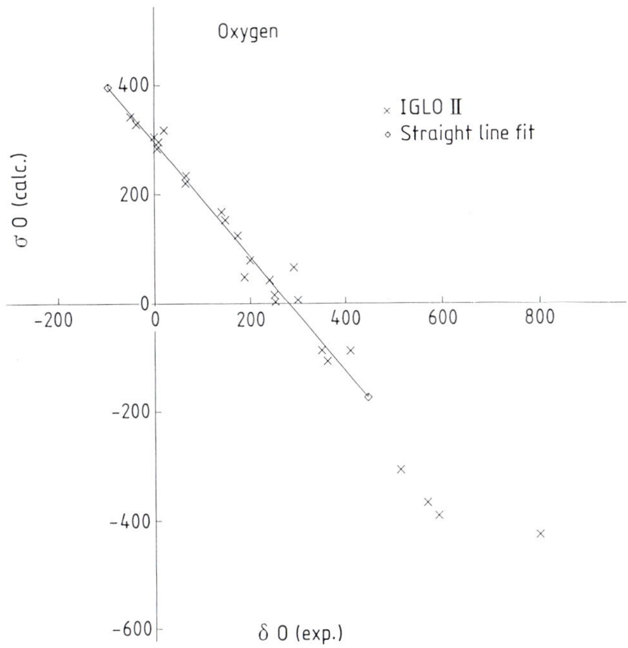

Fig. 18. Theoretical vs experimental oxygen shifts. See also Table 32. The straight line fit (the more paramagnetic systems as SO_2, F_2O etc. excluded) yields a slope of -1.04 ± 0.05 and an intercept of 298.3 ± 9.5 with 70.1 as maximum and 26.9 as mean deviation

good) is found in SF_2 and in H_2CS (for both of which experimental results are not yet available). The LMO contributions to $\sigma(^{33}S)$ are analyzed in Table 34. The overall inspection of Table 33 indicates the substantial variation of the L-shell contributions (like for $\sigma(^{29}Si)$ or $\sigma(^{31}P)$), the deshielding contribution of all bonds, with stronger contributions of the SO semipolar or the formal NS triple bonds than of single SX bonds, but the largest contribution from the 'genuine' double bond in H_2CS, and the dramatic changes of the lone-pair contributions. The strong deshielding in SF_2 and SO_2 is dominated by the lone-pair contributions. The difference between $\sigma(^{33}S)$ in SF_2, SF_4 and SF_6 is obviously related to the fact that there are two lone pairs in SF_2, one in SF_4 and none in SF_6. Similar trends are found in the series SF_2, OSF_2, or NSF, NSF_3. The high-field values of H_2S, OCS and thiirane arise because the deshielding contributions are all rather small in absolute value. The difference between COS and CS_2 is not easy to understand. Most striking are the differences between F and Cl substitution, as is seen from the comparison of SF_2 with SCl_2 or NSF with NSCl. In NSCl the NS bond has a strong positive (shielding) contribution, while the corresponding contribution in NSF is strongly negative (deshielding). The predicted results for NSF and NSCl are probably not very good since these molecule are candidates for strong correlation effects (although $\sigma(^{19}F)$ agrees well with experiment, see Sect. 3.10).

Table 33. Relative shifts $\delta(^{33}S)$[a]

	IGLO(II)	IGLO(III)	Experiment
H_2S	−156.2	−140.6	−171.0[b]
CH_3SH	−111.6		−125[c]
SF_2	1256.1	1317.3	
SF_4	370.1	407.4	
$F_3\mathit{S}SF$	370.2		
$F_3S\mathit{S}F$	611.9		
SF_6	107.3	147.5	155.4[b]
FSSF	522.8		
$\mathit{S}SF_2$	735.1	768.4	
$S\mathit{S}F_2$	213.6	248.2	
OSF_2	508.6	538.7	
$OS(CH_3)_2$	242.5	282.4	233 ± 20[d]; 325 ± 10[e]
$OSCl_2$	479.4	518.6	556[b]
O_2SF_2	294.8	324.4	292.2[b]
COS	−253.2	−248.6	−261.8[b]
CS_2	0.0	0.0	0.0
SO_2	862.9	883.1	706.9[b]
NSF	449.3	467.8	
NSCl	209.3	255.2	
NSF_3	129.0	167.9	
SCl_2	467.7	514.6	
Thiirane	−262.9	−261.8	−240[f]
Thiirane-S-oxide	133.7	166.9	120[f]
Thiirane-sulfone	227.4	255.6	245[f]
Thiophene	302.8	227.4	220.0[g]

[a]Referenced to CS_2; the calculated shielding $\sigma(^{33}S)$ of CS_2 is 516.0 and 536.1 ppm for basis II and III respectively; [b]Ref. 152; [c]Ref. 132; [d]Ref. 153; [e]Ref. 154; [f]Ref. 155; [g]Ref. 156

3.10 Fluorine and Chlorine NMR Shifts

The ^{19}F resonance is experimentally very important. Rather extensive IGLO data are available as well [23–25]. Some selected results are given in Table 35. As seen from Fig. 20, the agreement between theory (basis II) and experiment is satisfactory.

On the paramagnetic end of the $\sigma(^{19}F)$ scale one finds F_2 and on the diamagnetic end, somewhat surprisingly, the isovalence-electronic molecule ClF. It is worthwhile to have a detailed look at the MO contributions to $\sigma(^{19}F)$ in these two molecules and in HF.

The comparison of HF, F_2 and ClF in Table 36 first reveals that these three molecules differ very little as far as the chemical shift $\sigma_{\parallel}(^{19}F)$ for a field parallel to the molecular axis is concerned. This does not only hold for the total $\sigma_{\parallel}$, but even for the individual LMO contributions. For a field perpendicular to the molecular axis things are very different. For HF there is not a big difference between $\sigma_{\parallel}$ and $\sigma_{\perp}$ i.e. $\sigma(^{19}F)$ in HF is only slightly anisotropic. In F_2 $\sigma(^{19}F)$ is extremely anisotropic, but the anisotropy is dominated by a single MO

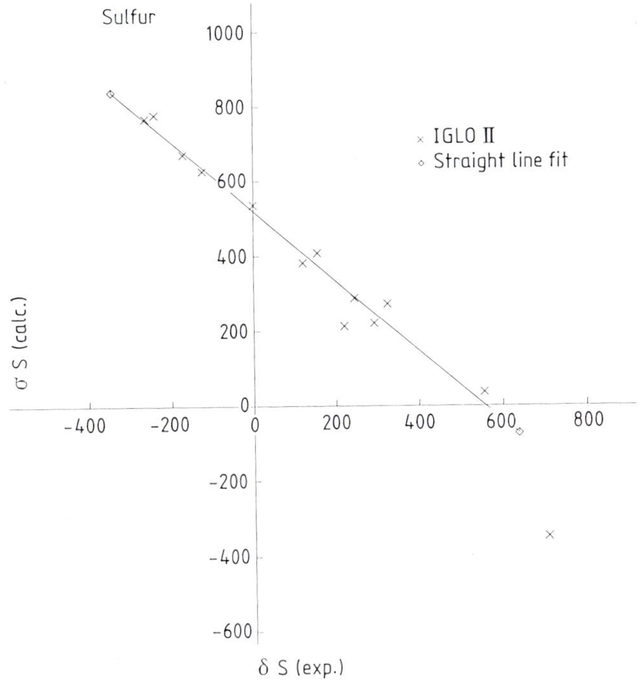

Fig. 19. Theoretical vs experimental sulphur shifts. See also Table 33. The straight line fit (SO_2 excluded) yields a slope of -0.93 ± 0.05 and an intercept of 520.6 ± 13.4 with 102.6 as maximum and 39.7 as mean deviation

contribution, namely that of the LMO nπ, which is -919.9 for $\sigma_{\perp}$ and 54.6 for $\sigma_{\parallel}$. This is the only LMO that by means of an angular momentum operator perpendicular to the bond can be 'rotated' to an MO that has strong overlap with the low-lying antibonding σ^*-MO. This is illustrated in Fig. 21 and very much in line with an argument that we have often used to explain paramagnetic currents and the chemical shifts induced by them.

The data on Table 36 are based on IGLO calculations in which the $\sigma-\pi$-separation is maintained, i.e. in the localization procedure σ- and π-MOs are not mixed. This variant is meaningful only for linear (and to some extent planar) molecules. The final results are rather close to those for 'full' localization. For the interpretation of the result the $\sigma-\pi$-separation is preferable, because one sees clearly that only a single π type lone pair MO causes the large paramagnetic effect. In full localization this would be shared among the three equivalent lone pairs.

The current density plotted in Fig. 22 clearly shows paramagnetic ring currents around either of the F-atoms. The big puzzle arises with ClF. There again the nπ-MOs, now both those of F and Cl, have large contributions in absolute values, but their sign is opposite to what one would have expected.

Table 34. LMO contributions to $\sigma(^{33}S)$[a]

	L	SH	SX	S^+X^-[b]	lp[c]	Other	σ	$\Delta\sigma$
H_2S	302.4	−23.2			−80.8		664.7	271.3
HSSH	295.6	−35.6	−27.2		−90.6[d]	−9.0	612.9	301.9
H_3CSH	296.2	−24.8	−38.4		−85.4	−5.0	627.6	327.2
HOSH	257.9	−96.1	−51.2		−250.9[d]	−8.0	171.0	841.7
FSH	228.8	−140.0	−45.4		−396.2	−6.0	−184.7	1371.3
F_2S	180.1		−89.5		−642.5	−26.6	−740.1	1281.7
F_4S	289.2		−119.1/−87.3[e]		−247.5	−53.5	145.9	245.7
F_6S	320.5		−72.1			−49.5	408.7	0.0
SCl_2	245.0		−72.1		−288.2	−46.3	48.3	251.5
S_2F_4	248.3	−51.1[f]	−239.2		−259.5	−34.0	−36.8	582.3
SSF_2	271.5		−94.6	−453.9[g]	−366.7	−23.5	−219.1	860.7
OSF_2	271.5		−107.7	−271.2[g]	−322.5	−25.4	7.4	575.2
$OS(CH_3)_2$	276.3		−109.4	−143.7[g]	−196.2	−14.3	273.5	331.7
$OSCl_2$	263.3		−127.7	−226.4[g]	−263.8	−51.5	36.6	575.3
O_2SF_2	292.2		−102.2	−210.3[g]		−16.9	221.2	431.1
OCS	330.7		−30.2		−85.2	−16.5	769.2	405.0
SCS	311.6		−284.6		−29.0	−52.3	516.0	840.3
SO_2	241.7		−395.1[h]		−432.0	−48.2	−346.9	577.8
SO_3	280.8		−263.2[h]			−11.9	49.8	−19.0
NSCl	297.3		−269.6	237.4	−330.3	−184.7	306.7	2879.2
NSF	276.7		−134.5	−146.0	−323.1	−176.8	66.7	2168.1
CH_2S	179.8		−369.9		−1053.2	−49.4	−722.3	2268.9
Thiirane	313.0		39.3		−91.4	−0.1	778.9	683.0
Thiophene	273.8		134.6		−119.1	−33.6	302.8	226.6

[a]The K-shell contribution is always 570.3, abbreviations of the LMOs as in Table 4 and 7; [b]Contribution of the semipolar bonds; [c]The semipolar bonded atoms bear one lone pair only; [d]Average of the two symmetry non-equivalent lone pairs; [e]Ax./eq.; [f]SF; [g]Sum of S^+X^- and lp(X); [h]Contribution of each "double bond"

As seen in Fig. 23 there is, in fact, a diamagnetic current around F and a paramagnetic one around Cl. One sees from Table 36 that $\sigma(^{35}Cl)$ behaves as expected, with very strong antishielding contribution of the $n\pi'$-MO of Cl.

There is no simple explanation for the unusual behaviour of ClF. Here a diamagnetic current is induced by those terms that usually give rise to paramagnetic currents, i.e. the imaginary part of the wave function and small energy denominators. A possible key towards an understanding of this analogy is that in this case the 'σ^{p0}-terms' as discussed in Sect. 2.9 are here unusually large.

We have, of course, tried to find molecules that share this anomaly with ClF. We found a similar though less pronounced behaviour for HSF, and as expected BrF resembles ClF as well. However, from Fig. 24 one sees that $\sigma(^{19}F)$ for ClF fits into the general pattern of the normal XF_n molecules. While for X from the first row, F becomes monotonically more deshielded in the sequence $CF_4 \rightarrow F_2$, for X from the 2nd row the behaviour is non-monotonic with an inversion of the trend after PF_3. In the series H_3CF, H_2NF, HOF, F_2 and H_3SiF, H_2PF, HSF, ClF as is seen in Fig. 25, $\sigma(^{19}F)$ changes monotonically. It decreases in the first series and increases in the latter. It has been known for quite a while that substitution by heavy atoms creates a high field shift, that is

Table 35. Relative NMR shifts $\delta(^{19}F)$[a]

		IGLO(DZ)	IGLO(II)	IGLO(III)	Exp. (gas)	Experiment
HF			−187.6		−214.4[c]	
BF_3		−90.4	−132.0	−116.7	−131.6[c]	−126[e]; −131[e]
BF_4^-			−168.9			−138−−145[e]
NFH_2		−92.3	−195.9	−212.3		−156.6[f]
NF_2H		23.7	−20.6	−18.0	−5.1[h]	−11.2[g]
NF_3		122.6	101.4	116.2	145.3[c]	138.0–146.9[e]
NF_4^+			206.1	208.4		213.5[q]
N_2F_4			36.5			59.8[q]
cis-N_2F_2			176.8	168.9		151[r]
trans-N_2F_2			102.4	103.8		90[r]
F_2			378.2	373.1	428.4[c]	428.7[e]
CFH_3		−188.3	−242.9	−253.1	−275.4[c]	−271.9−−276[e]
CF_2H_2		−102.2	−135.6	−135.5	−143.5[d]	−143.6−−148[e]
CF_3H		−50.3	−80.6	−73.6	−78.5[c]	−78−−80.0[e]
CF_4		−31.1	−61.8	−50.3	−63.4[c]	−63.3[e]; −62[e]
CH_3CH_2F		−165.0	−200.5	−206.2		−213[e]
CH_3CHF_2		−96.7	−120.4		−110.7[d]	−110[e]
CH_3CF_3		−51.4	−74.5		−63.1[c]	−61.7−−65[e]
CF_3CF_3		−51.5	−87.5		−87.4[c]	−89.1[i]
CH_2CF_2		−29.4	−78.4	−73.1	−83.5[d]	−81.3[e]
trans-CFHCFH		−107.9	−182.6	−183.0		−186.3[e]
cis-CFHCFH		−89.0	−151.8	−155.6		−165.0[e]
CH_2CFH		−52.3	−109.1	−109.1		−113[e]; −114.0[e]
$CHFCF_2$	CF_2(E)	−69.2	−130.4	−124.4	−129.7[d]	−127[e]; −129.4[e]
	CF_2(Z)	−45.0	−95.2	−92.1	−101.8[d]	−103[e]; −101.6[e]
	CHF	−130.2	−200.8	−202.5	−208.6[d]	−185[e]
F_2CCF_2		−75.6	−131.2	−127.4	−135.1[d]	−135[e]
F_2CO		27.3	−21.5	−7.6	−26.0[c]	−23.0[k]
CF_3CN		−26.1	−56.5		−56.4[d]	
FCCH		−108.1	−201.1	−202.9		
FCCF		−162.4	−256.6	−256.3		
c-$C_2H_2CF_2$		−85.8	−104.3			
CF_2CCH_2		−77.7	−113.2			
C_6H_5F			−119.8			−116[e]
SiF_4			−159.1		−167.6[d]	−160[e]; −170[e]
CF_3SiH_3			−66.2			−56.1[n]
SiF_6^{2-}			−163.0			−127[e]
Si_2F_6			−119.4			−123[e]; −126[e]
SiH_3SiF_3			−115.9			−120.3[o]
CH_3SiF_3			−134.0			−135[e]
PF_3			−41.7		−32.7[c]	−33.7−−35[e]
P_2F_4			−120.6			−115[l]; −116[l]
OPF_3			−86.3			−92.8[e]
SPF_3			−45.5			−51[l]
PF_5			−67.6[b]			−76[e]
PH_2F_3			−44.1[b]		−54[e]	
	eq		−21.7			−30[e]
	ax		−90.2			−106[e]
PHF_4			−50.7[b]		−52[e]	−49[e]
PF_6^-			−93.7			−65[e]
CF_3PF_2	PF_2		−107.7			−106.7[g]
	CF_3		−80.8[b]			−84.3[g]
CH_3PF_2			−103.9			−92.3[e]; −92.9[e]
CF_3PH_2			−56.2[b]			−42[m]

(*Continued*)

Table 35. (*Continued*)

		IGLO(DZ)	IGLO(II)	IGLO(III)	Exp. (gas)	Experiment
CH_2ClPF_2			−107.4			−96.3[e]
ClF_3	eq		−8.4		19.7[d]	−4[e]
	ax		89.4		145.6[d]	116[e]
			56.8[b]			86[e]
ClF			−468.5	−476.1	−441.5[c]	−448[e]
SF_2			−190.1	−183.2	−166[p]	
SF_4	eq		9.3	27.8		37.03[s]
	ax		84.2	98.3		97.73[s]
			46.7[b]	63.1[b]	76.9[c]	
SF_3SF	SF		−224.3			−204.1[t]
	SF_3-1		−46.0			−26.3[t]
	SF_3-2		42.3			53.2[t]
	SF_3-3		1.4			5.7[t]
SF_6			62.4		56.1[c]	50[e]; 56.21[s]
FSSF			−180.6	−171.3		−122.5[e]
SSF_2			32.4	55.3		79.5[e]
OSF_2			56.0	70.9		72.9–77.2[e,s,u]
O_2SF_2			34.4	47.5	31.0[d]	33.5[g]
NSF			218.6	225.0		234[v]
NSF_3			79.8	88.4		63.6[w]

[a]Referenced to $CFCl_3$; the calculated shielding $\sigma(^{19}F)$ of $CFCl_3$ is 276.1, 203.1 and 220.9 ppm for DZ, basis II and basis III; [b]Average; [c]Ref. 157; the data, which were referred originally to SiF_4 were converted using $\delta(SiF_4) = -167.6$ ppm (Ref. 158); [d]Ref. 158; [e]Ref. 159; [f]Ref. 160; [g]Ref. 161; [h]Ref. 129; [i]Ref. 162; [k]Ref. 163; [l]Ref. 143; [m]Ref. 101; [n]Ref. 164; [o]Ref. 165; [p]Ref. 87; [q]Ref. 136; [r]Ref. 166; [s]Ref. 167; [t]Ref. 168; [u]Ref. 169; [v]Ref. 170; [w]Ref. 171

Table 36. LMO contributions (in $\sigma-\pi$-separation) to $\sigma(^{19}F)$ or $\sigma(^{35}Cl)$

		HF	F_2	Cl<u>F</u>	<u>Cl</u>F	Cl_2	HCl
∥	K	306.3	306.4	306.3	607.3	607.3	607.3
	L	—	—	—	439.3	439.2	439.1
	Bond	27.8	16.5	23.2	7.1	10.4	13.6
	$n\sigma$	43.4	48.8	46.0	30.6	29.5	27.8
	$n\pi(2\times)$	52.3	54.6	53.1	31.8	31.3	30.7
	$n\sigma'$	—	2.3	2.7	1.5	1.7	—
	$n\pi'(2\times)$	—	2.7	3.8	2.0	2.6	—
	Sum	481.7	488.6	493.3	1153.5	1156.2	1149.1
⊥	K	306.4	306.4	306.3	607.3	607.3	607.3
	L	—	—	—	233.3	355.2	368.0
	Bond	−4.4	12.3	76.8	−42.4	13.6	−61.4
	$n\sigma$	25.2	44.0	64.9	−4.6	18.0	−17.7
	$n\pi$	1.3	−915.6	196.3	−2002.9	−542.7	−117.8
	$n\bar{\pi}$	26.2	27.2	26.5	15.6	15.6	15.4
	$n\sigma'$	—	−9.8	−13.7	−15.6	−14.8	—
	$n\pi'$	—	44.8	116.0	−3.3	31.8	—
	$n\bar{\pi}'$	—	−0.4	0.3	−1.3	−1.2	—
	Sum	379.4	−491.2	772.4	−1213.6	482.6	794.0
	σ_{iso}	413.5	−164.6	679.4	−424.6	707.1	912.4
	$\Delta\sigma$	102.3	979.8	−279.1	2367.1	673.6	355.1

Table 37. LMO contributions to $\sigma(^{19}F)$[a]

	FX	lp	Other	σ	$\Delta\sigma$
HF	37.6	15.6	—	390.8	104.5
H_2BF	−17.7	−10.8[b]	−22.0	227.4	448.6
HBF_2	−18.9	3.1[b]	−16.3	277.5	157.8
BF_3	9.1	14.7[b]	−9.7	335.1	80.6
H_3CF	47.0	26.3	0.8	446.0	−60.3
H_2CF_2	19.9	5.1	−2.7	338.7	144.2
HCF_3	4.9	−5.1	12.2	283.8	122.7
CF_4	3.3	−9.6	15.8	264.9	162.5
Cl_3CF	5.3	−29.5	−20.2	203.1	132.6
H_2CCF_2	2.9	−4.7	−13.5	281.5	103.9
F_2CCF_2	20.4	4.0	−4.4	334.3	67.6
OCF_2	−15.4	−13.7	−25.1	224.6	211.4
HCCF	26.0	20.0	12.0	404.2	106.0
H_2NF	51.2	5.9	12.3	399.0	86.7
HNF_2	17.6	−34.8	4.1	223.7	259.2
NF_3	−4.6	−65.2	−13.5	92.6	378.8
HOF	44.9	−44.9	32.3	248.8	327.9
OF_2	13.1	−106.6	13.1	12.7	573.8
F_2	25.9	−177.9	26.4	−175.1	964.8
AlH_2F	−6.2	17.1[b]	5.2	333.5	63.2
AlF_3	19.0	30.9[b]	−3.9	402.3	88.9
H_3SiF	10.3	25.3	−1.9	392.0	−18.5
SiF_4	−2.1	22.7	−10.1	362.2	88.0
SiF_6^{2-}	12.2	20.0	−12.4	366.2	93.0
H_2PF	36.8	−40.2	16.9	480.7	192.0
PF_3	−29.7	−3.5	−28.3	244.8	177.7
PF_5 ax	−3.7	−4.7	−22.1	266.4	152.3
PF_5 eq	−6.4	−1.9	−20.8	273.6	142.0
PF_6^-	2.6	2.7	−20.2	296.8	154.1
OPF_3	−9.0	4.7	−22.1	289.4	141.1
SPF_3	−15.5	−2.2	−31.6	252.6	182.6
HSF	73.9	65.4	53.6	630.0	500.8
SF_2	26.8	16.5	10.5	393.2	763.4
SF_4 ax	−39.7	−36.8	−37.2	118.9	317.5
SF_4eq	−8.1	−26.1	−26.2	193.9	225.1
SF_6	−7.2	−42.0	−32.2	140.7	350.3
OSF_2	−33.5	−28.7	−39.6	147.2	294.9
O_2SF_2	−18.2	−26.8	−39.0	168.7	290.8
SSF_2	−13.6	−28.4	−36.9	170.8	222.7
FSSF	26.1	14.9	6.5	383.8	216.9
NSF	−91.6	−50.5	−78.7	−15.5	576.2
NSF_3	−24.4	−36.9	−47.9	123.3	375.8
ClF	90.7	68.1	70.3	671.6	−290.5
ClF_3 ax	−16.0	−50.8	−24.3	113.7	399.6
ClF_3 eq	36.4	−45.8	6.2	211.5	538.8

[a]The K-shell contribution is always 306.3, abbreviations of the LMOs as in Tables 4 and 7; [b]Two lone pairs only

called the 'heavy-atom effect', as it is e.g. observed in the series CH_4, CCl_4, ... CI_4. The generally accepted interpretation of the 'heavy-atom effect' is that it has its origin in relativistic corrections [35]. The change of $\sigma(^{19}F)$ from F_2 to ClF or from NF_3 to PF_3 is certainly no relativistic effect, since non-relativistic IGLO calculations are in good agreement with experiment. In a preliminary study we

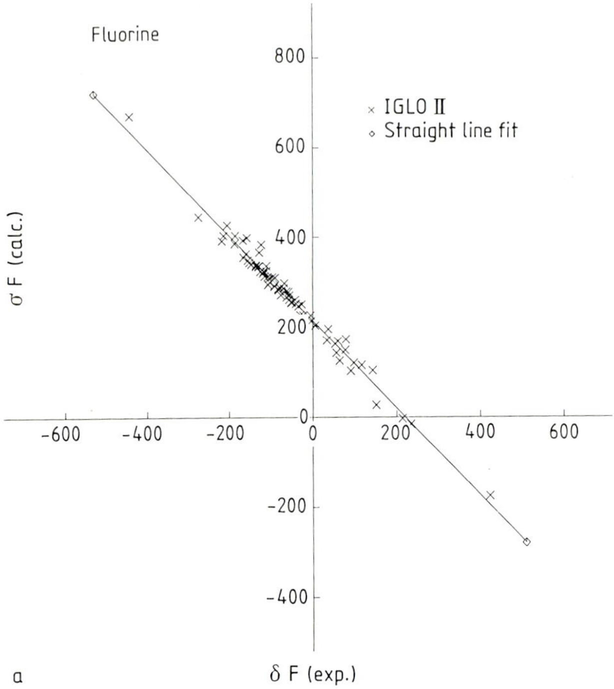
Fluorine
× IGLO II
◇ Straight line fit
σ F (calc.)
δ F (exp.)
800
600
400
200
0
-200
-400
-600
-400
-200
0
200
400
600
a

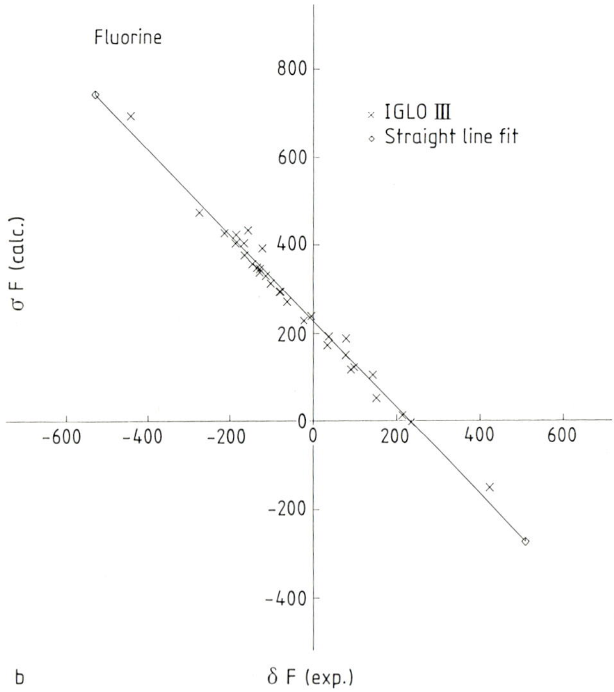
Fluorine
× IGLO III
◇ Straight line fit
σ F (calc.)
δ F (exp.)
800
600
400
200
0
-200
-400
-600
-400
-200
0
200
400
600
b

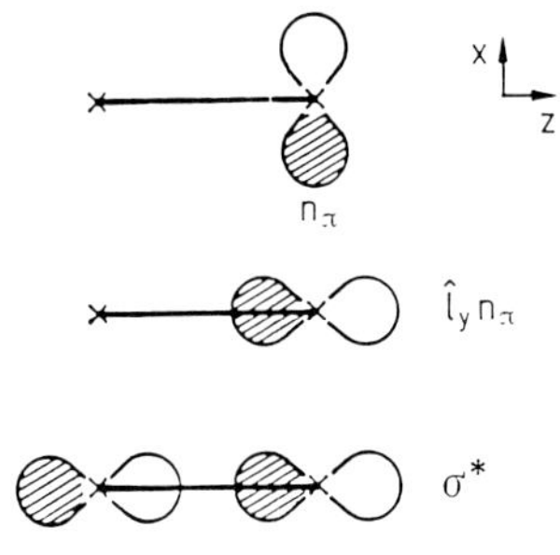

Fig. 21. Explanation of the paramagnetic current in F_2

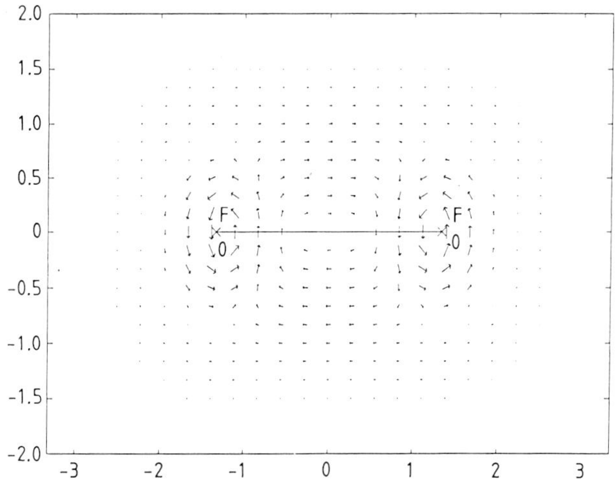

Fig. 22. Current density in F_2 for a field perpendicular to the bond. Coordinate scales in a_0

came to the conclusion that the heavy-atom effect has both nonrelativistic and relativistic origins. In 2nd row elements only non-relativistic terms are decisive, in the 4th row relativistic effects dominate definitely, while the 3rd row is somehow in between.

Higher shielding than in HF is found in organic fluorides like CH_3F and molecules like H_2PF, HSF and, of course ClF. Strong deshielding occurs in F_2, but also in hypervalent molecules such as SF_4, SF_6, in particular in NSF, but also in NF_3. PF_5 or OPF_3 behave differently. Concerning SF_2 the IGLO calculations were performed before the experimental study of Willner and

Fig. 20 a, b. Theoretical vs experimental fluorine shifts. See also Table 35. (**a**) Basis II data. The straight line fit yields a slope of -0.96 ± 0.02 and an intercept of 211.7 ± 2.1 with 54.1 as maximum and 16.4 as mean deviation. (**b**) Basis III data. The straight line fit yields a slope of -0.98 ± 0.02 and an intercept of 226.9 ± 4.1 with 52.6 as maximum and 21.2 as mean deviation

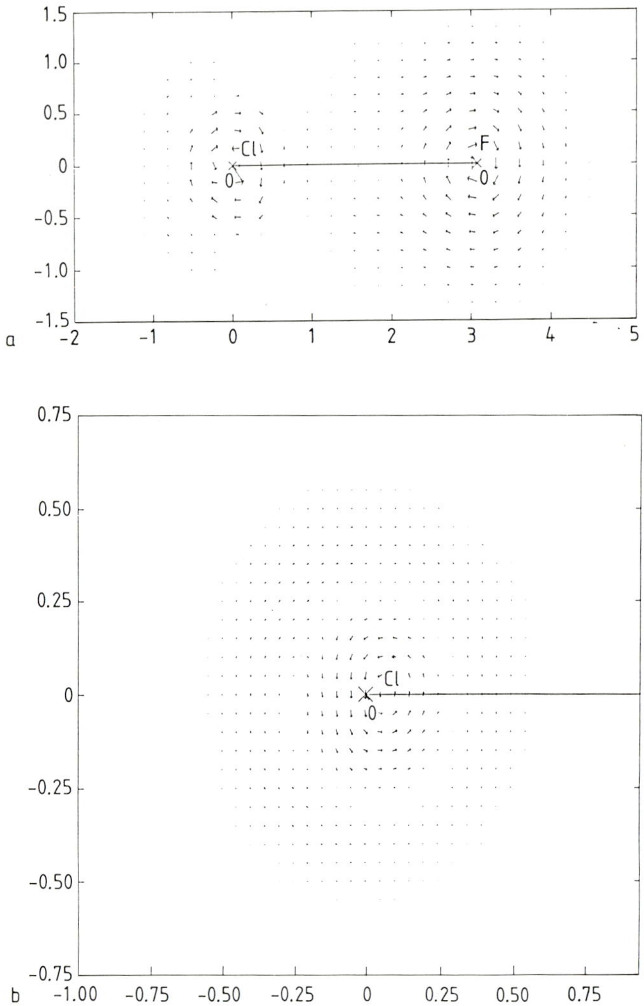

Fig. 23 a, b. Current density in ClF for a field perpendicular to the bond. Coordinate scales in a_0

Gombler [87]. The IGLO prediction was rather well confirmed, while a prediction from empirical rules was far off.

If one compares LMO contribution to $\sigma(^{19}F)$ one sees (Table 37) that the contributions of the lone pairs of F vary by far more than those of the XF bonds, but that it is not too easy to understand the trends in these shifts.

The amount of IGLO data on Cl shifts is by far smaller than on F shifts. A few results are collected in Table 38 and Fig. 26. The LMO analysis of $\sigma(^{35}Cl)$ in HCl, Cl_2 and FCl in Table 36 shows similar trends as the corresponding contribution to $\sigma(^{19}F)$—without an analogy to the unusual $\sigma(^{19}F)$ in ClF.

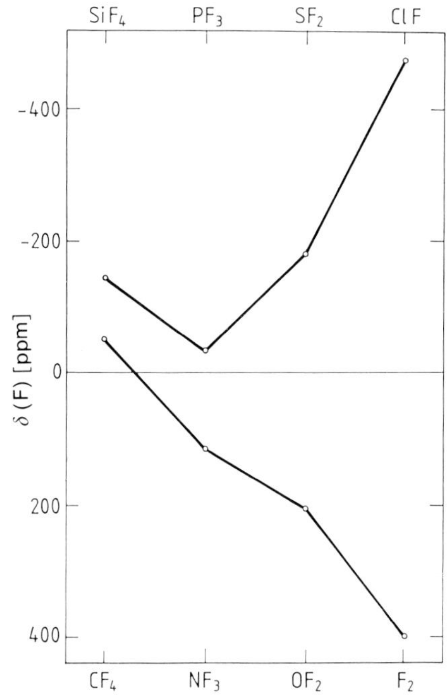

Fig. 24. $\delta(^{19}F)$ for XF_n molecules

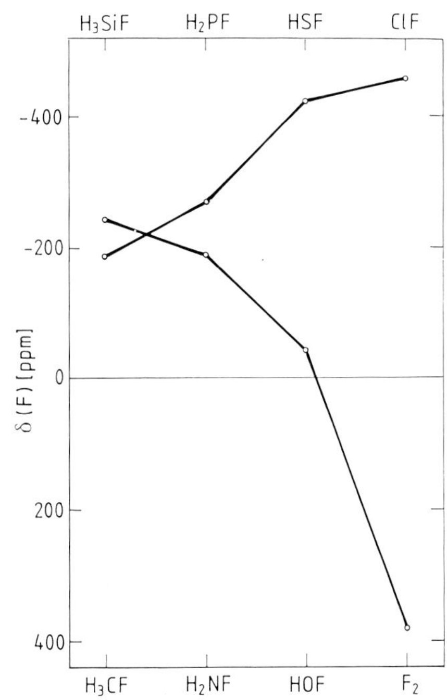

Fig. 25. $\delta(^{19}F)$ for FXH_{n-1} molecules

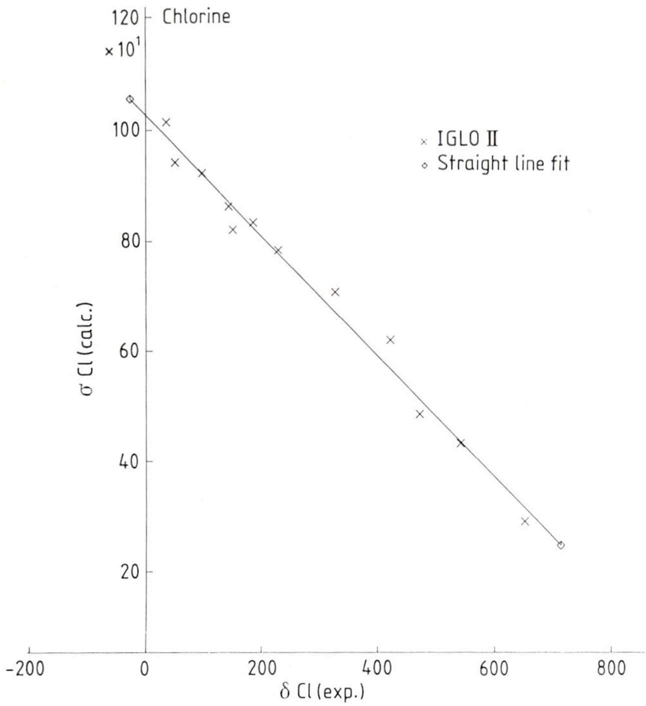

Fig. 26. Theoretical vs experimental chlorine shifts. See also Table 38. The straight line fit yields a slope of -1.10 ± 0.04 and an intercept of 972.6 ± 13.1 with 52.4 as maximum and 27.2 as mean deviation

Table 38. Relative shifts $\delta(^{35}Cl)$[a]

	IGLO(DZ)	IGLO(II)	IGLO(III)	Experiment[b]
CH_3Cl				33–50, 50
CH_2Cl_2	229	209		190–228, 228
$CHCl_3$	415	373		340–420, 420
CCl_4	638	569		500–560, 540
$CFCl_3$	538	503	498	466–471, 470
SiH_3Cl		−23		35
SiH_2Cl_2		69		97
$SiHCl_3$		129		143
$SiCl_4$		163		185
PCl_3		291		326–370, 326
SCl_2		171	182	150
$OSCl_2$		702	725	650–660, 650

[a]Referenced to Cl^-(aq.); CH_3Cl was used as a primary reference for the calculated data, a value of $\delta(CH_3Cl) = 50$ ppm was used for conversion, the calculated shielding of CH_3Cl is 975.2, 942.2 and 943.2 ppm for DZ, basis II and basis III; [b]Ref. 172

3.11 Concluding Remarks on Chemical Shifts

In principle, IGLO calculations are not limited to molecules containing elements of the 1st and the 2nd row. For higher rows the computational effort increases tremendously, as is generally the case for ab-initio calculations. Here not much experience has been gained on the question which basis sets are needed. So far only pilot calculations have been performed on compounds of Ge, As, Se and Br. They were encourageing, but it would be premature to include these preliminary results in the present review.

For elements of the higher rows relativistic effects will certainly become important [35] and it is probably not justified to neglect these.

As to correlations effects, which are present even in elements for the 1st and 2nd row there are arguments why they should be negligible in many cases. On the other hand it is also certain that for those molecules, which are formally of closed shell type, but which are not well described by a single configuration (like O_3, or cyclobutadiene etc.) the IGLO method in its present form cannot be adequate. Work is now in progress on the implementation of a multiconfiguration IGLO method.

Discrepancies between theory and experiment are not necessarily due to unjustified approximation in the calculation, they result more often from the fact that experiment and theory are not directly comparable. Theory treats a single molecule 'in space' at its equilibrium geometry. Gas phase experiments are closest to this situation, however, even in gas phase molecules are vibrating and rotating. To compare theory with experiment one should then either first calculate σ for the various rovibronic states and take a Boltzmann average, or one should extrapolate the measured results to $T = 0$. Then on the theoretical side only an averaging over the zero-point vibration is necessary, for which it is often sufficient to take σ at the r_z rather than the r_e-geometry.

Magnetic properties often depend sensitively on the geometry, and the results for r_z and r_e geometry may be quite different. For the example of F_2 IGLO calculations predict a change of $\sigma(^{19}F)$ by ~ 17 ppm. The experimental counterpart of this number, obtained indirectly is roughly twice as large [88]. The agreement is not too good, but the order of magnitude is certainly right.

We have, so far, not cared about isotope effects. Since isotopic molecules don't differ electronically, the main reason for isotope effects on σ is due to the mass dependence of the zero-point motion. Possibly the bulk effect can be accounted for by taking the different r_z-geometries. As C. Jameson has shown in her recent review [89] the decisive quantity is in fact the derivative of σ with respect to geometrical parameters, as they were e.g. calculated by the present authors [23], by Chesnut [37], Lazeretti et al. [32] and others.

Since IGLO results depend sensitively on the geometry, a good choice of the latter is important. If reliable experimental geometries are available these are usually better than geometries optimized at the SCF level. However, 'experimental' geometries are often based on unjustified assumptions, and in case of doubt it is recommended to use geometries optimized at a sufficiently high ab-initio level. Depending on the case one may have to include polarization functions or even electron correlation effects in the geometry optimization.

The present IGLO program is limited to closed-shell states. This is no severe restriction, because most experiments are done one closed-shell states as well. However, the MCSCF–IGLO program now under development will also allow IGLO calculations for open-shell states, e.g. free radicals.

Table 39. Magnetic susceptibilities[a] (with signs reversed)

Molecule	IGLO (DZ)	IGLO (II)	IGLO (III)	Experiment
BF_3	31.2	29.1	28.1	26.9[b]
CH_4	20.8	19.4	19.0	18.7[c]; 17.4[d]
C_2H_6	35.9	31.1	30.0	27.4[c]; 26.8[d]
C_3H_8	50.4	43.3	41.8	38.6–40.5[e]
n-C_4H_{10}	66.1	56.2		50.0; 50.3[e]
i-C_4H_{10}	64.6	56.0		50.5; 51.7[e]
n-C_5H_{12}		67.8		61.5–63.3[e]
i-C_5H_{12}		68.2		63.0–64.0[e]
neo-C_5H_{12}	79.6	70.0		63.0; 63.1[e]
C_2H_4	24.1	22.6		19.7[c]; 18.8[d]
c-C_3H_6	48.5	43.5		39.9[c]; 39.2[d]
C_6H_6	74.5	68.0		54.7–55.0[e]
CFH_3	23.5	20.2	19.4	17.8[f]
CF_2H_2	27.5	24.1	23.1	24.0[f]
CF_3H	33.3	30.0	28.9	30 ± 4[f]
CF_4	40.1	36.1	34.6	36 ± 5[f]
F_2CO	30.8	28.3	26.5	26.4[b]
FCCH	32.5	29.8	28.0	26.0[f]
FCCF	38.8	33.6	31.1	33 ± 4[f]
CH_3OH	29.6	22.6		21.4[g]
CH_3CHO	25.4	23.3		19.9[g]
CH_3CN	33.7	31.4	29.9	27.6[h]
CH_3NC	31.7	28.8	27.4	27.6[h]
$C_6H_5NH_2$	85.5	76.2		63.0[g]
C_6H_5CN	88.3	79.8		65.2[g]
OCS		40.4	37.2	32.4[i]; 35.71[k]
CS_2		54.1	48.4	42.2[i]; 44.94[k]
$Si(CH_3)_4$	95.2	81.4		74.9[l]
NH_3	17.3	17.4	17.5	18.0[i]
SiH_4	30.7	23.4		20.4 ± 0.8[d]
SiF_4		42.6		44 ± 7[m]
PH_3	30.5	28.0		26.6 ± 0.8; 26.2 ± 0.8[d]
SH_2		27.78	27.25	25.5[d]
SF_6		51.67	48.90	44[i]
OSF_2		32.34	30.22	27.1[h]
$OSCl_2$		54.75	50.59	44.3[n]; 45.2[h]
O_2SF_2		43.03	39.87	36.9[b]
SO_2		21.73	19.60	18.2[i]
SCl_2		58.52	55.84	47.9–49.4[h]

[a]Units are ppm cgs/mol; [b]Ref. 173; [c]Ref. 174; [d]Ref. 175; [e]Ref. 176; [f]Refs 177, 178; [g]Ref. 179; [h]Ref. 180; [i]Ref. 181; [k]Ref. 182; [l]Ref. 183; [m]χ_p from Ref. 180, χ_d calculated, see Ref. 23; [n]Ref. 184

Table 40. Magnetic susceptibility anisotropies[a] (with signs reversed)

		IGLO (II)	IGLO (III)	Experiment
NH_3	$\chi_\parallel - \chi_\perp$		-1.2	-1.3[b]
NF_3	$\chi_\parallel - \chi_\perp$	-2.77	-2.93	-3.0 ± 1.5[b]
CH_3F	$\chi_\parallel - \chi_\perp$	6.66	7.61	8.5 ± 0.6[b]; 8.2 ± 0.8[c]
CH_2F_2	$2\chi_{bb} - \chi_{aa} - \chi_{cc}$	-0.90	-0.73	-0.8 ± 0.4[b]; -0.58 ± 0.01[f]
	$2\chi_{aa} - \chi_{bb} - \chi_{cc}$	4.14	4.43	3.9 ± 0.5[b]; 4.16 ± 0.02[f]
CHF_3	$\chi_\parallel - \chi_\perp$	-1.31	-1.46	-1.2 ± 0.6[c]; -1.727[d]
$CHCl_3$	$\chi_\parallel - \chi_\perp$	6.2		8.0 ± 0.8[b]
CH_3CH_2F	$2\chi_{aa} - \chi_{bb} - \chi_{cc}$		6.03	6.6 ± 0.4[d]
	$2\chi_{bb} - \chi_{aa} - \chi_{cc}$		0.87	0.7 ± 0.5[d]
CH_3CHF_2	$2\chi_{aa} - \chi_{bb} - \chi_{cc}$		2.85	3.4 ± 0.7[d]
	$2\chi_{bb} - \chi_{aa} - \chi_{cc}$		2.19	1.4 ± 0.7[d]
CH_2CF_2	$2\chi_{bb} - \chi_{aa} - \chi_{cc}$	0.88	1.51	2.3 ± 0.6[b]
	$2\chi_{aa} - \chi_{bb} - \chi_{cc}$	-8.72	-8.45	-7.7 ± 0.5
cis-CHFCHF	$2\chi_{cc} - \chi_{bb} - \chi_{aa}$	-3.68	-4.86	-5.7 ± 0.4[b]
	$2\chi_{bb} - \chi_{cc} - \chi_{aa}$	-1.85	0.31	1.4 ± 0.3
CH_2CHF	$2\chi_{cc} - \chi_{bb} - \chi_{aa}$		1.04	0.8 ± 0.2[b]
	$2\chi_{bb} - \chi_{cc} - \chi_{aa}$		-10.78	-9.6 ± 0.2
F_2CO	$2\chi_{bb} - \chi_{aa} - \chi_{cc}$	-2.79	-1.11	-1.6 ± 0.9[d,e]
	$2\chi_{aa} - \chi_{bb} - \chi_{cc}$	8.22	6.96	5.3 ± 0.9[d]
H_2CO	$2\chi_{xx} - \chi_{yy} - \chi_{zz}$	-25.7		-25.5 ± 0.5[b]
	$2\chi_{yy} - \chi_{xx} - \chi_{zz}$	1.1		3.9 ± 0.3[b]
HCCF	$\chi_\parallel - \chi_\perp$	3.79	4.37	5.2 ± 0.7[b]
HCCCl	$\chi_\parallel - \chi_\perp$	5.4		9.3 ± 0.5[b]
$CHFCF_2$	$2\chi_{aa} - \chi_{bb} - \chi_{cc}$		-8.70	-7.7 ± 0.5[b]
	$2\chi_{cc} - \chi_{bb} - \chi_{aa}$		3.50	4.2 ± 0.2[b]
	$2\chi_{bb} - \chi_{cc} - \chi_{aa}$		-8.10	-7.7 ± 0.3[b]
CH_3CN	$\chi_\parallel - \chi_\perp$	9.50	9.52	10.5 ± 0.5[d]
CH_3NC	$\chi_\parallel - \chi_\perp$	12.36	12.98	13.5 ± 0.7[d]
HCO_2H	$2\chi_{aa} - \chi_{bb} - \chi_{cc}$	-5.5		-3.4 ± 0.5[b]
	$2\chi_{cc} - \chi_{aa} - \chi_{bb}$	15.7		12.8 ± 0.5[b]
HCO_2CH_3	$2\chi_{cc} - \chi_{aa} - \chi_{bb}$	13.6		11.0 ± 0.9[b]
	$2\chi_{bb} - \chi_{aa} - \chi_{cc}$	-3.3		-3.2 ± 0.9[b]
OCS	$\chi_\parallel - \chi_\perp$	5.0	7.8	9.3 ± 0.1[b]
CH_3SiH_3	$\chi_\parallel - \chi_\perp$	2.3		2.4 ± 0.2[b]
F_2O	$2\chi_{xx} - \chi_{yy} - \chi_{zz}$	12.6		8.8 ± 1.4[b]
	$2\chi_{yy} - \chi_{xx} - \chi_{zz}$	2.6		4.4 ± 0.7[b]
PH_3	$\chi_\parallel - \chi_\perp$	-3.24	-3.13[g]	-2.7 ± 0.8[b]

[a] Units are ppm cgs/mol.; [b] Ref. 20; [c] Ref. 6f; [d] Ref. 180; [e] There seems to be sign error in Ref. 20; [f] Ref. 185; [g] Basis II on hydrogen

3.12 Magnetic Susceptibilities

For all molecules for which IGLO calculations have been performed, the magnetic susceptibility tensors χ have automatically been calculated, and many results for χ have been published [1,4,22–26,42,83,84]. It is one of the advantages of the IGLO method that it furnishes good susceptibilities together with the chemical shifts.

Generally DZ calculations overestimate $|\chi|$, and increase of the basis size reduces $|\chi|$. On the whole one finds that results obtained with basis II are, for the isotropic part, systematically larger than the experimental values by

~10–20%. The discrepancy is reduced to ~5% if experimental gas phase values are used for comparison, which indicates that there is a strong influence of intermolecular forces on the susceptibilities. If experimental anisotropies of the susceptibilities are known, they usually agree rather well the IGLO results.

Some results for susceptibilities are collected in Table 39. Many more are found in the original references [1, 4, 22–26, 42, 83, 84]. A comparison of computed and experimental anisotropies $\Delta\chi$ is given in Table 40.

3.13 Computer Times

In Table 41 computer times for some molecules are given. As in all SCF-type methods they are roughly proportional to the fourth power of the number of basis functions. Additionally the IGLO times depend on the number of nuclei and of occupied MOs. Very recently parts of the IGLO program have been speeded up especially for systems with a larger number of nuclei and occupied MOs (the times given here are, however, for an older version of the program). The time needed for the IGLO part of the program is about one to twice the time needed for the SCF calculation including the integral evaluation. The program used here is highly optimized for a CYBER 205, but it has also been adapted to other computers. Calculations on a CONVEX C210 are slower by a factor of 2, on a CONVEX C100 by 5–6, a factor of about 35 is found for a comparison with a SUN 386i250, of about 20 for a DEC station 3200 and of about 10 for a SGI 4D20.

Table 41. Computer times[a]

Molecule/basis	N[b]	ZGM[c]	SCF	IGLO
CH_4/DZ	18	<1	<1	1
CH_4/II	46	8	4	9
CH_4/III	75	78	36	69
C_3H_8/DZ	46	6	4	26
C_3H_8/II	114	246	116	394
C_3H_8/III	185	1581	175	2034
$C_4H_4^{2+}$/DZ	48	9	4	30
$C_4H_4^{2+}$/II	112	274	68	294
$C_4Me_4^{2+}$/DZ	104	80	60	493
SiF_4/II	123	620	212	726
TMS/II	195	1320	619	3067
SF_2/II	79	71	30	125
SF_2/III	114	331	112	492
SF_6/II	167	948	277	1632
Pyridine/II	162	969	534	1308
Phosphabenzene/II	175	1180	651	1817
Phenylacetylene/II	212	2473	993	4663
Guanine/II	252	7570	3360	10960

[a]CPU times in seconds for the CYBER 205 (two pipes); [b]number of basis functions; [c]vectorized two electron integral program Ref. 28

Acknowledgements

The material presented in this review is in part the result of a cooperation with various groups as documented in the list of references. We are especially indebted to the very first cooperation with J. Michl, D.M. Grant, J. Facelli et al. on chemical shift tensors and the most recent one with P.v.R. Schleyer and his coworkers M. Bühl, M. Bremer and R. Herges mainly on carbocations, lithium organic compounds, boranes and carboranes. This cooperation had a large feedback on the write-up of this review and some unpublished results from the Schleyer group have been included. In addition to quoted private communications we further thank W. Koch, J. Sauer and Y. Apeloig for unpublished data. We have also profited from valuable comments or suggestions by V. Staemmler, P. Lazzeretti, Ch. van Wüllen, F. Reichel, W. Gombler and M. Willert-Porada. We thank H. Günther for the invitation to write this review and for valuable comments. Most calculations were performed on the CYBER 205 of the computer center of the Ruhr-Universität Bochum. The coorperative attitude of its staff is gratefully acknowledged. The authors thank R. Fink and R. Buchkremer for assistance in some of the IGLO calculations and T. van Mourik for her contribution to the current density plots. Some of the drawings were done with the help of B. Weinert and O. Adamczak. Mrs. U. Antoinette did an excellent job of typing and textediting the manuscript. Financial assistance by 'Deutsche Forschungsgemeinschaft' and 'Fonds der Chemischen Industrie' was very helpful.

4 References

1. Schindler M (1987) J Am Chem Soc 109: 1020
2. See e.g. Hehre WJ, Radom L, Schleyer PvR, Pople JA (1986) Ab initio molecular orbital theory, Wiley, New York
3. Kutzelnigg W (1980) Isr J Chem 19: 193
4. Schindler M, Kutzelnigg W (1982) J Chem Phys 76: 1919
5. Kutzelnigg W (1989) Theochem 202: 11
6a. Davies DW (1967) The theory of electric and magnetic properties of molecules, Wiley, New York
b. Memory JD (1968) Quantum theory of magnetic resonance parameters, McGraw Hill, New York
c. Lipscomb WN (1966) Adv Magn Reson 2: 137
d. Musher JI (1966) Adv Magn Reson 2: 177
e. Dichtfield R (1972) In: MTP Int Rev Sci, Phys Chem vol 2
f. Appleman BR, Dailey BP (1974) Advan Magn Reson 7: 231
g. Slichter CP (1978) Principles of Magnetic Resonace, 2nd edn Springer, Berlin Heidelberg New York; see also the fundamental paper: Ramsey NF (1950) Phys Rev 78: 699
7. Hegström RA, Lipscomb WN (1966) J Chem Phys 45: 2378; (1968) Revs Mod Phys 40: 354
8. Epstein ST (1965) J Chem Phys 42: 2897; (1973) J Chem Phys 58: 1592; (1980) Isr J Chem 19: 154
9. Langevin P (1905) J Phys 4: 678
10. Dalgarno A, Stewart AL (1956) Proc Roy Soc (London) A 238: 276; (1958) Proc Roy Soc 247: 245

11. Dalgarno A (1959) Proc Roy Soc (London) A 251: 282
12. Stevens RM, Pitzer RM, Lipscomp W (1963) J Chem Phys 38: 550; Stevens RM, Lipscomb WN (1964) J Chem Phys 40: 2238; (1964) J Chem Phys 41: 184; Kolker HJ, Karplus M (1964) J Chem Phys 41: 1259
13. Hameka H (1958) Mol Phys 1: 203; (1962) Rev Mod Phys 34: 87; Zeroka D, Hameka HF (1966) J Chem Phys 45: 300; Ditchfield R (1972) J Chem Phys 56: 5688; (1976) J Chem Phys 65: 3123; (1974) Mol Phys 27: 789
14. London F (1937) Naturwiss 15: 187; (1937) J Phys Rad 8: 397; (1937) CR Acad Sci (Paris) 205: 28
15. K Wolinski, JF Hinton, P Pulay (in press) J Am Chem Soc
16. Foster JM, Boys SF (1960) Rev Mod Phys 32: 296, 303, 305
17a. Okinski A, Sadlej AJ (1972) Acta Phys Polon A 42: 709
b. Maestro M, Moccia R (1975) Mol Phys 29: 81
c. Levy B, Ridard J (1981) Mol Phys 44: 1099
d. Hansen AE, Bouman TD (1985) J Chem Phys 82: 5035
18. Pulay P (1980) Chem Phys Lett 73: 393
19. Schneider RF (1968) J Chem Phys 48: 4905; Buckingham AD, Malm SM (1971) Mol Phys 22: 1127; Haeberlen U (1976) High resolution NMR in solids, In: Waugh JS (ed) Adv in Magnetic Resonance suppl 1, Academic, New York; Robert JB, Wiesenfeld L (1982) Phys Rep 86: 363
20. Flygare WH (1974) Chem Rev 74: 653
21. Huzinaga S (1971) Approximate atomic wave functions, University of Alberta, Edmonton, Alberta
22. Schindler M, Kutzelnigg W (1983) Mol Phys 48: 781
23. Fleischer U, Schindler M, Kutzelnigg W, (1987) J Chem Phys 86: 6337
24. Fleischer U, Schindler M (1988) Chem Phys 120: 103
25. Schindler M (1988) J Chem Phys 88: 7638
26. Schindler M, Kutzelnigg W (1983) J Am Chem Soc 105: 1360
27. Wüllen Cv, Fleischer U, Kutzelnigg W (to be published)
28. Meier U (1984) Diplomarbeit Bochum; Kutzelnigg W, Schindler M, Klopper WM, Meier U, Wallmeier H (1986) Lecture Notes in Chemistry 44: Springer, Berlin Heidelberg New York, p 55
29. Jameson CJ, Buckingham AD (1979) J Phys Chem 83: 3366; (1980) J Chem Phys 73: 5684
30. Sadlej A (1981) J Chem Phys 75: 320
31. Lazzeretti P (1978) Chem Phys Lett 59: 261; Iwai M, Saika A (1982) J Chem Phys 77: 1951; Iwai M, Saika A (1983) Int J Quantum Chem 24: 623
32. Jaszunski M (1978) Theor Chim Acta 48: 323; Daborn GT, Handy NC (1983) Mol Phys 49: 1277; Lazzeretti P, Zanasi R, Sadlej AJ, Raynes WT (1987) Mol Phys 62: 605
33. Daborn GT, Ferguson WI, Handy NC (1980) Chem Phys 50: 253; Daborn GT, Handy NC (1981) Chem Phys Lett 81: 201; see also Ref. 50
34. Vauthier GC, Comean M, Odiot S, Fliszar S (1988) Can J Chem 66: 1781
35. Pyykkö P, Göring A, Rösch N (1987) Mol Phys 61: 195
36. Lamanna UT, Guidotti C, Arrighini GP (1977) J Chem Phys 67: 604; Lazzeretti P, Zanasi R (1978) J Chem Phys 68: 832, 1523; Keil F, Ahlrichs R (1979) J Chem Phys 71: 2671; Höller R, Lischka H (1981) Mol Phys 41: 1017, 1041; (1981) Chem Phys Lett 84: 94; Weller T, Meiler W, Pfeifer H, Lischka H, Höller R (1983) Chem Phys Lett 95: 599; Fronzoni G, Galasso V (1986) Chem Phys 103: 29; Galasso V, Fronzoni G (1986) J Chem Phys 84: 3215; Ridard J, Levy B, Millie P (1978) Mol Phys 36: 1025
37. Ribas-Prado F, Giessner-Prettre C, Pullman B, Daudey JP (1979) J Am Chem Soc 101: 1737; Giessner-Prettre C, Pullman B (1982) J Am Chem Soc 104: 70; McMichael Rohlfing C, Allen LC, Dichtfield R (1984) Chem Phys 87: 9; Chesnut DB, Foley CK (1986) J Chem Phys 84: 852; Chesnut DB (1986) Chem Phys 110: 419; Wazer JRv, Ewig CS, Dichtfield R (1989) J Phys Chem 93: 2222
38a. Bouman TD, Hansen AE (1988) Chem Phys Lett 149: 510
b. Facelli JC, Grant DM, Bouman TD, Hansen AE (1990) J Comp Chem 11: 32
39. Nuclear magnetic resonance. A specialist's periodical report. The Royal Society of Chemistry London, vol 1 (1972) to vol 18 (1989)
40. Chesnut DB (1989) Annu Rep NMR Spectroscopy 50
41. Jameson AK, Jameson CJ (1987) Chem Phys Lett 134: 461
42. Schindler M (1987) J Am Chem Soc 109: 5950
43a. Wasylishen RE, Mooibroek S, Macdonald JB (1984) J Chem Phys 81: 1057

b. Jameson CJ, Jameson AK, Burrell PM (1980) J Chem Phys 73: 6013
c. Jameson CJ, Jameson AK, Honarbaksh J (1984) J Chem Phys 81: 5266
d. Alei M Jr, Florin AE, Litchman WM, O'Brien JF (1982) J Chem Phys 76: 152
44. Harmony MD, Laurie VW, Kuczkowski RL, Schwendemann RH, Ramsey DA, Lovas FJ, Lafferty WJ, Maki AG (1979) J Phys Chem Ref Data 8: 619
45. Hellwege KH, Hellwege AM (eds) (1987) Landolt-Börnstein, 'Zahlenwerte and Funktionen'. New Series II/15, Springer, Berlin Heidelberg New York
46. Günther H (1983) NMR-Spektroskopie, Georg Thieme, Stuttgart
47. Röhse R (priv. commun.)
48. Quinn WE, Bauer JM, Latourette JT, Ramsey NF (1958) Phys Rev 112: 1929
49. Cook DB, Davies AM, Raynes WT (1971) Mol Phys 21: 113; Raynes WT, Davies AM, Cook DB (1971) Mol Phys 21: 123; Zeroka D (1973) J Chem Phys 54: 3835; Ditchfield R (1981) Chem Phys 63: 185
50. Corcoran CT, Hirschfelder, JO (1980) J Chem Phys 72: 1524
51. Musher JI (1965) J Chem Phys 43: 4081; (1967) 46: 1219; see also Ref. 6
52. Lazzeretti P, Zanasi R (1981) Chem Phys Lett 80: 533; (1983) Chem Phys Lett 100: 67; Lazzeretti P, Rossi E, Zanasi R (1982) J Chem Phys 77: 3129
53. Haigh CW, Mallion RB (1980) Prog Nucl Magn Reson Spectrosc 13: 304; (1989) Croat Chem Acta 62: 1
54. McMichael Rohlfing C, Allen MC, Ditchfield R (1983) J Chem Phys 79: 4958
55. Günther H, Moskau D, Bast P, Schmalz D (1987) Angew Chem 99: 1242; (1987) Angew Chem Int Ed Engl 26: 1212
56. Malli G, Fraga S (1966) Theor Chim Acta 5: 275
57. Deverall C (1969) Prog Nucl Magn Reson Spectrosc 4: 278
58a. Benn R, Lehmkuhl H, Mehler K, Rufinska A (1984) Angew Chem 96: 521; (1984) Angew Chem Int Ed Engl 23: 534
b. Benn R, Rufinska A, Lehmkuhl M, Janssen E, Krüger C (1983) Angew Chem 95: 808; (1983) Angew Chem Int Ed Engl 22: 779
59. Fink R, Schindler M (unpublished)
60a. Schleyer PvR (private communication)
b. Bühl M, Schleyer PvR (in press) Angew Chem
c. Bühl M, Schleyer PvR (in press) in Olah GA, Wode K, Williams (Eds), Electron Defecient Boron and Carbon Clusters, Wiley
d. Schleyer PvR, Bühl M, Fleischer U, Koch W (1990) Inorg Chem 29: 153
61. Breitmaier E, Voelter W (1987) Carbon-13 NMR Spektroscopie, Verlag Chemie, Weinheim; Kalinowski H-O, Berger S, Braun S (1984) ^{13}C-NMR-Spectroscopy, Georg Thieme Verlag, Stuttgart
62. Roth WR, Lennartz HW, Doering WvE, Dolbier WR, Schmidhausen JC (1988) J Am Chem Soc 110: 1823
63. Orendt AM, Facelli JC, Grant DM, Michl J, Walker FH, Dailey WP, Waddell ST, Wiberg KB, Schindler M, Kutzelnigg W (1985) Theoret Chim Acta 68: 421
64. Beeler AJ, Orendt AM, Grant DM, Cutts PW, Michl J, Zilm KW, Downing JW, Facelli JC, Schindler M, Kutzelnigg W (1984) J Am Chem Soc 106: 7672
65. Christl M, Herbert R (1979) Org Magn Reson 12: 150; (1979) Chem Ber 112: 2020; Christl M, Freund S (1985) Chem Ber 118: 979; Christl M, Herzog C (1986) Chem Ber 118: 3067
66. Sander W, Gleiter R (1985) Chem Ber 118: 2548
67. Schindler M (to be published)
68a. Murray RW, Jeyaraman R (1985) J Org Chem 50: 2847
b. Adam W, Curci R, Edwards JO (1989) Acc Chem Res 22: 205
69. Cremer D, Schindler M (1987) Chem Phys Lett 133: 293
70. Schleyer PvR, Koch W, Liu B, Fleischer U (1989) J Chem Soc Chem Commun 1098; Bremer M, Schötz K, Schleyer PvR, Fleischer U, Schindler M, Kutzelnigg W, Koch W, Pulay P (1989) Angew Chem 101: 1063; (1989) Angew Chem Int Ed Engl 28: 1042; Schleyer PvR, de M Carneiro JW, Koch W, Raghavachari K (1989) J Am Chem Soc 111: 5475
71a. Saunders M, Kates MR (1983) J Am Chem Soc 105: 3571
b. Saunders M, Kates MR, Wiberg KB, Pratt W (1977) J Am Chem Soc 99: 8072
72. Saunders M, Laidig KE, Wiberg KB, Schleyer PvR (1988) J Am Chem Soc 110: 7652
73. Bremer M, Schleyer PvR, Schötz K, Kausch M, Schindler M (1987) Angew Chem 99: 795; (1987) Angew Chem Int Ed 26: 702
74. Bremer M, Schleyer PvR, Fleischer U (1989) J Am Chem Soc 111: 1147

75. Herges R, Schindler M (to be published)
76. R Herges (priv commun)
77. Herges R (1988) J Am Chem Soc 110: 7764
78. Jameson CJ, Jameson AK (1988) Chem Phys Lett 149: 300
79. Tossell JA, Lazzeretti P (1986) J Chem Phys 84: 369; (1986) Chem Phys Lett 132: 464; (1986) J Phys B 19: 3217
80. Raabe G, Michl J (1985) Chem Rev 85: 419
81. Philipsborn WV, Müller R (1986) Angew Chem 98: 381; (1986) Angew Chem Int Ed Engl 25: 383
82. Martin GJ, Martin ML, Guesnard J-P (1981) In: NMR Basic Principles and Progress, vol 18, Springer, Berlin Heidelberg New York
83. Schindler M (1988) Magn Reson Chem 26: 394
84. Schindler M (1988) J Am Chem Soc 110: 6623
85. Eshtiagh-Hosseini H, Kroto HW, Nixon JF, Brownstein S, Morton JR, Preston KF (1979) J Chem Soc Chem Commun 653
86. Solomon IJ, Keith JN, Kacharek AJ, Ranep JK (1986) J Am Chem Soc 90: 5408
87. Gombler W, Schaeps J, Willner (1990) Inorg Chem in press
88. Jameson CJ (1977) J Chem Phys 66: 4977
89. Jameson CJ (in press) In: Buncel/Jones (eds) Isotopes in the physical and biomedical sciences. Isotopic Applications in NMR Studies, Elsevier, Amsterdam
90. Schneider WG, Bernstein HJ, Pople JA (1958) J Phys Chem 28: 601
91. Emsley JW, Feeney J, Sutcliffe LH (1968) High Resolution NMR Spectroscopy, vol 2, Pergamon, New York
92. Cavanaugh JB, Dailey BP (1961) J Chem Phys 34: 1099
93. Petrakis L, Sederholm CH (1961) J Chem Phys 35: 1174
94. Raynes WT, Raza MA (1969) Mol Phys 17: 157
95. Sadtler Standard Spectra, NMR Spectra, Sadtler Research Inc., 1976
96. Chauvel JP, True NS (1985) Chem Phys 95: 435
97. Jackmann LM, Sternhell S (1969) Applications of NMR spectroscopy in organic chemistry, 2nd edn, Pergamon, New York
98. Litchman WM, Grant DM (1968) J Am Chem Soc 90: 1400
99. Shapiro BL, Mohrmann LE (1980) J Phys Chem Ref Data 39: 499
100. Hunter BK, Reeves LW (1968) Can J Chem 46: 1399
101. Dyer J, Lee J (1970) J Chem Soc B 409
102. Treichel PM, Goodrich RA, Pierce SB (1967) J Am Chem Soc 89: 2017
103. Hindman JC, Svirmickas A, Appelman EH (1972) J Chem Phys 57: 42
104. Nöth H, Wrackmeyer B (1978) In: NMR Basis Principles and Progress, vol 14, Springer, Berlin Heidelberg New York
105. Onak TP, Landesman HL, Williams RE, Shapiro I (1959) J Phys Chem 63: 1533
106. Onak TP, Drake RP, Dunks GB (1964) Inorg Chem 3: 1686
107. Onak TP, Gohart FJ, Williams RE (1963) J Am Chem Soc 85: 3378
108. Todd LJ, Siedle AR, Bodner GM, Kahl SB, Hieley JP (1976) J Magn Res 23: 301
109. Jackowski K, Raynes WT (1977) J Chem Research (S) 66
110. Ven LJMvd, Haan JWd (1978) J Chem Soc Chem Commun 94
111. Jackowski K, Raynes WT (1977) Mol Phys 34: 465
112. Tiffon B, Doulet J-P (1976) Can J Chem 54: 2045
113. Gay ID, Kriz JF (1978) J Phys Chem 82: 319
114. Somayajulu GK, Kennedy JR, Vickrey TM, Zwolinski BR (1979) J Magn Reson 33: 559
115. Beierbeck H, Saunders JK (1980) Can J Chem 58: 1258
116. Duncan TM (1987) J Phys Chem Ref Data 16: 4268
117. Ganapathy S, Chacko VP, Bryant RG (1984) J Magn Reson 99: 5026
118. Olah GA, Donovan DJ (1977) J Am Chem Soc 99: 5026
119. Falkenberg-Andersen C, Ranaganayakulu K, Schmitz LR, Sorensen TS (1984) J Am Chem Soc 106: 178
120. Myhre PC, Krugen JD, Mammond BL, Lou MS (1984) J Am Chem Soc 106: 6079
121. Olah GA, Prakash GKS, Arvanaghi M, Anet FAL (1982) J Am Chem Soc 104: 7105
122. Schmitz LR, Sorensen TS (1980) J Am Chem Soc 102: 1645
123. Marsmann HC (1981) In: NMR Basic Principles and Progress, vol 17, Springer, Berlin Heidelberg New York

124. Beckers H, Bürger H, Eujen R (1985) J Fluorine Chem 27: 461
125. Levy GC, Lichter R (1979) Nitrogen-15 nuclear magnetic resonance spectroscopy, Wiley, New York
126. Duthaler RO, Roberts JD (1978) J Am Chem Soc 100: 4969
127. Staedel W, Philipsborn Wv (1981) Org Magn Reson 15: 106
128. Witanowski M, Stefaniak L, Webb GA (1981) Annu Rep NMR Spectrosc 11B: 18
129. Jovanovic MV (1984) Spectrochim Acta 40A: 637
130. Willert-Porada M (priv. commun.)
131. Mason J, Bronswijk Wv (1969) J Chem Soc Chem Commun 357
132. Mason J (ed) (1987) Multinuclear NMR, Plenum, New York
133. Haas A, Rieland M (1988) Chimica 42: 67
134. Buchanan GW, Stothers JB (1982) Can J Chem 60: 787
135. Burger MI, Dhawan D, Fiat D (1982) Org Magn Reson 20: 184
136. Gonella NC, Nakanishi H, Holtwick JB, Horowitz DS, Kanamori K, Leonard NJ, Roberts JD (1983) J Am Chem Soc 105: 2050
137. Mason J, Christie KO (1983) Inorg Chem 22: 1849
138. Gorenstein DG (1983) Progr Nucl Magn Reson Spectrosc 16: 1
139. Seel F, Rudolph K, Budenz R (1965) Z Anorg Allg Chem 341: 196
140. Nixon JF, Schmutzler R (1964) Spectrochim Acta 20: 1835
141. Burg AB (1977) Inorg Nucl Chem Letters 13: 199
142. Crutchfield MM, Dungan CH, Letcher JM, Mark V, Wazer JRv (1967) P^{31} Nuclear magnetic resonance, John Wiley, New York
143. Horn H-G (1972) Chem Ztg 96: 666
144. Fritz G, Schaefer H (1974) Z Anorg Allg Chem 409: 137
145. Heckmann G, Fluck E (1972) Z Chem 23: 175
146. Appel R, Knoll F, Rupert I (1981) Angew Chem 93: 771; (1981) Angew Chem Int Ed Engl 20: 731
147. Rösch W, Vogelbacher U, Allspach P, Regitz M (1986) J Organomet Chem 306: 39
148. Karaghiosoff K, Schmidpeter A (1988) Phosphorus Sulfur 36: 217
149. Maatescu GD, Benedikt GM (1979) J Am Chem Soc 101: 3959
150. Kintzinger JP (1981) In: NMR Basis Principles and Progress, vol 17, Springer, Berlin Heidelberg New York
151. Sugawara T, Kawada Y, Katch M, Iwamura M (1979) Bull Chem Soc Jpn 52: 3391
152. Wasylishen RF, Connor C, Friedrich JO (1984) Can J Chem 62: 1181
153. Rodger C, Sheppard N, McFarlane C, McFarlane W (1978) In: NMR and the periodic table, Academic, London
154. Annunziata A, Barbarella G (1984) Org Magn Reson 22: 250
155. Barbarella G, Gongini A, Chatgilialoglu C, Rossini S, Tugnoli V (1987) J Org Chem 52: 3857
156. Drakenberg T, Forsén S (1983) In: The multinuclear approach to NMR spectroscopy, Reidel, Dordrecht
157. Jamesson CJ, Jamesson AK, Burrell PM (1980) J Chem Phys 75: 6013
158. Jamesson CJ, Jamesson AK, Honarbakhsh J (1984) J Chem Phys 81: 5266
159. Emsley JW, Phillips L (1971) Progr Nucl Magn Reson Spectrosc 7: 1
160. Nass R (1986) Dissertation Bortmund (FRG)
161. Wray V (1980) Annu Rep NMR Spectroscopy 10B: 1
162. Dungan CH, van Wazer JR (1970) Compilation of reported F-19 NMR chemical shifts, Wiley, New York
163. Gombler W (1981) Spectrochim Acta 37a: 57
164. Sharp KG, Coyle TD (1971) J Flourine Chem 1: 249
165. Drake JE, Westwood RPC (1971) J Chem Soc (A) 3300
166. Moy D, Young AR (1965) J Am Chem Soc 87: 1889
167. Spring CA, True NS (1983) J Am Chem Soc 105: 7231
168. Seel F, Budenz R, Gombler W (1970) Chem Ber 103: 1701
169. Gombler W, Seel F (1974) J Flourine Chem 4: 333
170. Appel R, Lassmann E (1971) Chem Ber 104: 2246
171. Richert H, Glemser O (1961) Z Anorg Allg Chem 307: 328
172. Barlos K, Kroner J, Nöth H, Wrackmeyer B (1978) Chem Ber 111: 1833; (1980) 113: 3716
173. Jameson CJ, Jameson AK, Wille, S (1981) J Chem Phys 74: 1613
174. Oldenziel JG Trappeniers NJ (1976) Physica (A) (Amsterdam) 82A: 565, 581
175. Barter C, Meisenheimer RG, Stevenson DD (1960) J Chem Phys 64: 1312

176. Hellwege KH, Hellwege AM (eds) (1967) Landolt-Börnstein, Zahlenwerte und Funktionen, II/10, Springer, Berlin Heidelberg New York
177. Singer SK, Monanty S (1983) Mol Phys 48: 1383
178. Appleman BR, Dailey BP (1974) Advan Magn Reson 7: 231
179. Haberditzl W (1987) In: Mulay LN, Bordeaus EA (eds) Theory and applications of molecular diamagnetism, Wiley, New York
180. Hellwege KH, Hellwege AM (eds) (1987) Landolt Börnstein, Zahlenwerte und Funktionen, New Series II/6, 14, Springer, Berlin Heidelberg New York
181. Handbook of Chemistry and Physics, 67th ed, CRC Press, 1987
182. Pocnan JM, Stone RG, Flygare WM (1969) J Chem Phys 51: 4278
183. Flygare WH (1974) Chem Rev 74: 653
184. Gierke TD, Flygare WH (1972) J Am Chem Soc 94: 7277
185. Kukolich SG Nelson AC (1972) J Chem Phys 56: 4446

Author Index Volumes 21–23